Instabilities and Nonequilibrium Structures VII & VIII

Nonlinear Phenomena and Complex Systems

VOLUME 8

The Centre for Nonlinear Physics and Complex Systems (CFNL), Santiago, Chile, and Kluwer Academic Publishers have established this series devoted to nonlinear phenomena and complex systems, which is one of the most fascinating fields of science today, to publish books that cover the essential concepts in this area, as well as the latest developments. As the number of scientists involved in the subject increases continually, so does the number of new questions and results. Nonlinear effects are essential to understand the behaviour of nature, and the methods and ideas introduced to treat them are increasingly used in new applications to a variety of problems ranging from physics to human sciences. Most of the books in this series will be about physical and mathematical aspects of nonlinear science, since these fields report the greatest activity.

The titles published in this series are listed at the end of this volume.

Instabilities and Nonequilibrium Structures VII & VIII

Edited by

Orazio Descalzi

Facultad de Ingeniería,
Universidad de los Andes,
Santiago, Chile

Javier Martínez

Instituto de Física,
Universidad Católica de Valparaíso,
Valparaíso, Chile

and

Enrique Tirapegui

Facultad de Ciencias Físicas y Matemáticas,
Universidad de Chile,
Santiago, Chile

KLUWER ACADEMIC PUBLISHERS
DORDRECHT / BOSTON / LONDON

A C.I.P. Catalogue record for this book is available from the Library of Congress.

ISBN 1-4020-1825-8

Published by Kluwer Academic Publishers,
P.O. Box 17, 3300 AA Dordrecht, The Netherlands.

Sold and distributed in North, Central and South America
by Kluwer Academic Publishers,
101 Philip Drive, Norwell, MA 02061, U.S.A.

In all other countries, sold and distributed
by Kluwer Academic Publishers,
P.O. Box 322, 3300 AH Dordrecht, The Netherlands.

Printed on acid-free paper

TABLE OF CONTENTS

FOREWORD

The contents of this book correspond to Sessions VII and VIII of the International Workshop on Instabilities and Nonequilibrium Structures which took place in Viña del Mar, Chile, in December 1997 and December 1999, respectively. We were not able to publish this book before and we apologize for this fact to the authors and participants of the meeting. We have made an effort to actualize the courses and articles which have been reviewed by the authors. Both Workshops were organized by Facultad de Ciencias Físicas y Matemáticas, Universidad de Chile, Instituto de Física of Universidad Católica de Valparaíso and Centro de Física No Lineal y Sistemas Complejos de Santiago. We are glad to acknowledge here the support of the Facultad de Ingeniería of Universidad de los Andes of Santiago which also be from now on one of the organizing Institutions of future Workshops.

Enrique Tirapegui

PREFACE

This book is divided in two parts. In **Part I** we have collected the courses given in Sessions VII and VIII of the Workshop and in **Part II** we include a selection of the invited Conferences and Seminars presented at both meetings. Three courses are related to the very important new developments in Bose-Einstein condensation: the first one by Robert Graham studies the classical dynamics of excitations of Bose condensates in anisotropic traps, the second by Marc Etienne Brachet is devoted to the bifurcations arising in attractive Bose-Einstein condensates and superfluid helium and the third course by André Verbeure is a pedagogical introduction to the subject with special emphasis on first principles and rigorous results. Part I is completed by two courses given by Michel Moreau: the first on diffusion limited reactions of particles with fluctuating activity and the second on the phase boundary dynamics in a one dimensional nonequilibrium lattice gas. The articles presented in Part II refer to a variety of subjects, both experimental and theoretical, of current interest in Nonlinear Physics.

Santiago, Marzo 2003

Orazio Descalzi
Javier Martínez
Enrique Tirapegui
(Eds.)

PART I

Courses

PART I

Diffusion-limited reactions of particles with fluctuating activity

by

O. Bénichou*, M. Moreau** and G. Oshanin**

* Physique de la Matière Condensée, Collège de France, 11 pl. Marcelin Berthelot, 75005 Paris, France

** Physique Théorique des liquides, Université Pierre et Marie Curie, 75252 Paris Cedex 05, France

Abstract

We study a diffusion-limited reaction, which consists in the annihilation of some species A by another species B. However, the B particles can only destroy A if they are in an activated form, whereas they also have an inert form which leaves B unchanged. Each form has a stochastic lifetime following an exponential law. We compute the reaction probability of a particle B, and deduce the kinetics of the reaction. It is shown that in one or two dimensions these fluctuations do not change the reaction kinetics, which differs from the ordinary, mean-field kinetics, as it is the case when there are no fluctuations. In three dimensions, the effective reaction constant, including the fluctuations, is computed. We prove that it is an increasing function of the relaxation frequency of the fluctuations.

O. Descalzi et al. (eds.), Instabilities and Nonequilibrium Structures VII & VIII, 5–22.
© 2004 *Kluwer Academic Publishers. Printed in the Netherlands.*

1. Introduction

Natural media are often subject to stochastic time fluctuations, due to their internal evolution or to their interaction with a changing environment. For instance, the microscopic description of an elementary reaction can involve a fluctuating interaction potential, taking into account the stochastic interaction of the reacting molecules with the surrounding fluid [1, 2, 3]. Another perturbation of a chemical reaction can be the random activation or desactivation of some reagent caused by to external factors, such as photons or solvent molecules. The existence of inert forms can also be due to the complexity of the molecules structures: some geometrical configurations may prevent the active parts of the reacting species to be in proper contact, and thus inhibit the reaction, whereas stochastic changes in the molecules geometries can allow them to react. Many cases of such geometrical limitations can be found for reactions in biomembranes [4] or for the ligand binding to proteins [5], for example. The effects of a fluctuating activity can be specially important for diffusion-limited reactions. They have been studied in the limiting case of fluctuations without time correlation [6]. However, a more realistic description of the fluctuations generally includes finite correlation times, which can have important and unsuspected effects on the reaction constants, as shown on other examples [2,3,7].

The purpose of this article is to consider the memory effects of an activation-desactivation process on the kinetics of a diffusion-limited annihilation reaction, using a discrete space and time description. After presenting our model, in Section 3 we shall review some classical results on diffusion in d-dimensions, and the main conclusions on the limiting case of uncorrelated fluctuations. The full model will be treated in Section 4.

2. Model

We consider two kinds of particles, A and B, located on a d-dimensional regular lattice including N sites. Initially, the respective numbers of particles A and B are $N_A(0)$ and $N_B(0)$, and they are randomly distributed among the N lattice sites, each site containing at most one particle. Particles A are immobile, whereas each particle B performs an independent, homogeneous discrete-time random walk on all sites of the lattice (including the sites occupied by other particles). Particles B fluctuate between two internal states: an inert form B^0 (state 0) and an activated form B^1 (state 1). For each particle B, the waiting times T_0 in state 0 and T_1 in state 1 are independent stochastic variables following exponential laws with the respective relaxation frequencies λ_0 and λ_1:

$$P(T_i > t) = \exp(-\lambda_i t) \quad \text{for } i = 0, 1. \tag{1}$$

The fluctuations between the activated and inert forms of B can be represented as a chemical equation

$$B^0 \leftrightarrow B^1 \tag{2}$$

Which, as mentioned in te introduction, can be the passage between two isomers or the thermal excitation-desexcitation of molecules due to their interaction with a random environment. A number of alternative interpretations of the fluctuations between the inert and activated forms of B can be found easily. For instance, B can be a predator with fluctuating aggressivity. An important point is that the waiting times T^0 and T^1 are independent of all prior events, such as the motions of B particles or their possible encounters with particles A. Furthermore, particles B do not interact with each other, and particles B^0 do not interact with particles A: if for instance a particle B meets another B, each one follows its motion without any change. Similarly, if an inert particle B^0 meets a A, both remain unchanged. On the contrary, if an active particle B^1 meets a A, particle A is destroyed instantaneously while B^1 is unchanged, which can be formally represented by the reaction

$$A + B^l \rightarrow B^l \qquad (3)$$

the kinetics of reaction (3) being determined by the random walk performed by B on the lattice, since the reaction is reaction-controlled. However, it has been recognized from a long time [8-11] that the evolution of such a reaction is not correctly described by the usual, mean-field laws of chemical kinetics, at least for low dimensions: if the concentration b^l of B^l is kept constant, the concentration of A should decrease exponentially like $exp(-k\ b^l\ t)$, which is not the case in 1 or 2 dimensions, as it is recalled further on. Thus a complete, stochastic theory is needed.

It is clear that the average number of A particles at time t, $<N_A(t)>$, is the sum of the probability for each particle A to survive up to t. If all positions are equivalent (which amounts to neglecting boundary effects), we have

$$<N_A(t)> = N_A(0)\ \psi(t) \qquad (4)$$

$\psi(t)$ being the probability for a given A to survive up to time t (at least), which is clearly independent of the number and positions of the other A. Thus, we will now consider the survival probability of one particle A submitted to the action of $N_B(0) = N_B$ particles B.

It is well known [12] that the so-called "random telegraph" dichotomous process, described by the stochastic laws (1), leads to the following conditional probability $p(j, t+\tau\ /\ i, t)$ to find the system in state j at time $t+\tau$, knowing that it is in state i at time t:

$$p(j, t+\tau\ /\ i, t) \equiv p_{ji}(\tau) = (\delta_{ji} - \alpha_i)\ e^{-\lambda\tau} + \alpha_j \qquad (5)$$

where   is the Kronecker symbol and

$$\lambda = \lambda_0 + \lambda_1$$

$$\alpha_0 = \lambda_1\ /\ \lambda\ ; \quad \alpha_1 = \lambda_0\ /\ \lambda$$

$\qquad (5')$

In the limiting cases of very large or very small relaxation frequency λ, the previous annihilation problem was addressed by different authors [6, 8-11]. Their results will be summarized in Section **3**.

Following most authors [6, 8-12], we will consider discrete time processes: particles B perform discrete random-walks with the same time step, which is taken as time unit. Then they can only move or interact with A at integer times 0, 1, 2... We assume that the particles B do not interact with A at time 0: a B particle cannot destroy A at time t = 0, whatever may be its position or its internal state. Our main propose will be to compute the probability $\Psi(n)$ that A survives from time 0 up to time n (at least) and related quantities.

3. Limit cases

3.1. Mobile particles with permanent, full activity

In this case, particles B are always active: particle A is destroyed as soon as a particle B reaches it. The probability $\Psi(n)$ that particle A survives up to time n can be computed in the thermodynamic limit, when the number N_B of particles B as well as the number N of available lattice sites tend to infinity in such a way that he concentration N_B / N of particles B tends to a finite limit b:

$$b = \lim_{n \to \infty} N_B / N$$

It can be shown [6] that

$$\Psi(n) = \exp(- bD(n)) \tag{6}$$

Here $D(n)$ is the expectation value of the number of distinct sites visited by a n-step random walk, the expectation being taken over all n-steps trajectories starting from the origin. $D(n)$ can be computed exactly [13]. When $n \to \infty$, its asymptotic behavior depends on the dimensionality and on the type of the lattice. For d-dimensional Polya random walks, we have

$$\text{if } d = 1, \quad D(n) = (8n/\pi)^{1/2} + O(n^{-1/2}) \tag{7.a}$$

$$\text{if } d = 2, \quad D(n) = \pi C\, n/\log n + O\,(n/\log^2 n) \tag{7.b}$$

$$\text{if } d = 3, \quad D(n) = n\, S + O(n^{-1/2}) \tag{7.c}$$

In (7.b), C is a constant depending on the lattice. For instance, $C = 4 \times 3^{-3/2}$, 1 and $2 \times 3^{-1/2}$ for hexagonal, square and triangular lattices, respectively.

In (7.c), S is *the probability that a B particle never returns to its initial position*. This probability has a well-know, finite value in 3 dimensions, depending on the type of the lattice [13], whereas it vanishes in 1 or 2 dimensions.

It is seen that the ordinary, mean-field kinetics only applies for reaction (3) in 3 dimensions, since then, Eqs.(6) and (7.b) imply $\Psi(n) = \exp(- S b n)$, which corresponds to a chemical constant

$$k = S \tag{8}$$

On the contrary, $\Psi(n)$ does not decay exponentially in 1 or 2 dimensions, so that the mean-field theory breaks down.

3.2. Mobile Particles with permanent, partial activity

This case occurs when the respective relaxation frequencies λ_0 and λ_1 of the inert or activated states of B both tend to infinity, with finite values of the stationary probabilities of these states, respectively $\alpha_0 = \lambda_1 / (\lambda_0 + \lambda_1)$ and $\alpha_1 = \lambda_0 / (\lambda_0 + \lambda_1)$. This means that each time a particle B meets particle A, it has the probability α_0 to be inactive, independently of past events.

This problem has been analyzed in ref.[6]. It has been shown that the survival probability of particle A is given by

$$\Psi(n) = \exp(- b \underline{D}(n)) \tag{9}$$

where quantity $\underline{D}(n)$, which replaces $D(n)$ of Eq(6), also depends on the lattice. Detailed formulas are given and derived in [6]. The most important results are

(i) *in one and two dimensions:* if $n \to \infty$, $\underline{D}(n) \sim D(n)$ given by Eq.(7.a)-(7.b). Thus the survival probability of A is essentially the same as in the case of permanent full activity, described in Section **3.1**, and in a first approximation, it is *independent* of the probability of the inert state α_0. However, $\underline{D}(n) - D(n)$ is approximately a

constant depending on α_0, which tends to $-\infty$ when $\alpha_0 \to 1$, *i.e.* when the particles B become almost completely inactive. When $\alpha_0 = 1$, the survival probability $\Psi(n)$ of A is obviously infinite, but if $\alpha_0 < 1$, no matter how small is the probability $\alpha_1 = 1 - \alpha_0$ that a particle B is active, $\Psi(n)$ obeys the same laws given by Eq(7.a) and (7.b) when $n \to \infty$. This conclusion, which can seem surprising, is related to the fact that in 1 and 2 dimensions, each particle B reaches A with probability 1, whatever may be its initial position, and its return probability to A is 1. Thus, for very large n, the same particle B has most probably visited A many times, and the probability that it was each time inactive is very low if $\alpha_0 < 1$.

(ii) *in dimension $d > 2$*, it is found [6] that

$$\underline{D}(n) = \frac{\alpha_1 S}{\alpha_1 + \alpha_0 \ S} \ n + O(n^{1/2}) \tag{10}$$

S being, as in (7.c), the probability that a particle B never returns to its initial position.

From (10) it is seen that one can actually define a reaction constant $\underline{k}$ such that

$$\frac{1}{\underline{k}} = \frac{1}{\partial \underline{D}(n)/\partial n} = \frac{1}{S} + \frac{\alpha_0}{\alpha_1} \tag{11}$$

which is an example of the inverse addition law that often applies for the chemical rate constants [1,2,14-17], and more generally, for the transmission probability of "chain processes" [16].

4. Mobile particles with fluctuating activity

4.1. Survival probability of a particle A with N_B particles B.

We now consider the complete model presented in Section2: the mobile particles B fluctuate between the inactive state 0 and the activated state 1 according to the probability law (1) with a finite relaxation frequency λ. If a particle B reaches A in state 0 at time t, and returns to A at a further time $t + \tau$, the probability that it is then in state i (i = 0 or 1) is p_{i0} (τ) given by Eq(2). Thus in order to compute the probability that A is annihilated by a particle B at time n, it is necessary to consider the complete trajectory of B from $t = 0$ to $t = n$.

Let us number the N_B particles B with the index j and denote them B_j, j= 1,2,...N_B. Let $\{\Gamma_j\}$ j = 1, 2,...N_B be their trajectories from $t = 0$ to $t = n$. Considering the trajectory Γ_i of a particle B_i which visits A m times between times $t = 0$ (excluded) and n (included), we denote $t_i^1, t_i^2, ... t_i^m$ the successive times of these visits

$$0 < t_i^1 < t_i^2 < ... < t_i^m \leq n$$

If particle B_i does not visit A between 0 and n, we take m = 0.

Given the trajectories $\{\Gamma_i\}$, the conditional probability that particle B_i is in state 0 at all its visits to A after time 0 and up to time n is, if there is at least one visit to A (m ≥ 1)

$$P_i(n/\Gamma_i) = \alpha_0 \prod_{h=1}^{m-1} p_{00}(\tau_i^h) \tag{12}$$

with

$$\tau_i^h = t_i^{h+1} - t_i^h \quad (i=1,...m-1) \tag{13}$$

and

$$p_{00}(\tau) = \alpha_0 + \alpha_1 e^{-\lambda \tau} \tag{14}$$

In (11), we assumed that at it first visit to A, the probability that particle B_i is inert is identical to its asymptotic value α_0. For all further visits, this probability is computed by using formula (2) repeatedly.

If the trajectory does not visit A (m=0), the right-hand side of Eq(11) should be replaced by 1.

Because the particles B are independent of each other, the conditional probability that particle A survives at time n is

$$\psi(n/\{\Gamma_j\}) = \prod_{i=1}^{N_B} P_i(n/\Gamma_i) \tag{15}$$

and the total survival probability of particle A at time n is

$$\psi(n) = <\psi(n/\{\Gamma_j\})>$$

the average being taken on all possible trajectories from t = 0 to t = n. Since all particles B are identical, we have

$$\psi(n) = <P_1(n/\Gamma)>^{N_B} \tag{16}$$

the average being taken on all possible trajectories Γ of one particle B.

Let us call M the initial position of this particle B, and Γ_M a trajectory starting from M at time 0. One can write

$$<P_1(n/\Gamma)> = <<P_1(n/\Gamma_M)>_{\Gamma_M}>_M \tag{17}$$

Here $<P_1(n/\Gamma_M)>_{\Gamma_M}$ denotes the (conditional) average on all trajectories *starting from* M. The final average $< >_M$ is taken on all possible initial positions M. Assuming that the probability of the initial position is uniformly distributed among the N available sites, we have

$$<<P_1(n/\Gamma_M)>_{\Gamma_M}>_M = \frac{1}{N}\sum_M <P_1(n/\Gamma_M)>_{\Gamma_M}$$

and Eq(16) can be written

$$\psi(n) = (1 - \frac{1}{N}\sum_M (1 - <P_1(n/\Gamma_M)>_{\Gamma_M}>))^{N_B} \tag{18}$$

Assuming that, in the thermodynamic limit, N $\to \infty$ and $N_B \to \infty$, with

$$N_B/N \to b$$

we see from (18) that in this limit the survival probability of A at time n is

$$\psi(n) \approx \exp(-b\sum_M(1 - <P_1(n/\Gamma_M)>_{\Gamma_M}>) \tag{19}$$

Thus, the survival probability $\psi(n)$ is simply related to the survival probability of A in presence of one particle B only, starting from M, which is $<P_1(n/\Gamma_M)>_{\Gamma_M}$, or to the quantity

$$1 - <P_1(n/\Gamma_M)>_{\Gamma_M}$$

which is the probability that B, starting from M, destroys A at some time t $\leq$ n.

Similar results were obtained [6] in the case of fluctuations without memory, mentioned in Section **3.3** : clearly, they do not depend on the internal evolution of particles B along their trajectories.

If particles B are always activated, $1 - <P_1(n/\Gamma_M)>_{\Gamma_M}$ is just the probability $L_M(n)$ that B, starting from M, reaches A before or at time n, and

$$D(n) = \sum_M L_M(n) \qquad (20)$$

is *the average number of distinct sites visited by a B particle during a n-steps trajectory* [6,13].

We shall now study the survival probability and related quantities in presence of one fluctuating B particle.

4.2. First arrival time of one activated particle B and reaction probability

We consider the first arrival time T_M at A *after t=0* of a given particle B starting from M at time 0, and the first arrival time $\underline{T}_M$ at A after t=0 of some particle B *under its activated form*, knowing that it starts from M at time 0 with the stationary distribution $\{\alpha_i\}$ for the internal states i, i= 0 or 1.

Let us define the probability $P_M(n)$ that B reaches A for the first time at time n

$$P_M(n) = \text{Proba.}(T_M = n) \qquad (21)$$

and the probability $Q_M(n)$ that B has never reached A at time n

$$Q_M(n) = \text{Proba.}(T_M > n) \qquad (22)$$

with the obvious relation

$$P_M(n) = Q_M(n-1) - Q_M(n) \quad (n > 0) \qquad (23)$$

We also define the similar probabilities for the first arrival at A of B under its activated form

$$\underline{P}_M(n) = \text{Proba.}(\underline{T}_M = n) \tag{24}$$

and

$$\underline{Q}_M(n) = <P_1(n/\Gamma_M)>_{\Gamma_M} = \text{Proba.}(\underline{T}_M > n) \tag{25}$$

which is a simpler notation for the probability $<P_1(n/\Gamma_M)>_{\Gamma_M}$ in Eq(17). We have

$$\underline{P}_M(n) = \underline{Q}_M(n\text{-}1) - \underline{Q}_M(n) \quad (n > 0) \tag{26}$$

The probabilities $P_M(n)$ and $Q_M(n)$ are known for most random walks and networks in d dimensions [13]. In order to relate them to $\underline{P}_M(n)$ and $\underline{Q}_M(n)$, we introduce their Laplace tranforms. $F(n)$ being some function of the integer n, its Laplace transform $\hat{F}(s)$ will be defined by

$$\hat{F}(s) = \sum_{n=0}^{\infty} F(n) e^{-sn} \tag{27}$$

Let us consider a trajectory Γ_M starting from M at time 0 and reaching A at the successive times $t_i^1, t_i^2, \ldots t_i^m$ between times 0 and n, with, according to the conventions of Section **4.1**

$$0 = t_i^0 < t_i^1 < t_i^2 < \ldots < t_i^m \le n$$

and

$$\tau_i^h = t_i^{h+1} - t_i^h \ge 1 \quad (i=0,\ldots m\text{-}1)$$

$$\tag{28}$$

$$\tau_i^m = n - t_i^{m-1} \ge 0$$

Using formulas (11) and (13), it is seen that the probability that B never reaches A under its activated form up to time n is

$$\underline{Q}_M(n) = <P_1(n/\Gamma_M)>_{\Gamma_M} =$$

$$Q_M(n) + \sum_{m\ge 1} \sum_{\tau 0 + \tau 1 + \ldots + \tau m = n} \alpha_0 \, P_M(\tau_0) \, p_{00}(\tau_1) P_A(\tau_1) \ldots p_{00}(\tau_{m-1}) P_A(\tau_{m-1}) \, Q_A(\tau_m) \tag{29}$$

with $\tau_i \ge 1$ for $i = 0,1,\ldots m\text{-}1$ and $\tau_m \ge 0$, and

$$p_{00}(\tau) = \alpha_0 + \alpha_1 \, e^{-\lambda\tau}$$

Thus the Laplace tranform of $\underline{Q}_M(n)$ is, using definition (26) and writing $P_M(0) = 0$

$$\hat{\underline{Q}}_M(s) = \hat{Q}_M(s) + \sum_{m \geq 1} \alpha_0 \hat{P}_M(s) \, [\alpha_0 \hat{P}_A(s) + \alpha_1 \hat{P}_A(s+\lambda)]^{m-1} \, \hat{Q}_A(s)$$

or

$$\hat{\underline{Q}}_M(s) = \hat{Q}_M(s) + \alpha_0 \hat{P}_M(s) \, [1 - \alpha_0 \hat{P}_A(s) - \alpha_1 \hat{P}_A(s+\lambda)]^{-1} \, \hat{Q}_A(s) \tag{30}$$

Now, by Eq(26) we have

$$\hat{P}_M(s) = e^{-s} \sum_{n \geq 1} Q_M(n-1) \, e^{-s(n-1)} - \sum_{n \geq 1} Q_M(n) \, e^{-sn} \ = 1 - (1-e^{-s}) \hat{Q}_M(s) \tag{31}$$

Furthermore, since Q_A and P_A refers to the return of particle B to A, starting from A, we can drop the index A if we assume that the medium is homogeneous, and write

$$\hat{Q}_A(s) \equiv \hat{Q}(s)$$

$$\hat{P}_A(s) \equiv \hat{P}(s) = 1 - (1 - e^{-s}) \, \hat{Q}(s)$$

and similar relations for $\hat{\underline{P}}_M(s), \hat{\underline{Q}}_M(s), \hat{\underline{P}}_A(s) \equiv \hat{\underline{P}}(s)$ and $\hat{\underline{Q}}_A(s) \equiv \hat{\underline{Q}}(s)$.

Using relation (31) repeatedly, it is easily found from Eq(30) that

$$\hat{\underline{P}}_M(s) = \hat{P}_M(s) \, [1 + \frac{\alpha_0}{\alpha_1} \frac{1 - \hat{P}(s)}{1 - \hat{P}(s+\lambda)}]^{-1} \tag{32}$$

which is the main result of this Section.

We notice that

$$\hat{P}(0) = \sum_n P(n) \equiv R \equiv 1 - S \tag{33}$$

is the probability that a B particle will ever return to its initial position, whereas S is the probability that it will never returns. Similarly, $P_M \equiv \hat{P}_M(0)$ and $\underline{P}_M \equiv \hat{\underline{P}}_M(0)$ are the probability that B, starting from M, will ever reach A, and the probability that it will ever destroy A, respectively. $\underline{P}_M$ is the *reaction (or annihilation) probability*.

Taking s = 0 in Eq(32), we see that

(i) *in one and 2 dimensions,* where $R = P_M = 1$, (32) implies $\underline{P}_M = P_M = 1$: the reaction occurs with probability 1, independently of the internal fluctuations of particle B.

(ii) *in three (or more) dimensions,* then $P_M = R = 1 - S$ [13], and (32) can be written

$$\frac{P_M}{\underline{P}_M} = 1 + \frac{\alpha_0}{\alpha_1} \frac{S}{1 - \hat{P}(\lambda)} \tag{34}$$

It should be noticed that if $\alpha_0 = 0$, (34) yields $\underline{P}_M = P_M$, as it should be, since the particle B is always activated then.

If the relaxation frequency λ tends to 0, which correspond to a frozen situation (the internal state of particle B does not change with time), $\hat{P}(\lambda) \rightarrow S$ and we obtain

$$\frac{P_M}{\underline{P}_M} = 1 + \frac{\alpha_0}{\alpha_1} = \frac{1}{\alpha_1} \tag{35}$$

so that we have $\underline{P}_M = \alpha_1 P_M$, which is again obvious since if B is activated the annihilation will occur at the first encounter of B with A

If on the contrary we consider the case of infinite relaxation frequencies λ_0 and λ_1, (which corresponds to particles B with permanent, partial activity mentioned in Section **3.3**), $\lambda \rightarrow \infty$, and from its definition, $\hat{P}(\lambda) \rightarrow 0$. Then

$$\frac{P_M}{\underline{P}_M} = 1 + \frac{\alpha_0}{\alpha_1} S \tag{35}$$

Furthermore, from its definition $\hat{P}(\lambda)$ is a decreasing function of λ, so that Eq(34) shows that $\underline{P}_M$ *is an increasing function of the internal relaxation frequency* : when λ increases from 0 to infinity, the reaction probability $\underline{P}_M$ increases from $\alpha_1 P_M$ to $\alpha_1 (\alpha_1 + \alpha_0 S)^{-1} P_M$. It should be pointed out that this result is independent of the number of dimensions d $\geq$ 3 and of the type of the lattice. A similar conclusion was obtain in a very different model of reaction with a fluctuating reaction potential [3].

4.3. Reaction kinetics

The kinetics of the annihilation reaction is defined by the survival probability $\psi(n)$, which is given by formulas (19) and (25)

$$\psi(n) \;=\; \exp(\,\text{-}\, b\, \underline{D}(n)) \tag{36}$$

with

$$\underline{D}(n) \;=\; \sum_{M}(1 - \underline{Q}_{M}(n)) \tag{37}$$

The reaction "constant" $\underline{k}(n)$ can be defined as

$$\underline{k}(n) = -\frac{1}{b}\frac{\partial}{\partial n}\ln\psi(n) = \underline{D}(n) - \underline{D}(n-1) \tag{38}$$

Using (37) and (26), we obtain

$$\underline{k}(n) = \sum_{M}(\underline{Q}_{M}(n-1) - \underline{Q}_{M}(n)) = \sum_{M}\underline{P}_{M}(n) \tag{39}$$

The Laplace transform of $k(n)$ is, by Eq(32)

$$\underline{\hat{k}}(s) = \hat{k}(s)\,[1 + \frac{\alpha_{0}}{\alpha_{1}}\,\frac{1-\hat{P}(s)}{1-\hat{P}(s+\lambda)}]^{-1} \tag{40}$$

where $\hat{k}(s)$ is the Laplace transform of the reaction constant $k(n)$ in the absence of internal fluctuations ($\alpha_{0} = 1$) given by

$$k(n) = \sum_{M} P_{M}(n) = D(n) - D(n-1) \tag{41}$$

which, according to the interpretation of Eq(20), is the average number of new distinct sites visited at the nth step of the random walk, or the probability that at its nth step the B particle visits a site which was never visited previously.

In one or two dimensions, it has already been remarked in Section **3** that $\hat{P}(s) \to 1$ if $s \to 0$, so that $\underline{\hat{k}}(s) \approx \hat{k}(s)$ if $s \to 0$. Thus $\underline{k}(n)$ *has the same asymptotic behavior as* $k(n)$ when $n \to \infty$. Neither k nor $\underline{k}$ tend to a a finite limit when $n \to \infty$, and the ordinary law of chemical kinetics do not apply.

In three (or more) dimensions, $1 - \hat{P}(s) \to S$ if $s \to 0,$ and $k(n) \to S$ if $n \to \infty.$
Then $\hat{k}(s) \sim S\, s^{-1}$ when $s \to 0,$ so that by (41)

$$\hat{\underline{k}}(s) \sim S\, s^{-1}\,[1 + \frac{\alpha_0}{\alpha_1}\,\frac{S}{1 - \hat{P}(\lambda)}]^{-1}$$

which implies that if $n \to \infty.$

$$k(n) \to \underline{k} = S\,[1 + \frac{\alpha_0}{\alpha_1}\,\frac{S}{1 - \hat{P}(\lambda)}]^{-1} \tag{42}$$

which can be written under the form

$$\frac{1}{\underline{k}} = \frac{1}{S} + \frac{\alpha_0}{\alpha_1}\,\frac{1}{1 - \hat{P}(\lambda)} \tag{43}$$

We recall that in (43), $\hat{P}(s)$ in the Laplace transform of probability that the first return of B to its initial position takes place at step n exactly (which is independent from the initial position in an infinite homogeneous medium), and $S = 1 - \hat{P}(0)$ is the total probability that B never returns to its initial position.

When $\lambda \to \infty$ we recover the inverse addition law (11) for the rate constants which was obtained in Section **3.2** for particle B with permanent partial activity.

As it was observed for the reaction probability, (43) shows that *the reaction constant* $\underline{k}$ *is an increasing function of the relaxation frequency* λ : when λ increases from 0 to ∞, $\underline{k}$ increases from $\alpha_1 S$ to $\alpha_1 S (\alpha_1 + \alpha_0 S)^{-1}.$

This result may be understood intuitively, since if a particle B reaches A under inert form, it has a large probability to return to A after a few steps. Then, if the relaxation frequency of B is low, A will have a high probability to be still inert at these returns and the reaction rate will be low, whereas increasing the relaxation frequency should increase the probability that the reaction is completed.

5. Conclusion

We have studied the effect of the random activation and desactivation of a reagent on the kinetics of a diffusion controlled annihilation reaction. Although rather specific, this mechanism can also mimic other elementary reactions, or represent interaction processes in different fields, such as population dynamics. This problem is thus of interest in many circumstances.

It has been shown that the activation-desactivation process of the "scavengers", or mobile molecules B, does not amounts to replacing the concentration b of these molecules by the average concentration $\alpha_1 b$ of their activated form, except in the trivial case of a zero relaxation frequency λ, corresponding to an infinite correlation time. We have proved that decreasing this correlation time always increases the overall reaction constant $\underline{k}$, which was already the case in very different models of fluctuating reactions.

However, our conclusions are obviously limited by the restrictive conditions adopted in this work, and further advances are necessary to address realistic problems. First, more general reactions should be considered, such as reactions producing new species, or auto-catalytic reactions. Second, spatial correlations could be taken into account as well as time correlations. Finally, it should be possible to extend the present theory to a time continuous description.. Such developments are in progress.

References

[1]   For the general kinetic theory of chemical reactions, see for instance:
R. Kapral, "Kinetic theory of chemical reactions in liquids", Adv. Chem. Phys. **48**, 71 (1981)
J Hynes, "The theory of reactions in solution" in "Theory of Chemical Reaction Dynamics" vol **4**, ed. M. Baer CRC Press (1985)
P. Hanggi, P. Talkner and M. Borkovec "Reaction rate theory: Fifty years after Kramers", Rev. Mod. Phys. **62**, 252 (1990)

[2]   M. Moreau, B. Gaveau, M. Frankowicz and A. Perera, "Diffusion and reaction in a model fluctuating medium" in "Far from Equilibrium Dynamics of Chemical Systems", eds. A. Popielawski and J . Gorecki, World Scientific, Singapore (1993)
C.R. Doering and J.C. Gadoua, Phys. Rev. Lett. **69**, 2318 (1993)

[3]   B. Gaveau and M. Moreau, Phys. Lett. A **192**, 364 (1994)
B. Gaveau and M. Moreau, Internat. J. Bifurcat. Chaos **4**, 1297 (1994)
B. Gaveau and M. Moreau, Phys. Lett. A **211**, 331 (1996)

[4]   M. O Caceres, CE. Budde and M. A. Re, Phys. Rev. E **52**, 3462 (1995)

[5]   M. F. Perutz and F.S Mathews, J. Mol. Biology **21**, 199 (1966)

[6]   O. Bénichou, M. Moreau and G. Oshanin, Phys. Rev. E **61**, 3388 (2001)

[7]   P. Pechukas and P. Hanggi, Phys. Rev. Lett. **73**, 2772 (1994)

[8]   A Blumen, G. Zumofen and J. Klafter, Phys. Rev. B **30**, 5379 (1984)
A Blumen, J. Klafter and G. Zumofen, in Optical Spectroscopy of Glasses, ed. I. Zschokke, Reidel Publ, Dordrecht (1986)

[9]   S F. Burlatsky and A. A. Ovchninnikov, Sv. Phys. JETP **65**, 908 (1987)

[10]   A. Szabo, R Zwanzig and N. Agmon, Phys. Rev Lett. **61**, 2496 (1989)

[11]   S Redner and K Kang, J. Phys. A 1, 451 (1984)

[12]   see for instance
C. W. Gardiner, "Handbook for Stochastic Methods" Springer, Berlin (1983)

[13]   B. D. Hughes, Random Walks and Random Environments, Oxford Science
       Publishers, Oxford (1995)

[14]   F. Collins and G. Kimball, J. Colloid Sc. **4**, 425 (1949)

[15]   S.F. Burlatsky, A.A. Ovchinnikov and G. Oshanin, Sov. Phys. JETP **68**, 1153
       (1989)

[16]   B. Gaveau, J. Hynes, R. Kapral and M. Moreau,, J. Stat. Phys. **65**, 879 and
       895 (1989)

[17]   G. Oshanin, M. Moreau and S. F. Burlatsky, Adv. Colloid and Interface Sc.
       49, 118 (1994)

Classical Dynamics of Excitations of Bose Condensates in Anisotropic Traps

Robert Graham

Fachbereich Physik, Universität-Gesamthochschule Essen,
45117 Essen, Germany

This lecture discusses some aspects of the dynamics of the collective and single-particle excitations at zero temperature of Bose-Einstein condensates of alkali-vapors in magnetic traps. We shall discuss those aspects which can be understood by taking the short-wavelength or 'eikonal' limit of the excitations. Trapped Bose-Einstein condensates can be excited experimentally either directly via periodic modulations of the trap potential or by scattering light off the condensate. My discussion here will closely follow some theoretical work published in [1–3] that has recently been done in collaboration with Andras Csordas and Peter Szepfalusy at the Research Institute for solid State Physics and Optics in Budapest, Hungary and with Martin Fliesser at the University of Essen, Germany.

I. INTRODUCTION

The achievement of Bose-Einstein condensation of clouds of magnetically trapped alkali atoms [4–6] by evaporative cooling to temperatures in the 100 nano-Kelvin regime has revived the interest in the physics of weakly interacting Bose-condensed gases. As is well-known the bulk properties of such gases are rather well described by ideal gases of quasi-particles, first derived microscopically by Bogoliubov [7], which are collisionless phonons at long wavelength and approach free particles at short wavelength. In the case of trapped condensates of repelling atoms, to which we confine our discussion here, the quasiparticle description retains its usefulness. However, trapped condensates are spatially inhomogeneous. They form around the minima of the trapping potential and can form rather sharp surfaces. The thickness of the surface layer, given by the healing length [8], can become very small compared to the radius of the condensate. Quasiparticles may therefore leave the condensate and reenter it after being reflected back by the trapping potential. Therefore their dynamics are more complex than in a homogeneous system. In Bogoliubov's approach single quasiparticles are described by wavefunctions, which are solutions of a set of linear wave equations. This is in complete analogy to the description of single particles in quantum mechanics by Schrödinger's equation. We know that it is extremely fruitful to examine the classical limit of the Schrödinger equation, which gives all of classical physics. In the same spirit we can examine the classical limit of the wave equations for the quasiparticles. In spatially homogeneous condensates this gives a simple theory of free quasiparticles with conserved momentum which are distinguished from free particles merely by their unusual relation between energy and momentum, and hence also between velocity and momentum.

In the case of trapped inhomogeneous condensates recent work has shown [1] that the classical limit of the dynamics of quasiparticles becomes more interesting because they experience forces both from the trap and from the condensate. In the case of isotropic traps, and hence also isotropic condensates, angular momentum conservation ensures the integrability of the quasiparticle dynamics. This can be used to construct WKB solutions of Bogoliubov's wave equations for this case [9]. In the experimentally more relevant case of axially symmetric traps integrability of the quasiparticle dynamics is lost and numerical studies [2,1] indeed have shown a generally mixed phase-space in this case. Detailed analytical work was performed for quasiparticle energies E much smaller than the chemical potential μ. Here two integrable limits were identified: (i) the phonon limit, where the phonon-like quasiparticle is confined to the interior of the condensate and is specularly reflected back when it strikes the surface, and (ii) a surface-particle limit, where the motion of the single-atom-like quasiparticle consists of rapid small-amplitude oscillations between the repelling main bulk of the condensate and the potential wall of the trap and a slow secular motion along the surface of the condensate. The numerical examination of the quasiparticle dynamics at larger energies (and for the trap anisotropy of the experiment [4]) revealed [1] a strong chaotic component at $E = \mu$ and a curious 'quasi-integrable' regime at $E \gg \mu$, where the dynamics in phase-space clusters around the tori corresponding to single atom motion in the trapping potential, however with small-scale chaos superimposed.

In subsequent work [3] this large energy regime was studied more closely, both analytically and numerically. To discuss the onset of chaos coming from large energies a classical perturbation theory for the quasiparticle dynamics at large energy was developed. The influence of the thickness of the surface on the appearance of chaos was also examined. In the Thomas-Fermi approximation [8] this thickness is neglected, leading to a spatially discontinuous effective force on the quasiparticle. This violates assumptions of the KAM theorem and turns out to be the reason that noticeable small-scale chaos survives even at very large energies, where the system should be close to the integrable limit of independent atoms in the trap. Simulations with different boundary layers substantiated this hypothesis.

O. Descalzi et al. (eds.), Instabilities and Nonequilibrium Structures VII & VIII, 23–41.
© 2004 *Kluwer Academic Publishers. Printed in the Netherlands.*

24

For a condensate with boundary layer the transition to chaos as the energy is lowered to values comparable with the chemical potential was studied in [3].

In the present lecture I intend to give an overview of this work on the *classical* limit of the dynamics of the excitations of trapped Bose-Einstein condensates. A lot of numerical [13] and analytical [14] work has also been done on the quantum mechanics of the excitations, in paticular also analytical work on the hydrodynamic regime in anisotropic traps [2,16,19,20] which I shall not be able to discuss here, however.

II. DESCRIPTION OF A WEAKLY INTERACTING BOSE-EINSTEIN CONDENSED GAS

The experimental realization of Bose-condensates of atoms harmonically bound in magnetic traps [4–6] call for a space-dependent version of Bogoliubov's theory, or some modification thereof. Such a theory proceeds by splitting the field operator $\hat{\psi}(\boldsymbol{x})$ and its adjoint in a c-number part $\psi_0(\boldsymbol{x})$ and a residual operator $\hat{\varphi}(\boldsymbol{x})$,

$$\hat{\psi}(\boldsymbol{x}) = \psi_0(\boldsymbol{x}) + \hat{\varphi}(\boldsymbol{x}) \tag{1}$$

and an accompanying decomposition of the Hamiltonian in terms of 0, 1, 2, 3, 4 order in $\hat{\varphi}$, $\hat{\varphi}^+$. The term of 1 order in $\hat{\varphi}$, $\hat{\varphi}^+$ is made to vanish by choosing $\psi_0(\boldsymbol{x})$ to satisfy the time-independent Gross-Pitaevskii equation [10], which at low temperatures, takes the form

$$-\frac{\hbar^2}{2m}\nabla^2\psi_0(\boldsymbol{x}) + (U(\boldsymbol{x}) - \mu)\psi_0(\boldsymbol{x}) + V_0|\psi_0(\boldsymbol{x})|^2\psi_0(\boldsymbol{x}) = 0\,, \tag{2}$$

with the normalization $\int |\psi_0|^2 d^3x = N_0$. Here

$$U(\boldsymbol{x}) = \frac{m}{2}(\omega_x^2 x^2 + \omega_y^2 y^2 + \omega_z^2 z^2) \tag{3}$$

is the generally anisotropic harmonic trap potential,

$$V_0 = \frac{4\pi\hbar^2 a}{m} \tag{4}$$

is the strength of the pseudo-potential replacing the true two-particle potential at low energies, with the s-wave scattering length a, which is here assumed to be positive.

For $(N_0 a/d_0) \gg 1$, where $d_0 = \sqrt{\hbar/m\bar{\omega}}$, $\bar{\omega} = (\omega_x\omega_y\omega_z)^{1/3}$, the solution to the Gross-Pitaevskii equation can be determined in the Thomas-Fermi approximation [8] by neglecting the kinetic-energy term

$$|\psi_0|^2 = \frac{\mu - U(\boldsymbol{x})}{V_0}\Theta(\mu - U(\boldsymbol{x}))\,. \tag{5}$$

In the following we shall choose ψ_0 as real and positive, which is without restriction of generality as long as we confine our attention to a non-rotating condensate at rest.. The chemical potential is determined from the normalization. The next step is the diagonalization of that part of H, which is a quadratic form in $\hat{\varphi}$, $\hat{\varphi}^+$, by a Bogoliubov transformation to quasi-particles

$$\hat{\varphi}(\boldsymbol{x}) = \sum_j \left(u_j(\boldsymbol{x})\hat{\alpha}_j + v_j^*(\boldsymbol{x})\hat{\alpha}_j^+\right) \tag{6}$$

with

$$\int d^3x \left(|u_j(\boldsymbol{x})|^2 - |v_j(\boldsymbol{x})|^2\right) = 1 \tag{7}$$

and

$$[\alpha_j, \alpha_{j'}] = 0 = [\alpha_j^+, \alpha_{j'}^+]$$
$$[\alpha_j, \alpha_{j'}^+] = \delta_{jj'}\,. \tag{8}$$

The second-order part of H is diagonalized by this transformation, if $U_j(\boldsymbol{x})$ and $V_j(\boldsymbol{x})$ satisfy the Bogoliubov-Degennes-Fetter equations [12]

$$\begin{pmatrix} \hat{H}_{\mathrm{HF}} & +K(\boldsymbol{x}) \\ +K(\boldsymbol{x}) & \hat{H}_{\mathrm{HF}} \end{pmatrix} \begin{pmatrix} u_j \\ v_j \end{pmatrix} = E_j \begin{pmatrix} u_j \\ -v_j \end{pmatrix} \tag{9}$$

with the Hartree-Fock Hamiltonian

$$\hat{H}_{\mathrm{HF}} = -\frac{\hbar^2}{2m}\nabla^2 + U(\boldsymbol{x}) - \mu + 2V_0|\psi_0(\boldsymbol{x})|^2 \tag{10}$$

and the coupling term

$$K(\boldsymbol{x}) = V_0|\psi_0(\boldsymbol{x})|^2 \tag{11}$$

between the two components $u_j(\boldsymbol{x})$, $v_j(\boldsymbol{x})$ of a quasi-particle wave-function. Because of the different signs of the u_j, v_j components on the right-hand side, they play the role of particle and anti-particle components of the complete wave-function. As the equations are symmetric under the particle-antiparticle transformation $E_j \to -E_j$, $u_j \to v_j^*$, $v_j \to u_j^*$ we may define E_j to be non-negative without restriction of generality. Various numerical [13] and approximate analytical [14] treatments of these equations are available in the literature.

In the present paper we wish to study the classical and semi-classical limit of the center-of-mass motion of the quasi-particles. In order to discuss the dynamics rather than the eigenstates of the quasi-particles, it is useful to introduce time-dependent wave functions via

$$\begin{pmatrix} u(t) \\ v(t) \end{pmatrix} = \sum_j c_j \begin{pmatrix} u_j \\ v_j \end{pmatrix} e^{-iE_j t/\hbar} \tag{12}$$

with arbitrary coefficients c_j. They satisfy the time-dependent Schrödinger equation

$$i\hbar\frac{\partial}{\partial t}\begin{pmatrix} u(t) \\ -v(t) \end{pmatrix} = \begin{pmatrix} \hat{H}_{\mathrm{HF}} & +K \\ +K & \hat{H}_{\mathrm{HF}} \end{pmatrix} \begin{pmatrix} u(t) \\ v(t) \end{pmatrix}. \tag{13}$$

For large energies E_j, $E_j \gg \mu$, the classical motion can be interpreted as the center of mass-motion of quasi-particle wave packets. For small energies E_j, $E_j \ll \mu$, such a straightforward physical interpretation of the classical quasi-particle dynamics is no longer possible. However, even in this regime, there is still a close mathematical relation between the classical and the quantum dynamics, as the classical trajectories are the characteristics of the quantum mechanical wave-equation. This is made explicit by the derivation of the classical dynamics as a limit of the Schrödinger equation via the Hamilton-Jacobi equation. The Hamilton-Jacobi equation corresponding to eq. (13) is obtained by the asymptotic ansatz for $\hbar \to 0$

$$\begin{pmatrix} u(\boldsymbol{x},t) \\ v(\boldsymbol{x},t) \end{pmatrix} \simeq \begin{pmatrix} a_0(\boldsymbol{x},t) & + & 0(\hbar) \\ b_0(\boldsymbol{x},t) & + & 0(\hbar) \end{pmatrix} e^{iS(\boldsymbol{x},t)/\hbar} \tag{14}$$

with $\int d^3x(|a_0(\boldsymbol{x},t)|^2 - |b_0(\boldsymbol{x},t)|^2) = 1$. It reduces eq. (13) to the form, to zeroth order in $\hbar$,

$$\begin{pmatrix} \epsilon_{\mathrm{HF}} + \frac{\partial S}{\partial t} & +K \\ +K & \epsilon_{\mathrm{HF}} - \frac{\partial S}{\partial t} \end{pmatrix} \begin{pmatrix} a_0 \\ b_0 \end{pmatrix} = 0. \tag{15}$$

Here

$$\epsilon_{\mathrm{HF}} = \frac{p^2}{2m} + U(\boldsymbol{x}) - \mu + 2V_0|\psi_0(\boldsymbol{x})|^2. \tag{16}$$

We may restrict to $-E = \frac{\partial S}{\partial t} \le 0$ in accordance with our restriction on E. To first order in $\hbar$ we obtain

$$\frac{\partial}{\partial t}\begin{pmatrix} a_0 \\ -b_0 \end{pmatrix} + \frac{1}{2m}\boldsymbol{\nabla}\cdot\left((\boldsymbol{\nabla}S)\begin{pmatrix} a_0 \\ b_0 \end{pmatrix}\right) + \frac{1}{2m}\boldsymbol{\nabla}S\cdot\boldsymbol{\nabla}\begin{pmatrix} a_0 \\ b_0 \end{pmatrix} = \frac{i}{\hbar}\begin{pmatrix} \epsilon_{\mathrm{HF}} + \frac{\partial S}{\partial t} & +K \\ +K & \epsilon_{\mathrm{HF}} - \frac{\partial S}{\partial t} \end{pmatrix} \begin{pmatrix} a_1 \\ b_1 \end{pmatrix}. \tag{17}$$

Here $\begin{pmatrix} a_1 \\ b_1 \end{pmatrix}$ are the $0(\hbar)$-components of the amplitudes in (14). These will exist, and the expansion will be well-defined, only if the left-hand side of (17) is orthogonal on the kernel $\begin{pmatrix} a_0 \\ b_0 \end{pmatrix}$ of the matrix in (15), which also appears on the right-hand side of (17). This condition gives rise to the conservation law

$$\frac{\partial}{\partial t}(|a_0|^2 - |b_0|^2) + \frac{1}{2m}\boldsymbol{\nabla}\cdot((|a_0|^2 + |b_0|^2)\boldsymbol{\nabla}S) = 0 \tag{18}$$

which ensures that the normalization condition

$$\int d^3x(|a_0|^2 - |b_0|^2) = 1$$

is consistent with the classical dynamics and represents the classical limit of the continuity equation following from (18) [9]. The zeroth order equation has a nontrivial solution only if the determinant condition

$$\left(\frac{\partial S}{\partial t}\right)^2 = \epsilon_{\mathrm{HF}}^2 - K^2 \tag{19}$$

is satisfied, which, observing our sign convention for $\partial S/\partial t$, gives the time-dependent Hamilton-Jacobi equation

$$\frac{\partial S(\boldsymbol{x},t)}{\partial t} + H\left(\boldsymbol{x}, \frac{\partial S}{\partial \boldsymbol{x}}\right) = 0 \tag{20}$$

with the classical Hamiltonian

$$H(\boldsymbol{x},\boldsymbol{p}) = \sqrt{\epsilon_{\mathrm{HF}}^2(\boldsymbol{x},\boldsymbol{p}) - K(\boldsymbol{x})^2} \,. \tag{21}$$

The time-independent Hamilton Jacobi equation results from the separation

$$S(\boldsymbol{x},t) = S(\boldsymbol{x}) - Et \tag{22}$$

and reads

$$H(\boldsymbol{x}, \frac{\partial S}{\partial \boldsymbol{x}}) = E \,. \tag{23}$$

In the following sections we analyze the classical dynamics described by the Hamiltonian (21,23).

III. CLASSICAL QUASI-PARTICLE DYNAMICS

For the case of isotropic harmonic traps angular momentum is conserved and the quasi-particle dynamics is integrable and separable in spherical coordinates. This case is discussed in [9], where it is made the basis of a semiclassical quantization procedure. In the following we concentrate on the analysis of the case of anisotropic harmonic traps in the limit where the Thomas-Fermi approximation applies. In the present section we shall assume cylindrical symmetry of the trap

$$U(\boldsymbol{x}) = \frac{m\omega_0^2}{2}(x^2 + y^2) + \frac{m\omega_z^2}{2}z^2 \tag{24}$$

In the experiment [4] $\omega_z > \omega_0$, namely, $(\omega_z/\omega_0)^2 \approx 8$. As the parameter denoting the anisotropy of the potential we introduce ϵ by $\epsilon^2 = 1 - (\omega_0/\omega_z)^2$, which is the numerical excentricity of the Thomas-Fermi surface $\mu = U(\boldsymbol{x})$, a rotational symmetric ellipsoid. This two-dimensional surface is the boundary of the condensate.

Our problem has a characteristic energy, namely the chemical potential. Thus, the second relevant parameter of the classical motion is the ratio E/μ. We note that measuring the energy in units of μ, coordinates, momenta and time in units of

$$r_0 = \sqrt{\frac{2\mu}{m\omega_0^2}} \quad , \quad p_0 = \sqrt{2m\mu} \quad , \quad t_0 = \omega_0^{-1} \tag{25}$$

respectively, the dimensionless Hamiltonian can be put in a form, which depends only on the anisotropy parameter ϵ. This shows that condensates with the same anisotropy but with different chemical potential behave similarly in the classical description, if the physical quantities are scaled appropriately.

In the isotropic case $\epsilon = 0$ the classical dynamics is completely integrable. As three independent constants of motion we can choose the energy E, the modulus of the angular momentum and its z-component L_z. As we keep rotational symmetry around the z-axis in the anisotropic case $\epsilon \neq 0$ the z-component of the angular momentum L_z and of course the energy are still conserved quantities, whereas the total angular momentum is not conserved. Thus, in the following we shall investigate the classical behaviour of this three degrees of freedom system depending on the two constants of motion E and L_z, and we address the question wether the dynamics are integrable or chaotic.

Let us introduce the usual cylindrical coordinates $\rho = \sqrt{x^2 + y^2}$, z and ϕ. Because of the rotational symmetry around the z-axes the angle ϕ is a cyclic variable. In cylindrical variables the Hamiltonian has merely two degrees of freedom ρ and z, L_z just enters as a parameter.

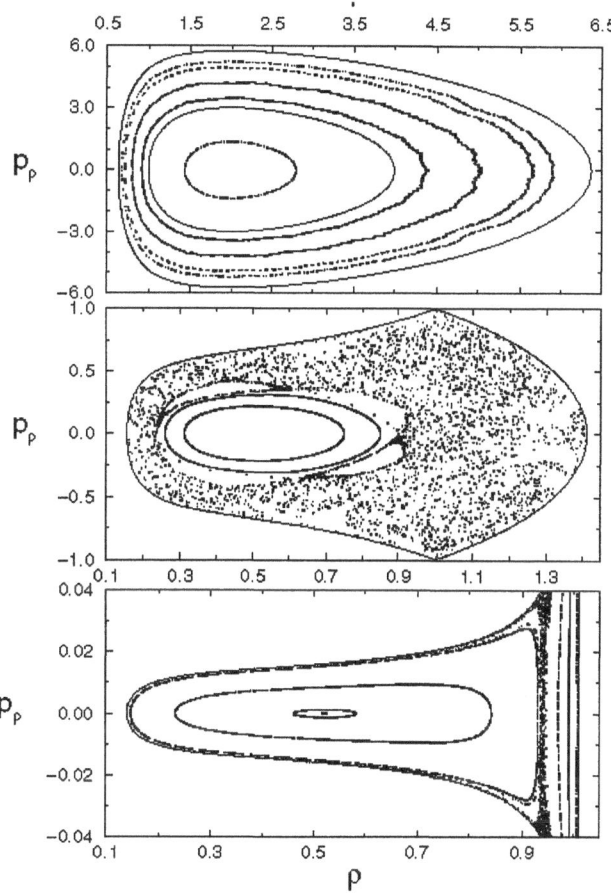

FIG. 1. Poincaré sections of the dynamics of the Bogoliubov Hamiltonian (21) in cylindrical coordinates for the different energies (from top to bottom) $E/\mu = 40$ (a), 1 (b) and 0.02 (c). The cut is taken at $z = 0$ and diplayed in the variables ρ, p_ρ in units of $(2\mu/m\omega_0^2)^{1/2}$, $(2m\mu)^{1/2}$, respectively. The anisotropy is chosen as $\omega_z/\omega_0 = \sqrt{8}$, the angular momentum was fixed as $\omega_0 L_z/E = 0.2$. (From [1])

28

Certain conditions have to be satisfied, as can be seen from the Hamiltonian, in the region outside the condensate. For $E > \mu$ the condition $E + \mu > \omega_0 L_z$ has to be guaranteed, for $E < \mu$ we must have $E > (\omega_0 L_z)^2/4\mu$.

The dynamics of this two dimensional system we can visualize by Poincaré cuts, see Fig.1. For different energies we observe different dynamical behaviour. For $E > (\omega_0 L_z)^2/4\mu > \mu$ two different kinds of trajectories can occur typically. If the repulsive effective potential in ρ-direction due to the angular momentum L_z is strong enough, the particle cannot enter the condensate and is only moving in the harmonic potential of the external trap. The motion in an anisotropic harmonic potential is completely integrable, as a third constant of motion we can choose the energy in the z-degree of freedom $E_z = p_z^2/2m + m\omega_z z^2/2$. These trajectories, which are not perturbed by the condensate, can be seen as the integrable tori around the fixed point of the Poincaré map in the centre of Fig.1a, which is the periodic orbit moving only in z and ϕ directions. If the particle enters the condensate, E_z is no more a conserved quantity. Nevertheless for energies large compared to the chemical potential also those trajectories are still quite similar to unperturbed motion. Typically the trajectories are confined to thin stochastic layers separated by each other by integrable tori. No Arnold diffusion occurs, as usually for a system of two, not of three degrees of freedom. At high energies the system behaves quasi integrable. The influence of the condensate can be taken as a small perturbation to the motion in the external potential.

For energies in the range $10 > E/\mu > 0.1$ (Fig.1b) we typically observe a mixed phase space. The fixed point is now inside the condensate, but does not loose its stability. The detailed structure depends on the parameters chosen. Already for small anisotropy ($\epsilon^2 = 0.2$) a relevant part of phase space can be chaotic. This shows that for energies of the order of the chemical potential the isotropic case with its integrable dynamics is an exceptional rather than a typical situation. If $E < \mu$ all trajectories move inside and outside the condensate.

For energies small compared to the chemical potential $E < 0.1\mu$ (Fig.1c) the chaotic part of phase space decreases again and is restricted to a thin layer separating and surrounding two regular islands, corresponding to two stable fixed points separated by an unstable one. Most orbits seem to lie on integrable tori. This suggests that the system has an integrable regime in the limit of small energies.

This limit corresponds to the hydrodynamical regime [11] investigated in several contexts. In a bulk case, when there is no external potential $U(\mathbf{x})$ the lowest lying excitations are phonons with linear wave-number dependence.

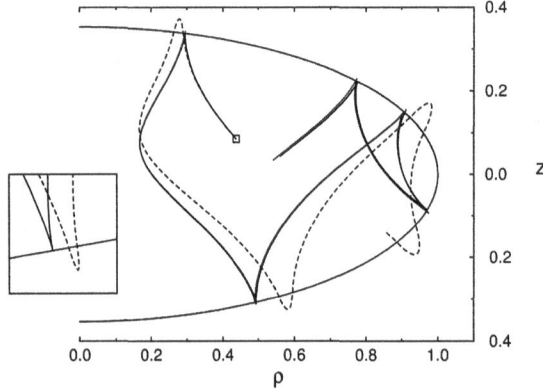

FIG. 2. Trajectories in coordinate space of the Bogoliubov dynamics of (21) starting from the same point and in the same direction for different energies $E/\mu = 0.1$ (dashed line), 0.01 (solid) and 10^{-6}(dotted). (From [1])

Numerically we have found that tending with the energy to zero, keeping μ fixed the range of the classical motion outside the condensate for trajectories starting inside is getting smaller and smaller and in the limit the motion is confined to the region inside of the Thomas-Fermi surface. Starting trajectories from the same point inside the condensate under the same direction and changing only the modulus of Cartesian-momentum we have found that they differ from each other only in a thin region near the boundary whose width scales with the energy. Lowering the modulus of the initial momentum to zero they tend to a well-defined limiting trajectory. This can be clearly seen in Fig.2. In the isotropic case this is the limit $E/\mu, \omega_0 L/\mu \to 0$, keeping the ratio L/E fixed. In the following section this 'hydrodynamic regime' will be studied in detail for anisotropic traps. However, it will turn out that in traps there exists a second low-energy regime, which for isotropic traps is defined by $E/\mu \to 0$ with $E - (\omega_0 L)^2/4\mu \ll E$, where the quasi-particles are single-particle like excitations confined to a narrow layer around the surface of the condensate. This low-energy Hartree-Fock regime will be discussed in detail in section IV, also for anisotropic traps, together with the usual high-energy $(E \gg \mu)$ Hartree-Fock regime.

IV. QUASI-PARTICLE DYNAMICS IN THE HYDRODYNAMIC REGIME

A. Hydrodynamic Hamiltonian

If there exist limiting trajectories for different initial conditions there should exist limiting dynamics described by some limiting Hamilton-function. Inside the condensate the Bogoliubov Hamiltonian can be written as

$$H(\boldsymbol{p}, \boldsymbol{x}) = \sqrt{\epsilon_{kin}(\boldsymbol{p})(\epsilon_{kin}(\boldsymbol{p}) + 2K(\boldsymbol{x}))}, \tag{26}$$

where $\epsilon_{kin}(\boldsymbol{p}) = \boldsymbol{p}^2/2m$. For small energies $K(\boldsymbol{x})$ is much bigger than ϵ_{kin} everywhere except in a small region near the boundary. This suggests that the approximant of the Hamilton function (26) can be obtained by neglecting the kinetic energy square

$$H_{hyd}(\boldsymbol{p}, \boldsymbol{x}) = \sqrt{2\,\epsilon_{kin}(\boldsymbol{p})\,K(\boldsymbol{x})} \tag{27}$$

for describing the motion in the hydrodynamical regime. This approximate Hamiltonian is in accordance with the bulk case, when $K(\boldsymbol{x}) = \mu$ should be taken in (27) in order to obtain the linear phonon spectra from the Bogoliubov dispersion relation.

This Hamiltonian is meaningful only inside the condensate and only near the boundary of the condensate the full Hamiltonian (26) differs from this approximate one. On the Thomas-Fermi surface the full Hamilton function gives definite values for the Cartesian momenta, whereas according to H_{hyd} they become infinite. Following the trajectories of H_{hyd} in the isotropic case the angular momentum conservation requires that the tangential component of the momentum remains finite even though the absolute value of the momentum diverges like $K^{-1/2}$. Therefore each trajectory hits the boundary orthogonally and is reflected back orthogonally without change in the tagential component of the momentum. As this local rule is independent of the global symmetry of the trap potential it must hold also in the anisotropic case.

The Hamiltonian (27) has some further unusual features. It is not of the usual form but is a homogeneous first order function of the momenta. The consequence is that with the same initial value $\boldsymbol{x}(t = 0)$ and with the same direction of the initial momenta the orbit $\boldsymbol{x}(t)$ is the same independently of the energy. A constraint follows from the canonical equations of motion, namely

$$m\dot{\boldsymbol{x}} \cdot \dot{\boldsymbol{x}} = \mu - U(\boldsymbol{x}), \tag{28}$$

relating the velocities and the coordinates. Thus one cannot choose the initial point and the velocity independently. Furthermore, due to this constraint one cannot express the three velocities in terms of the momenta, i.e., one cannot do the inverse Legendre transformation in the usual way to derive the Lagrangian. From (28) it is clearly seen that despite of the divergence of momenta on the boundary of the condensate the velocities even tend to zero here.

B. Anisotropic case with cylindrical symmetry

The case of a trap with axial or cylindrical symmetry is the experimentally most relevant one. In Poincaré cuts of the full dynamics we have seen regular behaviour for small energies. Therefore one can expect that the classical motion given by the approximate Hamiltonian is fully integrable. To show this let us introduce new coordinates, namely the cylindrical elliptical coordinates ξ, η given by

$$\rho = \sigma\sqrt{(\xi^2 + 1)(1 - \eta^2)} \quad , \quad z = \sigma\xi\eta \tag{29}$$

which are orthogonal coordinates. Surfaces of constant ξ are confocal ellipsoids with foci at a distance σ in ρ direction, surfaces of constant η are confocal hyperboloids with the same foci. For σ, the parameter of the transformation, we take the foci of the Thomas-Fermi ellipsoid, $\sigma = \epsilon(2\mu/m\omega_0^2)^{1/2}$ for $\omega_z > \omega_0$. For $\omega_0 > \omega_z$ one has to change $\xi^2 + 1$

to $\xi^2 - 1$ and take $\sigma = \epsilon(2\mu/m\omega_z^2)^{1/2}$. In the following we consider only the first case (29), in the second case the analysis proceeds similarly. ξ can take any value in the range $[0, (1/\epsilon^2 - 1)^{1/2}]$. The limiting case $\xi = (1/\epsilon^2 - 1)^{1/2}$ describes the Thomas-Fermi ellipsoid. η can be in the range $[-1, 1]$. Making the point transformation from cylindrical to cylindrical elliptical coordinates the momenta transform as

$$p_\rho = \frac{1}{\sigma} \frac{1}{\xi^2 + \eta^2} \sqrt{(\xi^2 + 1)(1 - \eta^2)}(\xi p_\xi - \eta p_\eta) \quad ,$$

$$p_z = \frac{1}{\sigma} \frac{1}{\xi^2 + \eta^2}((\xi^2 + 1)\eta p_\xi + (1 - \eta^2)\xi p_\eta) \quad . \tag{30}$$

The Hamiltonian (27) in cylindrical elliptical coordinates is

$$H_{hyd}^2 = \frac{\omega_z^2}{2\epsilon^2} \frac{(1 - \epsilon^2(\xi^2 + 1))(1 - \epsilon^2(1 - \eta^2))}{\xi^2 + \eta^2}$$

$$\times \left((\xi^2 + 1)p_\xi^2 + (1 - \eta^2)p_\eta^2 + (\frac{1}{1 - \eta^2} - \frac{1}{\xi^2 + 1})p_\phi^2 \right). \tag{31}$$

Taking the energy E and $p_\phi = L_z$ as constants one can write down the Hamilton-Jacobi equation for ξ and η, which is separable in these coordinates. Thus the problem is fully integrable. Introducing a separation constant $B > 0$ the two separated Hamilton-Jacobi equations are

$$(\xi^2 + 1)\left(\frac{dS_\xi}{d\xi}\right)^2 - \frac{L_z^2}{\xi^2 + 1} - \frac{2E^2}{\omega_z^2} \frac{1}{1 - \epsilon^2(\xi^2 + 1)} = -B$$

$$(1 - \eta^2)\left(\frac{dS_\eta}{d\eta}\right)^2 + \frac{L_z^2}{1 - \eta^2} + \frac{2E^2}{\omega_z^2} \frac{1}{1 - \epsilon^2(1 - \eta^2)} = B \quad . \tag{32}$$

Combining these two equations one gets for the separation constant B the phase-space function

$$B = \frac{1}{\epsilon^2} \frac{1}{\xi^2 + \eta^2} \left[(1 - \epsilon^2(\xi^2 + 1))(\xi^2 + 1)p_\xi^2 \right.$$

$$+ (1 - \epsilon^2(1 - \eta^2))(1 - \eta^2)p_\eta^2 + \left(\frac{1}{1 - \eta^2} - \frac{1}{\xi^2 + 1}\right)p_\phi^2 \Bigg]$$

$$= \frac{\sigma^2}{\epsilon^2} \left[p_x^2 + p_y^2 + (1 - \epsilon^2)p_z^2 \right] - (xp_x + yp_y + zp_z)^2 \quad . \tag{33}$$

in elliptical and cartesian coordinates respectively. This is the third independent constant of motion in addition to the energy E and L_z. This can be checked directly, using the equations of motion for the time derivatives of B. Similarly to the isotropic case conservation of E and B require that trajectories hit the boundary orthogonally, because the momenta there diverge. In the isotropic limit $\sigma \to 0$ the elliptical coordinates become singular, and therefore it is more instructive to see this limit in cartesian coordinates. In this limit σ/ϵ is the Thomas-Fermi radius, and B has the simple meaning

$$B = \frac{2E^2}{\omega_0^2} + L^2 \quad . \tag{34}$$

The existence of the three independent constants of motion E, L_z and B explains the integrable motion generated by H_{hyd} and therefore the almost integrable situation found numerically in the motion generated by the total Hamiltonian (26) in the small energy and small angular momentum region. We notice that two kinds of trajectories can occur in this regime. From (32) we can determine the turning points in ξ and η. In ξ-direction all the trajectories reach the Thomas-Fermi surface and are reflected back there. If the condition

$$B > B^* = \frac{2E^2}{\omega_0^2} + L_z^2 \,, \tag{35}$$

is satisfied, there is an inner turning point in ξ-direction and η takes a range $[-\eta_{max}, \eta_{max}]$. For an example see Fig.3a.

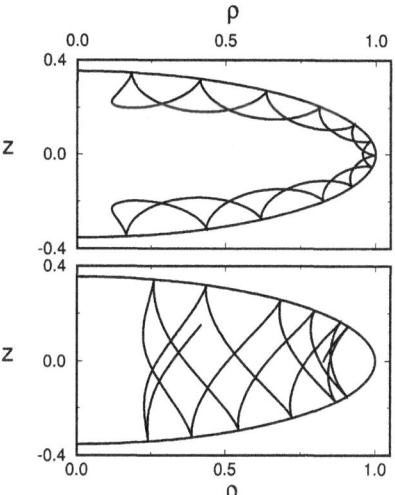

FIG. 3. Trajectory of the hydrodynamic Hamiltonian (27) for $B > B^*$(a) and for $B < B^*$(b). z, ρ are plotted in units of $(2\mu/m\omega_0^2)^{1/2}$. (From [1])

For $B < B^*$ however there are further turning points in η-direction, the motion being confined between two hyperbolas with ξ values extending to zero, which can be seen in Fig.3b. This kind of trajectory only occurs in the anisotropic system. $B = B^*$ is the separatrix between these two types of motion. As usual this separatrix is structurally unstable against small nonintegrable perturbations of the integrable motion in the hydrodynamic limit. It plays a crucial role for the appearance of chaos in the Bogoliubov Hamiltonian as the energy is increased from values very small compared to μ, because it is destroyed and replaced by a chaotic separatrix layer, which is very narrow at first, but grows in width as the energy is increased. In Fig.1c two regular islands corresponding to the two kinds of trajectories and the chaotic separatrix layer between them can be seen.

C. The Hartree-Fock dynamics

Another limiting case of the Bogoliubov description of quasiparticles (9) consists in neglecting the hole-component $V_j(\boldsymbol{x})$ in the field operator $\hat{\varphi}(\boldsymbol{x})$. The remaining component $U_j(\boldsymbol{x})$ is then described by the Hartree-Fock Hamiltonian (10). The interaction between particles is taken into account by the potential $K(\boldsymbol{x})$, describing the mean interaction of one particle with all the other particles. Restricting ourselves to $T = 0$ all those other particles are in the condensate. In the homogenous systems this approach simply results in a shift of the dispersion relation of noninteracting particles by the chemical potential μ. For spatially homogeneous Bose condensates and also Bose condensates in traps such a description can be applied for energies larger than the mean interaction energy given by μ. However, in traps there is even a regime for energies smaller than μ where the Hartree-Fock approximation applies [17], namely in the case when the kinetic energy $\epsilon_{kin}(\boldsymbol{p})$ is large compared to the *local* mean interaction energy $K(\boldsymbol{x})$. This can be satisfied in a layer around the surface of the Bose condensate where $K(\boldsymbol{x})$ is very small.

Using the Thomas-Fermi approximation for the wave function (5) the Bogoliubov Hamiltonian and the Hartree-Fock Hamiltonian coincide outside the condensate. Inside the condensate, if the kinetic energy $\epsilon_{kin}(\boldsymbol{p})$ is much larger than the potential term $K(\boldsymbol{x})$, an expansion of the Bogoliubov Hamiltonian (26) to first order in $K(\boldsymbol{x})$ just gives the Hartree-Fock Hamiltonian

$$H_{\text{HF}} = \frac{p^2}{2m} + |U(\boldsymbol{x}) - \mu|, \tag{36}$$

which is therefore valid, for $\epsilon_{kin} \gg K(\boldsymbol{x})$, inside and outside the condensate.

We now want to investigate the classical dynamics of this Hartree-Fock Hamiltonian. As the isotropic problem is completely integrable we immediately turn to the classical dynamics of the anisotropic, but axially symmetric case in the trap potential (24) and consider it as a system with two degrees of freedom. The conserved angular momentum around the symmetry axes L_z enters only as a parameter. Again we investigate the dynamics by Poincaré cuts, now taken on the Thomas-Fermi surface $\xi = (1/\epsilon^2 - 1)^{1/2}$ and parametrized by the second elliptical coordinate η and its conjugate momentum p_η. For energies much larger than the chemical potential the interaction with the condensate is only a small perturbation to the integrable motion in the harmonic trap and we observe quasi-regular behaviour. In this limit the Bogoliubov description of quasi-particles reduces to the Hartree-Fock description, the condition $\epsilon_{kin}(\boldsymbol{p}) \gg K(\boldsymbol{x})$ being fulfilled for all trajectories, and the classical motions generated by both Hamiltonians are essentially the same. Trajectories not entering the condensate are even identical, since here the two descriptions fully coincide.

For energies in the approximate range $10 > E/\mu > 0.1$ one finds a mixed phase space again [1]. However, for energies much smaller than the chemical potential $E < 0.1\mu$ one finds a second regime of regular behaviour [1]. Particles are confined to the sharp potential channel near the Thomas-Fermi surface. The width of this potential channel scales as E/μ. Roughly the particles spend the same time outside and inside the condensate. We look at the problem in elliptical coordinates (29), and choose for concreteness again the case $\omega_z > \omega_0$. The oscillations in ξ orthogonal to the Thomas-Fermi surface $\xi = (1/\epsilon^2 - 1)^{1/2}$ are much faster than the oscillations in η-direction along the channel. This suggests to make an adiabatic approximation in which the action-integral $I_\xi = (2\pi)^{-1} \oint p_\xi d\xi$ over one full cycle in ξ at fixed η, p_η emerges as an adiabatic constant for the motion. Evaluating this adiabatic invariant for $E/\mu \ll 1$ we get as a function of η, p_η

$$I_\xi = \frac{4\mu}{3\pi\omega_z} \frac{1}{\sqrt{1 - \epsilon^2(1 - \eta^2)}} \left(\frac{E}{\mu} - \frac{1 - \eta^2}{1 - \epsilon^2(1 - \eta^2)} (\frac{\omega_0 p_\eta}{2\mu})^2 - \frac{1}{1 - \eta^2} (\frac{\omega_0 L_z}{2\mu})^2 \right)^{3/2} . \tag{37}$$

This new adiabatically conserved quantity which emerges in the low-energy limit of the Hartree-Fock dynamics, is the cause of integrability in that limit.

V. ONSET OF CHAOS COMING FROM LARGE QUASIPARTICLE-ENERGIES

A. Perturbation theory for large energy

The single-particle interaction energy with the condensate is of the order of the chemical potential μ and is only a small perturbation for energies $E \gg \mu$. Expanding to second order the Hamiltonian (21) takes the form

$$H = \epsilon_0(\boldsymbol{x}, \boldsymbol{p}) + 2V_0 |\psi_0(\boldsymbol{x})|^2 - \frac{V_0^2}{2} \frac{|\psi_0(\boldsymbol{x})|^4}{\epsilon_0(\boldsymbol{x}, \boldsymbol{p})} \tag{38}$$

with

$$\epsilon_0 = \frac{\boldsymbol{p}^2}{2m} + \frac{m}{2}\omega_0^2(x^2 + y^2 + \lambda z^2) - \mu . \tag{39}$$

In the following we shall restrict our discussion to the case of vanishing axial angular momentum $L_z = 0$. In this case the dynamics is restricted to a plane containing the z-axis, which can be taken as the (x, z)-plane $y \equiv 0$, $p_y = 0$. The action angle variables of the unperturbed harmonic motion in the trap are

$$x = \sqrt{\frac{2I_x}{m\omega_0}} \sin \theta_x \quad , \quad p_x = \sqrt{2m\omega_0 I_x} \cos \theta_x$$

$$z = \sqrt{\frac{2I_z}{m\sqrt{\lambda}\omega_0}} \sin \theta_z \quad , \quad p_z = \sqrt{2m\sqrt{\lambda}\omega_0 I_z} \cos \theta_z \tag{40}$$

with

$$\epsilon_0 = \omega_0 I_x + \sqrt{\lambda}\omega_0 I_z - \mu . \tag{41}$$

To express the perturbed Hamiltonian in action-angle variables we need the Fourier coefficients of the condensate density and its square

$$\bar{\rho}_{\ell n}(\mathrm{I}_x, \mathrm{I}_z) = \frac{1}{(2\pi)^2} \int_0^{2\pi} d\theta_x \int_0^{2\pi} d\theta_z e^{-i(\ell\theta_x + n\theta_z)} |\psi_0(x, 0, z)|^2$$

(42)

$$\bar{\rho^2}_{\ell n}(\mathrm{I}_x, \mathrm{I}_z) = \frac{1}{(2\pi)^2} \int_0^{2\pi} d\theta_x \int_0^{2\pi} d\theta_z e^{-i(\ell\theta_x + n\theta_z)} |\psi_0(x, 0, z)|^4 .$$

The canonical transformation $(\theta, \mathrm{I}) \to (\phi, J)$ with

$$\mathrm{I}_x = J_x - \frac{2V_0}{\omega_0} \sum_{\ell,n} \frac{\ell}{\ell + \sqrt{\lambda}n} \bar{\rho}_{\ell,n}(J_x, J_z) e^{i(\ell\theta_x + n\theta_z)}$$

(43)

$$\phi_x = \theta_x - \frac{2V_0}{\omega_0} \sum_{\ell,n} \frac{1}{i(\ell + \sqrt{\lambda}n)} \frac{\partial \bar{\rho}_{\ell,n}(J_x, J_z)}{\partial J_x} e^{i(\ell\theta_x + n\theta_z)}$$

and analogous for I_z, ϕ_z removes the angle-dependence in the first order perturbation term and we are left with the second-order Hamiltonian

$$H = \omega_0(J_x + \sqrt{\lambda}J_z) - \mu + 2V_0\bar{\rho}_{00}(J_x, J_z) + \frac{V_0^2}{\omega_0} \sum_{\ell,n} K_{\ell,n}(J_x, J_z) e^{i(\ell\phi_x + n\phi_z)}$$

(44)

where

$$K_{\ell,n} = -\frac{\bar{\rho^2}_{\ell,n}}{2(J_x + \sqrt{\lambda}J_z) - 2\mu/\omega_0} - 4\sum_{p,r} \left(p \frac{\partial \bar{\rho}_{\ell-p,n-r}}{\partial J_x} + r \frac{\partial \bar{\rho}_{\ell-p,n-r}}{\partial J_z} \right) \frac{\bar{\rho}_{p,r}}{p + r\sqrt{\lambda}} .$$

(45)

Eq. (44) differs from (38) only by terms of higher than second order.

B. Resonances and their overlap

For irrational $\sqrt{\lambda}$ there are no resonances to zeroth order in the interaction, but in first order isolated resonances $\Omega_x(J_x, J_z)P + \Omega_z(J_x, J_z)R = 0$ appear with integers P, R and the perturbed frequencies

$$\Omega_x = \omega_0 + 2V_0 \frac{\partial \bar{\rho}_{00}}{\partial J_x}$$

(46)

$$\Omega_z = \sqrt{\lambda}\omega_0 + 2V_0 \frac{\partial \bar{\rho}_{00}}{\partial J_z} .$$

Their distances in frequency space have to be compared with their widths, given by [15]

$$\Delta\omega = 4\sqrt{\frac{V_0^3}{\omega_0} \frac{|K_{PR}(J_x, J_z)|}{M_{PR}(J_x, J_z)}}$$

(47)

where

$$\frac{1}{M_{PR}(J_x, J_z)} = \frac{\sum_{i,k=x}^z P_i \frac{\partial^2 \bar{\rho}_{00}}{\partial J_i \partial J_k} P_k}{P_x^2 + P_z^2}, \quad P_x = P, \quad P_z = R.$$

(48)

If resonances overlap somewhere in phase space Chirikov's resonance-overlap criterion [15] tells us that we should expect chaos there. Unfortunately, eqs. (44), (45) are too difficult to evaluate analytically for a realistic equilibrium solution $\psi_0(\boldsymbol{x})$ of the time-independent Gross-Pitaevskii equation. However, these equations are still useful to understand some qualitative features of numerical simulations of the complete Hamiltonian dynamics at large energies.

34

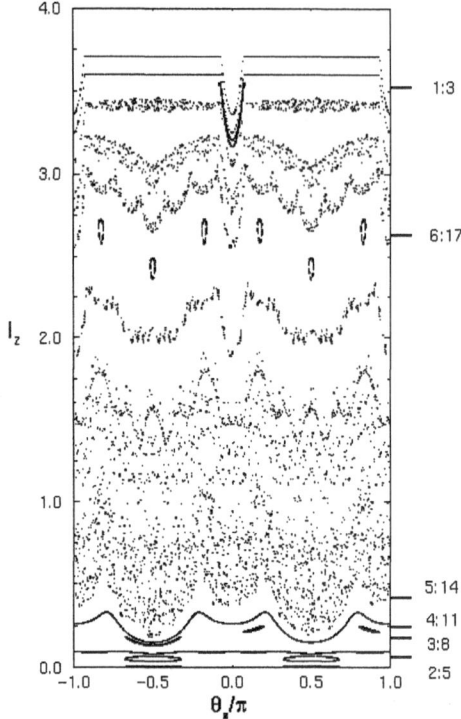

FIG. 4. Poincaré surface of section of the quasiparticle dynamics. The cut is taken at constant energy at $x = 0$, $p_x > 0$ for energy $E/\mu = 20$ and $\lambda = 8$. The cross-section is presented in the (θ_z, I_z)-plane, where $\sqrt{\lambda}\omega_0 I_z = p_z^2/2m + (m\lambda\omega_0^2/2)z^2$, $\arctan\theta_z = m\sqrt{\lambda}\omega_0 z/p_z$. The action variable I_z is given in units of $2\mu/\omega_0$. Frequency ratios $\Omega_x : \Omega_z$ of resonances are given on the right hand margin together with the resonance actions determined from perturbation theory. For resonances existing in doublets related by symmetries only one member of the doublet is shown for clarity. The 5:14 resonance could not be detected numerically. (From [3])

Let us discuss in particular the case of large condensates to which the Thomas-Fermi approximation (5) applies. As $|\psi_0(\boldsymbol{x})|^2$ is an even function of x, y, z the perturbation amplitudes are nonvanishing only for even ℓ, n. Therefore K_{PR} in (47) must be replaced by K_{2P2R}. The Thomas-Fermi approximation of the condensate density (5) has a discontinuous first-order derivative at the surface. This means that the Fourier coefficients $\bar{\rho}_{\ell,n}$ and $\bar{\rho^2}_{\ell,n}$ at large $|\ell|$, $|n|$ fall off like $|\ell|^{-2}$, $|n|^{-2}$ and $|\ell|^{-3}$, $|n|^{-3}$, respectively giving the (P, R)- resonances for large $|P|$, $|R|$ widths, which according to eq. (45), fall only like $|P|^{-1}$, $|R|^{-1}$. This estimate results from the second term in eq. (45), which is predicted to fall off only like $|\ell|^{-2}$, $|n|^{-2}$. On the other hand the number of large-order resonances $\Omega_x/\Omega_z = -R/P$ scales like $|P|$, $|R|$. Therefore, barring non-generic cases where the $|K_{PR}|$ are small for some exceptional reason, one expects large-order resonances to always overlap in the Thomas-Fermi approximation, i.e. the tori of the free harmonic oscillations in the trap will typically all be broken. By the same arguments a condensate density with

M smooth derivatives and a discontinuous $M + 1$ order derivative will give rise to resonance widths scaling like $|P|^{-\frac{M}{2}-1}$, $|R|^{-\frac{M}{2}-1}$ and tori with large $|P|$, $|R|$ can exist if $M > M_c = 2$. The critical order of smoothness for general Hamiltonian systems with f degrees of freedom was determined by Chirikov [15] as $M_c = 2f - 2$.

In Fig.4 a Poincaré surface of section of the dynamics at $E = 20\mu$, $L_z = 0$, taken at $x = 0$ and plotted in the (θ_z, I_z)-plane is shown for the experimentally realized [4] value $\lambda = 8$. The action variable I_z is plotted in units of $2\mu/\omega_0$. With the exception of tori at small I_z and at large $I_z \simeq (E+\mu)/\sqrt{\lambda}\omega_0$ all tori in Fig.4 are broken. The survival of the exceptional unbroken tori can be understood from the fact that for $I_z \to 0$ or $I_x \to 0$ (with $I_z \simeq (E+\mu)/\sqrt{\lambda}\omega_0$) all the coefficients $K_{\ell,n}$ approach 0 with the exception of $K_{\ell,0}$ or $K_{0,n}$, respectively, which cannot influence appreciably tori with frequency ratios $\Omega_x/\Omega_z \simeq 8^{-1/2}$.

In Fig.4 a number of resonances with frequency ratios in the neighborhood of $\Omega_x/\Omega_z = 8^{-1/2}$ can be discerned. Their frequency ratios are given on the right hand side of the graph. To understand the position where these resonances occur we consider $\bar{\rho}_{00}$ in the Thomas-Fermi approximation

$$\bar{\rho}_{00} = \frac{4}{\pi^2 V_0} \int_0^{\frac{\pi}{2}} d\phi_x \int_0^{\frac{\pi}{2}} d\phi_z \left(\mu - \omega_0 J_x \sin^2 \phi_x - \sqrt{\lambda}\omega_0 J_z \sin^2 \phi_z \right)$$
$$\theta \left(\mu - \omega_0 J_x \sin^2 \phi_x - \sqrt{\lambda}\omega_0 J_z \sin^2 \phi_z \right) \tag{49}$$

and evaluate the first-order frequency shifts

$$\Delta\omega_{x,z} = 2V_0 \frac{\partial \rho_{00}}{\partial J_{x,z}} \tag{50}$$

from the integrals

$$\Delta\omega_x = -\frac{8\omega_0}{\pi^2} \int_0^{\pi/2} d\phi_x \int_0^{\pi/2} d\phi_z \sin^2 \phi_x \theta(\mu - \omega_0 J_x \sin^2 \phi_x - \sqrt{\lambda}\omega_0 J_z \sin^2 \phi_z)$$
$$\tag{51}$$
$$\Delta\omega_z = -\frac{8\omega_0\sqrt{\lambda}}{\pi^2} \int_0^{\pi/2} d\phi_x \int_0^{\pi/2} d\phi_z \sin^2 \phi_z \theta(\mu - \omega_0 J_x \sin^2 \phi_x - \sqrt{\lambda}\omega_0 J_z \sin^2 \phi_z) \,.$$

It is manifest that $\Delta\omega_{x,z}$ are negative. Near the bottom of Fig.4 the action J_z is small, while $J_x \simeq (E+\mu)/\omega_0 - \sqrt{\lambda}J_z$ is large. The Heaviside function in eqs. (51) therefore restricts $\sin^2 \phi_x$ to small values of the order of $\mu/\omega_0 J_x = O(\mu/E)$ while $\sqrt{\lambda}\sin^2 \phi_z$ is not so restricted. Therefore $|\Delta\omega_x| \ll |\Delta\omega_z|$ near the bottom of Fig.4 and $\frac{\Omega_x}{\Omega_z} - \frac{1}{\sqrt{\lambda}} \simeq \frac{\Delta\omega_x}{\sqrt{\lambda}\omega_0} - \frac{\Delta\omega_z}{\omega_0\lambda} > 0$ holds there. In the upper parts of Fig.4 J_x is small while $J_z \simeq (E + \mu)/\sqrt{\lambda}\omega_0 - J_x/\sqrt{\lambda}$. The situation is therefore reversed and we obtain $\frac{\Omega_x}{\Omega_z} - \frac{1}{\sqrt{\lambda}} < 0$ by the same argument.

The periodic orbit $z = 0$, $p_z = 0$ which has $J_z = 0$ forms the lower border of the range of J_z. For this case the frequency shifts $\Delta\omega_x$, $\Delta\omega_z$ are easily evaluated from eq. (51) with the result

$$\Delta\omega_x = -O\left(\omega_0(\mu/\omega_0 J_x)^{3/2} \right)$$
$$\tag{52}$$
$$\Delta\omega_z = -\frac{2}{\pi}\sqrt{\frac{\lambda\omega_0\mu}{J_x}}$$

indicating that the fixed point is stable for $E \gg \mu$. The arguments of the previous paragraph indicate, furthermore, that $|\Delta\omega_x|$ is minimal for this case, while $|\Delta\omega_z|$ is maximal, leading to a maximal value of the ratio

$$\left(\frac{\Omega_x}{\Omega_z} \right)_{\max} \simeq \frac{1}{\sqrt{\lambda}} + \frac{2}{\pi}\sqrt{\frac{\mu}{\lambda\omega_0 J_x}} \simeq \frac{1}{\sqrt{\lambda}} \left(1 + \frac{2}{\pi}\sqrt{\frac{\mu}{E+\mu}} \right) \tag{53}$$

where we used $E + \mu \simeq \omega_0 J_x + \sqrt{\lambda}\omega_0 J_z$ with $J_z = 0$ in the last estimate. For $E/\mu = 20$ this gives $(\Omega_x/\Omega_z) \simeq 0.402$, which is just barely larger than the ratio 0.4 of the 2:5 resonance visible near the lower border of Fig.4. Similar arguments apply to the upper border of the range of J_z which is formed by the periodic orbit $x = 0$, $p_x = 0$ with vanishing J_x for which

$$\Delta\omega_x = -\frac{2}{\pi}\sqrt{\frac{\omega_0\mu}{\sqrt{\lambda}J_z}} \quad , \quad \Delta\omega_z = -\sqrt{\lambda\omega_0}O\left((\mu/\sqrt{\lambda}\omega_0 J_z)^{3/2} \right) \,. \tag{54}$$

Thus

$$\left(\frac{\Omega_x}{\Omega_z}\right)_{\min} \simeq \frac{1}{\sqrt{\lambda}}\left(1 - \frac{2}{\pi}\sqrt{\frac{\mu}{E+\mu}}\right) . \tag{55}$$

For $E/\mu = 20$ this yields $(\Omega_x/\Omega_z)_{\min} \simeq 0.304$ which is smaller than the ratio 0.333 of the 1:3 resonance visible near the upper border of Fig.4.

In general we can conclude that for fixed energy E we should expect to see the strongest resonances in the interval

$$\frac{1}{\sqrt{\lambda}}\left(1 - \frac{2}{\pi}\sqrt{\frac{\mu}{E+\mu}}\right) < \Omega_x/\Omega_z < \frac{1}{\sqrt{\lambda}}\left(1 + \frac{2}{\pi}\sqrt{\frac{\mu}{E+\mu}}\right) , \tag{56}$$

which are, for $\lambda = 8$, $E/\mu = 20$, the resonances 1:3, 2:5, 3:8, 4:11, 5:14, 6:17 and are indeed all seen in Fig.4 with the exception of the 5:14 resonance. In its place a chaotic region is seen. The resonance values of the actions following from the perturbation theory are indicated on the right-hand side of Fig.4. The 4:11 resonance predicted close to the missing 5:14 resonance is nearly completely destroyed and barely separated from the chaotic region replacing the 5:14 resonance.

C. Influence of the thickness of the surface-layer

Let us now consider the effect of the sharpness of the boundary layer of the condensate. Its thickness is given by the healing length [8]

$$\ell_H = \frac{\hbar}{\sqrt{2\mu m}} . \tag{57}$$

It enters as an independent parameter whose ratio to the radial Thomas-Fermi radius $\rho_{TF} = \sqrt{2\mu/m\omega_0^2}$ is determined by the value of the chemical potential, i.e. by the number of particles in the trap, $\ell_H/\rho_{TF} = \hbar\omega_0/2\mu$. As example we consider $\ell_H/\rho_{TF} = 0.1$. The condensate with boundary layer is modelled by joining

$$\begin{aligned}
|\psi_0(\boldsymbol{x})|^2 = &\frac{\mu - \frac{1}{2}m\omega_0^2 r^2}{V_0}\theta(\rho_{TF} - \ell_H - r) \\
&+ \exp(-a_0 - a_1(r - \rho_{TF} + \ell_H) - a_2(r - \rho_{TF} + \ell_H)^2 \\
&- a^3(r - \rho_{TF} + \ell_H)^3)\theta(r - \rho_{TF} + \ell_H)
\end{aligned} \tag{58}$$

continuously with continuous three derivatives at $r = \rho_{TF} - \ell_H$, thereby fixing $a_0, \ldots, a_3$. Here the abreviation $r = \sqrt{\rho^2 + \lambda z^2}$ is used. The degree of smoothness thereby introduced should be sufficient to see smooth KAM tori. The model condensate (58) is of course not a solution of the Gross-Pitaevskii equation, but does not differ qualitatively from solutions taking into account the boundary layer [17]. For our present purposes it is therefore perfectly acceptable while avoiding unnecessary complications. In Fig.5 we compare Poincaré surface of sections in the same format as in Fig.4 for the Thomas-Fermi approximation and the smooth model condensate with $l_H/\rho_{TF} = 0.1$ at energy $E/\mu = 100$ and $\lambda = 8$. It can be seen that with the smooth boundary layer included most of the broken tori of the Thomas-Fermi condensate seen in the upper part of Fig.5 become smooth, as seen in the lower part of Fig.5, however with ripples on the tori still indicating the presence of the narrow boundary layer. If the boundary layer is narrowed further (not shown) these ripples become stronger and develop sharp cusps. The chaotic band at small actions I_z survives even for very large quasiparticle energies E/μ. Resonances within this band have a comparatively large width for two reasons. For one the factors $1/M_{PR}$ are large, because I_z is not far above the value $I_z = \mu/\omega_0\sqrt{\lambda}$ where M_{PR}^{-1} peaks (with a logarithmic singularity in Thomas-Fermi approximation). For larger values of I_z, of the order of $E/\omega_0\sqrt{\lambda}$, like for the 6:17 resonance or values of $P : R$ close to $\sqrt{\lambda}$, $1/M_{PR}$ is much smaller by a factor of the order of $(\mu/E)^2$. Second, the interaction coefficient $|K_{PR}|$ has a resonantly enhanced contribution $-4(\partial\bar{\rho}_{00}/\partial J_z)\bar{\rho}_{2P2R}(P/R + \sqrt{\lambda})^{-1}$. The last factor in this expression favors resonances $P : R$ close to the ratio $\sqrt{\lambda}$.

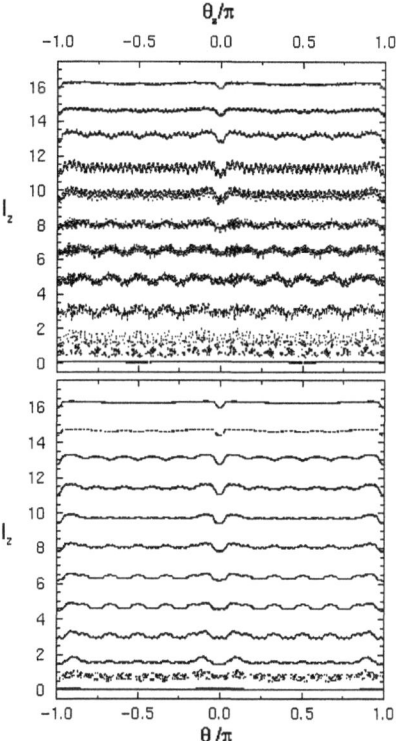

FIG. 5. Poincaré surface of sections as in Fig.4 for $E/\mu = 100$ for the dynamics with the condensate in Thomas-Fermi approximation (upper part) and for the condensate with boundary layer of thickness $l_H/\rho_{TF} = 0.1$, continuous up to and including the third derivative (lower part) (From [3])

The existence of the narrow boundary layer of the condensate furthermore leads to appreciable interaction coefficients $|K_{PR}|$ even for comparatively high-order resonances whose overlap may be responsible for the formation of the chaotic band. The requirement that *both* factors M_{PR}^{-1} and $|K_{2P2R}|$ must be large may explain the appearance of a band of actions I_z with appreciable resonance overlap somewhat above the value $I_z = \mu/\omega_0\sqrt{\lambda}$.

Dynamically the chaotic band is related to orbits in a layer of actions $I_z \gtrsim \mu/\sqrt{\lambda}/\omega_0$ which are just sufficient for the quasiparticle to hit or miss the condensate at random as it oscillates back and forth in x-direction. This mechanism is able to introduce sensitive dependence on initial conditions and exists, if only in a narrow band, even at very large energies. For just slightly smaller actions I_z the quasiparticle has to pass the condensate twice in each period of θ_x and the tori are smooth even in Thomas-Fermi approximation, as can be seen in Fig.4 for $E/\mu = 20$. A similar mechanism for a chaotic band should actually exist also for the oscillations in the z-direction, the short axis of the condensate-ellipsoid. However the instabilty in this case for action $I_x \gtrsim \mu/\sqrt{\lambda}/\omega_0$ seems to be much less pronounced (which is plausible, because the perturbation by the condensate should be weaker along the short axis) and is not visible in the numerical data.

38

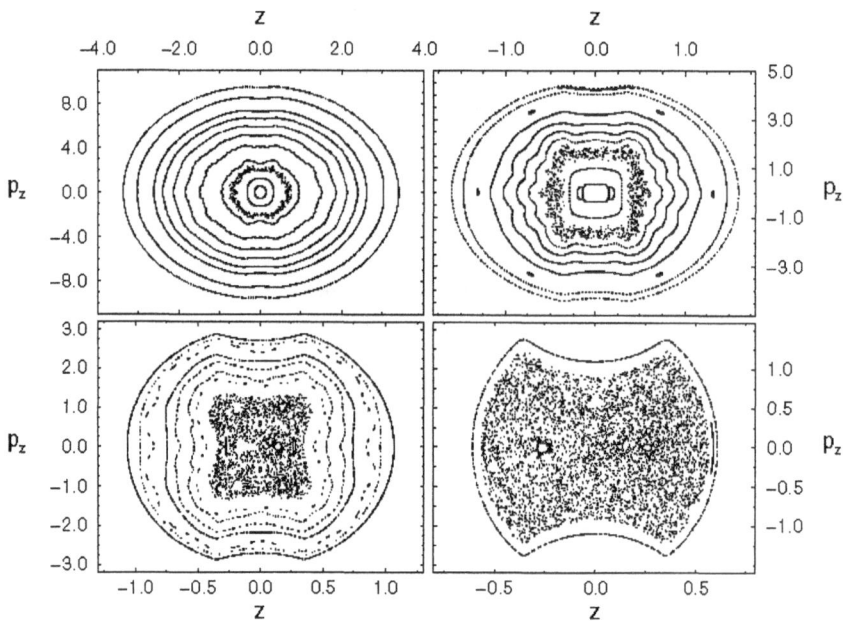

FIG. 6. Poincaré surface of sections as in Figs.4,5 but plotted in the (z, p_z)-plane. z, p_z are plotted in units $\sqrt{2\mu/m\omega_0^2}$, $\sqrt{2m\mu}$. The value of E/μ is 100 (upper left), 20 (upper right), 10 (lower left), and 2 (lower right). (From [3])

Finally we consider numerically the transition to chaos in the condensate with boundary layer $l_H/\rho_{TF} = 0.1$ as the quasiparticle energy is lowered. In Fig.6 the same Poincaré surface of section is plotted as in Figs.4,5 (and again for $\lambda = 8$) but in the (z, p_z)-plane rather than the (θ_z, I_z)-plane. The two plots in the upper row are for $E/\mu = 100$ and $E/\mu = 20$, those in the lower row are for $E = 10$ and $E = 2$ respectively. The plots for $E = 100$ and $E = 20$ can be compared with Fig.5 and Fig.4 respectively, but Fig.4 is, of course for vanishing l_H only. The slightly rounded Thomas-Fermi surface at $z/\rho_{TF} = \lambda^{-1/2}$ is visible in these graphs. Based on these and similar plots the transition to chaos can now be roughly described as follows: The chaotic band in a layer of small actions $I_z \gtrsim \mu/\sqrt{\lambda}\omega_0$ existing even at very large energies (see the plot for $E/\mu = 100$) becomes wider as the energy is lowered (see $E/\mu = 20$). Then, at a critical energy $E/\mu = 13.9$ which we discuss further below, the periodic orbit along the long axis of the condensate-ellipsoid becomes unstable and a second chaotic region in phase-space surrounding the unstable orbit $z = 0 = p_z$ is created. The inner and outer chaotic regions then rapidly grow together as the energy is lowered further (see $E/\mu = 10$) and finally fill up most of the accessible regions of phase-space (see the last plot with $E/\mu = 2$).

Apart from a continuous widening of the chaotic regions the main event in this transition to chaos is the appearance of the instability of the periodic orbit along the long axis of the condensate-ellipsoid, which for $\lambda > 0$ is the x-axis. In Thomas-Fermi approximation this orbit has the time-dependence for $|x| < \rho_{TF}$

$$x = \rho_{TF}\sqrt{E/\mu - 1}\sinh(\omega_0 t - \varphi^\circ)$$

$$p_x = m\omega_0\rho_{TF}\sqrt{E/\mu - 1}\cosh(\omega_0 t - \varphi^\circ)$$

(59)

and for $|x| > \rho_{TF}$

$$x = \rho_{TF} \sqrt{E/\mu + 1} \sin(\omega_0 t - \varphi'^{\circ})$$

$$p_x = m\omega_0 \rho_{TF} \sqrt{E/\mu + 1} \cosh(\omega_0 t - \varphi'^{\circ})$$

(60)

with φ° and φ'° suitably adjusted by continuity at $|x| = \rho_{TF}$. To determine the energy where this orbit looses its stability we can perform a linear stability analysis for small perturbations z, p_z away from this orbit, which satisfy

$$m\dot{z} = p_z, \quad \dot{p}_z = \lambda \left(1 - 2\theta(|x(t)| - \rho_{TF})\right) m\omega_0^2 z.$$

(61)

The growth of a perturbation of the periodic orbit is determined by the monodromy matrix $\mathbf{M}$ for a half-period $T/2$

$$\begin{pmatrix} z(T/2)/\rho_{TF} \\ p_z(T/2)/m\omega_0\rho_{TF} \end{pmatrix} = \mathbf{M} \begin{pmatrix} z^{(0)}/\rho_{TF} \\ p_z^{(0)}/m\omega_0\rho_{TF} \end{pmatrix}.$$

(62)

The Hamiltonian form of the dynamics ensures that Det $\mathbf{M} = 1$. The stability condition for the perturbations in z-direction then becomes

$$|Tr\mathbf{M}| \leq 2.$$

(63)

With a little algebra it is straight-forward to evaluate $\mathbf{M}$ and its trace thereby reducing (63) to

$$|\cos\left(\sqrt{\lambda}\omega_0 t_2\right)\cosh\left(\sqrt{\lambda}\omega_0 t_1\right)| \leq 1$$

(64)

Here t_1 and t_2 are the total lengths of the time-intervals during each half-period where $|x| < \rho_{TF}$ and $|x| > \rho_{TF}$, respectively. They are given by

$$t_1 = \frac{2}{\omega_0} \text{Artanh} \sqrt{\frac{\mu}{E}}, \qquad t_2 = \frac{2}{\omega_0} \text{Arctan} \sqrt{\frac{E}{\mu}}.$$

(65)

For $\lambda = 8$ stability is lost, according to the criterion (64) for $E/\mu = 14.4$ which is in reasonable agreement with the already quoted value $E/\mu = 13.9$ determined from the numerical simulation of the complete quasiparticle dynamics.

VI. CONCLUSION

The quasi-particle excitations are the basic constituents of the dynamical and thermodynamical properties of Bose condensates. In the present lecture we have discussed their dynamics for Bose condensates of atomic gases in traps in the classical limit. The two limiting types of excitations, collective modes and quasi-particle excitations consisting essentially of single atoms moving in a mean field correspond, in the classical limit, to particles and anti-particles of zero mass, moving 'relativistically' with the speed of sound, and to single atoms moving in the potential created by the trap and the Hartree-Fock potential energy of all other atoms. In spatially homogeneous (untrapped) condensates these two types of excitation strongly differ in energy E, the collective modes occurring at $E \ll \mu$, the single-particle modes at $E \gg \mu$. In the trapped condensates both types of excitations coexist, at least classically, at small energies $E \ll \mu$, and are instead spatially separated. The collective modes live inside the condensate, the single-particle modes at small energies in a narrow layer at the border.

One principal result we have discussed here is that the classical dynamics of both, the collective modes and the single-particle modes, become integrable in the limit $E/\mu \ll 1$. This has important consequences for the quantum dynamics as well: the integrability can be used there to separate the Schrödinger equation and to obtain not only the low-lying levels of the collective modes [2,16,19,20], but also of the single-particle modes. After quantization an energy gap reappears separating the collective modes with typical energies $\hbar\omega_0$ and the single-particle modes whose lowest levels have energies of the order $(\hbar\omega_0)^{2/3}\mu^{1/3}$ due to their close confinement in normal direction to the surface of the condensate. However, this energy difference is much smaller than, and has a different origin as the energy difference between both types of modes in homogeneous systems.

Another principal result we have discussed at length in this lecture is the *nonintegrabilty* of the classical dynamics of the quasi-particle excitations at intermediate energies $E \simeq \mu$. This applies to both, the full Bogoliubov dynamics and the limiting Hartree-Fock dynamics approximating it wherever the kinetic energy is large compared to the *local* mean interaction energy. Again this nonintegrability has a direct consequence also for the quantum dynamics, because it implies avoided crossings between quasi-particle levels as functions of the dimensionless interaction strength $N_0 a/d_0$ with $d_0 = \sqrt{\hbar/m\omega_0}$, if the energy and μ are comparable. Such avoided crossings have indeed been seen in numerically generated plots [18].

We have given a detailed discussion of the onset of chaos as the energy is lowered from the large energy regime. For quasiparticle energies large compared to the chemical potential a perturbative expansion becomes possible. The dynamics then looks like that of an anisotropic harmonically bound particle which is perturbed by the presence of a weakly repelling condensate localized around the center of the trap with a rather sharp surface. The appearance of an infinitely sharp surface in the Thomas-Fermi approximation leads to the break-up of all tori with the exception of those in the immediate phase-space neighborhood of the periodic orbits along the main axes of the ellipsoidal condensate. Therefore in this case chaos exists in phase-space for arbitrarily large energies E/μ. Still resonances can be identified even in this case and their actions are well described by perturbation theory. We have compared this non-smooth case to that of a condensate with a smooth (up to third derivatives) but narrow boundary layer. In real condensates the thickness of the boundary layer is determined by the two-particle interaction and the number of particles [8]. In that case smooth KAM tori exist at sufficiently large energies, but they are rippled by the influence of the boundary layer. Furthermore it turns out that an appreciable region of chaos in phase-space persists even to large energies. It is related to orbits along the long axis of the condensate-ellipsoid which are sufficiently perturbed in the direction of the short axis to sometimes miss and sometimes hit the condensate in a random way with sensitive dependence on small perturbations. Finally we have studied how large scale chaos appears as the energy is gradually lowered to values of the order of the chemical potential. The instability of the periodic orbit along the long axis of the ellipsoid was found to play a major role in this transition. It is connected with the appearance of a second inner chaotic region at lower energies which joins up with the chaotic band existing also at large energies.

ACKNOWLEDGEMENTS

The present lecture-notes are based on work [1–3] done jointly with András Csordás, Péter Szépfalusy, and Martin Fliesser, to whom I am gratefull for fruitful collaboration. This work has been supported by the Deutsche Forschungsgemeinschaft through the Sonderforschungsbereich 237 "Unordnung und große Fluktuationen".

[1] M. Fliesser, A. Csordás, R. Graham, P. Szépfalusy, *Phys. Rev* **A 56**, 4879 (1997)
[2] M. Fliesser, A. Csordás, R. Graham, P. Szépfalusy, *Phys. Rev.* **A 56**, R2533 (1997)
[3] M. Fliesser and R. Graham, *Physica* **D 131**, 141 (1999)
[4] M. H. Anderson, J. R. Ensher, M. R. Matthews, C. E. Wiemann, and E. A. Cornell, Science **269**, 198 (1995).
[5] C.C. Bradley, C.A. Sackett, J.J. Tollett, and R.G. Hulet, Phys. Rev. Lett. **75**, 1687 (1995).
[6] K.B. Davis, M.O. Mewes, M.R. Andrews, N.J. van Druten, D.D. Durfee, D.M. Kurn, and W. Ketterle, Phys. Rev. Lett. **75**, 3969 (1995).
[7] N. Bogoliubov, *J. Phys. USSR* **11**, 23 (1947)
[8] G. Baym and C. J. Pethick, *Phys. Rev. Lett.* **76**, 6 (1996)
[9] A. Csordás, R. Graham, P. Szépfalusy, *Phys. Rev* **A 56**, 5179 (1997)
[10] L. P. Pitaevskii, Zh. Eksp. Teor. Fiz. **40**, 646 (1961) [Sov. Phys. JETP **13**, 451 (1961)]; E. P. Gross, Nuovo Cimento **20**, 454 (1961); J. Math. Phys. **4**, 195 (1963).
[11] S. Stringari, Phys. Rev. Lett. **77**, 2360 (1996).
[12] A. L. Fetter, Ann. Phys. (N.Y.) **70**, 67 (1972).
[13] M. Edwards, P. A. Ruprecht and K. Burnett, R. J. Dodd and C. W. Clark, *Phys. Rev. Lett.* **77**, 1671 (1996); P. A. Ruprecht, M. Edwards, and K. Burnett, Phys. Rev. **A 54**, 4178 (1996); K. G. Singh and D. S. Rokhsar, *Phys. Rev. Lett.* **77**, 1667 (1996); L. You, W. Hoston, and M. Lewenstein, *Phys. Rev.* **A55**, R1581 (1997); A. Smerzi and S. Fantoni, *Phys. Rev. Lett.* **78**, 3589 (1997); B. D. Esry, *Phys. Rev.* **A55**, 1147 (1997).
[14] A. L. Fetter, *Phys. Rev.* **A53**, 4245 (1996); W.-C. Wu and A. Griffin, *Phys. Rev.* **A54**, 4204 (1996); V. M. Pérez-García, H. Michinel, J. I. Cirac, M. Lewenstein, P. Zoller, *Phys. Rev. Lett* **77**, 1520 (1996); Y. Castin and R. Dum, *Phys. Rev. Lett.* **77**, 5315 (1996); Yu. Kagan, E. L. Surkov, G. V. Shlyapnikov, *Phys. Rev.* **A54**, R1753 (1996); *Phys. Rev.* **A55**, R18 (1997); F. Dalfovo, C. Minniti, S. Stringari, L. P. Pitaevskii, Phys. Lett. **A227** 259 (1997).
[15] B.V. Chirikov, *Physics Reports* **52**, 263 (1979)
[16] P. Öhberg, E. L. Surkov, I. Tittonen, M. Wilkens, and G. V. Shlyapnikov, *Phys. Rev.* **A 56**, R3346 (1997).
[17] F. Dalfovo, L. Pitaevskii and S. Stringari, *Phys. Rev* **A 54**, 4213 (1996)

[18] L. You, W. Hoston, and M. Lewenstein, *Phys. Rev.* **A55**, R1581 (1997)

[19] A. Csordas and R. Graham, *Phys. Rev.* **A59**, 1477 (1999)

[20] M. Fliesser and R. Graham, *Phys. Rev.* **A59**, R27 (1999)

Bifurcations in Attractive Bose-Einstein Condensates and Superfluid Helium

Cristián Huepe[1], Caroline Nore[2], and Marc-Etienne Brachet[1]

[1] Laboratoire de Physique Statistique de l'Ecole Normale Supérieure associé au CNRS et aux Universités Paris 6 et 7, 24 rue Lhomond, 75005 Paris, France
[2] Université de Paris-Sud, LIMSI, Bâtiment 508, F-91403 Orsay Cedex, France

Abstract. We consider systems containing a Bose-Einstein condensate described by a macroscopic wave function that obeys a Nonlinear Schrödinger like equation (NLSE). Using a continuation method, we characterize the bifurcation of stationary states.

For attractive Bose condensates confined in isotropic potentials, we show the presence of an Hamiltonian saddle-node bifurcation where the stable (elliptic) and unstable (hyperbolic) solutions meet. The condensate decay rates corresponding to macroscopic quantum tunnelling, thermal fluctuations and inelastic collisions are determined. The influence of anisotropy on the bifurcation diagram is also characterised. This requires three-dimensional computations.

For two-dimensional superflows past a cylinder, we also find a saddle-node bifurcation. Through a secondary pitchfork bifurcation, the unstable branch generates one-vortex asymmetric fields that are the nucleation solutions. We characterize on the bifurcation diagram the influence of the ratio of the coherence length to the disc diameter. A study of the system's three-dimensional instabilities is carried out. We demonstrate that vortex stretching can be induced at subcritical velocities.

1 Introduction

The present paper is a review of some recent applications [1–5] of branch following methods [6] through which stationary solutions of nonlinear partial differential equations can be found and studied. We study two different physical systems, both described by the Nonlinear Schrödinger equation (NLSE): attractive Bose condensates and superflows past a cylinder. The dynamics of dilute Bose-Einstein condensates is accurately described by the NLSE, allowing direct quantitative comparison between theory and experiment [7]. The NLSE can also be considered to describe the dynamics of superfluid ^{4}He, at temperatures low enough for the normal fluid to be negligible.

The paper is organised as follows. In section 2 we describe the numerical tools that are needed to obtain the stationary solutions of the NLSE. Section 3 is devoted to Bose-Einstein condensates with attractive interactions. Such condensates are known to be metastable in spatially localised systems, provided that the number of condensed particles is below a critical value $\mathcal{N}_c$. In section 4, we study the stability of superflows past a cylinder and find the corresponding two-dimensional bifurcation diagram: a saddle-node bifurcation followed by a secondary pitchfork bifurcation. Three-dimensional vortex stretching, similar to that present in superfluid turbulence[8–10], is obtained at subcritical velocities.

O. Descalzi et al. (eds.), Instabilities and Nonequilibrium Structures VII & VIII, 43–68.
© 2004 *Kluwer Academic Publishers. Printed in the Netherlands.*

44

The implications for recent experiments in Bose-Einstein condensed gas [11] are discussed. Finally, section 5 is our conclusion.

2 Numerical branch following in NLSE

In this section we define and test the numerical tools needed to obtain the stationary solutions of the NLSE.

Consider the following action functional associated to the NLSE

$$\mathcal{A} = \int d\tilde{t} \left\{ \int d\tilde{\mathbf{x}} \frac{i}{2} \left(\bar{\psi} \frac{\partial \psi}{\partial \tilde{t}} - \psi \frac{\partial \bar{\psi}}{\partial \tilde{t}} \right) - \mathcal{F} \right\}, \tag{1}$$

where ψ is a complex field, $\bar{\psi}$ its conjugate and $\mathcal{F}$ is the energy of the system. Here, $\mathbf{x}$ and $\tilde{t}$ correspond to adequately adimensionalised space and time variables respectively.

The Euler-Lagrange equation corresponding to (1) leads to the NLSE in terms of the functional $\mathcal{F}$

$$\frac{\partial \psi}{\partial \tilde{t}} = -i \frac{\delta \mathcal{F}}{\delta \bar{\psi}}. \tag{2}$$

This equation obviously admits ψ_S as a stationary solutions if $\delta\mathcal{F}/\delta\psi|_{\psi=\psi_S} = 0$. Thus, stationary solutions of (2) are extrema of $\mathcal{F}$.

2.1 General formulation

In general, we are looking for an extremum of an energy functional $\mathcal{E}$ under some constraint $\mathcal{Q}[\psi] = cst$. The usual Lagrange multiplier trick consists in introducing a control parameter ν and, rather than solving for extrema of $\mathcal{E}[\psi]$, searching for extrema of the new functional $\mathcal{F}[\psi] = \mathcal{E}[\psi] - \nu \mathcal{Q}[\psi]$. We thus solve for

$$\left. \frac{\delta \mathcal{F}}{\delta \psi} \right|_{\nu = cst.} = 0. \tag{3}$$

We now turn to the precise definitions, corresponding to the two systems considered in this paper : (a) Bose-Einstein Condensates and (b) Superflows.

a- Bose-Einstein Condensates We consider a condensate of $\mathcal{N}$ particles of mass m and effective scattering length a in a radial confining harmonic potential $V(r) = m\omega^2 r^2/2$ [1]. Quantities are rescaled by the natural quantum harmonic oscillator units of time $\tau_0 = 1/\omega$ and length $L_0 = \sqrt{\hbar/m\omega}$, thus obtaining the adimensionalised variables $\tilde{t} = t/\tau_0$, $\tilde{\mathbf{x}} = \mathbf{x}/L_0$ and $\tilde{a} = 4\pi a/L_0$. The control parameter ν becomes in this context the chemical potential μ. The total number of particles in the condensate is therefore given by $\mathcal{Q} \equiv \mathcal{N}$. Functionals $\mathcal{F}$, $\mathcal{E}$ and $\mathcal{N}$ are given, in terms of rescaled variables, by

$$\mathcal{F} = \mathcal{E} - \mu \mathcal{N} \tag{4}$$

$$\mathcal{E} = \int d^3\tilde{\mathbf{x}} \left(\frac{1}{2}|\nabla_{\tilde{\mathbf{x}}}\psi|^2 + V(\tilde{\mathbf{x}})|\psi|^2 + \frac{\tilde{a}}{2}|\psi|^4 \right) \quad \cdot \tag{5}$$

$$\mathcal{N} = \int d^3\tilde{\mathbf{x}}|\psi|^2. \tag{6}$$

Two different situations are possible, depending on the sign of the (rescaled) effective scattering length $\tilde{a}$. When $\tilde{a}$ is positive the particles interact repulsively. A negative $\tilde{a}$ corresponds to an attractive interaction. The dynamical equation is

$$\frac{\partial\psi}{\partial\tilde{t}} = -i\frac{\delta\mathcal{F}}{\delta\bar{\psi}} = i\left[\frac{1}{2}\nabla_{\tilde{\mathbf{x}}}^2\psi - \frac{1}{2}|\tilde{\mathbf{x}}|^2\psi - \left(\tilde{a}|\psi|^2 - \mu\right)\psi\right]. \tag{7}$$

We will be interested in the solutions of $\delta\mathcal{F}/\delta\bar{\psi} = 0$, for a given value of μ. According to equation (3), these solutions are extrema of $\mathcal{E}$ at constant particle number $\mathcal{N}$.

b- Superflows In the problem of a superflow past an obstacle, $\mathcal{E}$ is the hydrodynamic energy and $\nu \equiv \mathbf{U}$ is the flow velocity with respect to the obstacle [2,4]. This implies that $\mathcal{Q} \equiv \mathcal{P}$ is the flow momentum. Functionals $\mathcal{F}$, $\mathcal{E}$ and $\mathcal{P}$ are given by the expressions

$$\mathcal{F} = \mathcal{E} - \mathcal{P} \cdot \mathbf{U} \tag{8}$$

$$\mathcal{E} = c^2 \int d^3x \left([-1 + V(\mathbf{x})]|\psi|^2 + \frac{1}{2}|\psi|^4 + \xi^2|\nabla\psi|^2 \right) \tag{9}$$

$$\mathcal{P} = \sqrt{2}c\xi \int d^3x \frac{i}{2} \left(\psi\nabla\bar{\psi} - \bar{\psi}\nabla\psi \right). \tag{10}$$

Here, c and ξ are the physical parameters characterising the superfluid. They correspond to the speed of sound (c) for a fluid with mean density $\rho_0 = 1$, and to the coherence length (ξ). The potential $V(\mathbf{x})$ is used to represent a cylindrical obstacle of diameter D. The NLSE reads

$$\frac{\partial\psi}{\partial t} = -\frac{i}{\sqrt{2}c\xi}\frac{\delta\mathcal{F}}{\delta\bar{\psi}} = i\frac{c}{\sqrt{2}\xi} \left([1 - V(\mathbf{x})]\psi - |\psi|^2\psi + \xi^2\nabla^2\psi \right) + \mathbf{U} \cdot \nabla\psi. \tag{11}$$

We will be interested in the solutions of $\delta\mathcal{F}/\delta\bar{\psi} = 0$, for a given value of $\mathbf{U}$. According to equation (3), these solutions are extrema of $\mathcal{E}$ at constant momentum $\mathcal{P}$.

2.2 Numerical method

When the extremum of $\mathcal{F}$ is a local *minimum*, the stationary solution ψ_S of (11) can be reached by a relaxation method. If the extremum is not a minimum, Newton's iterative method is used to solve for ψ_S.

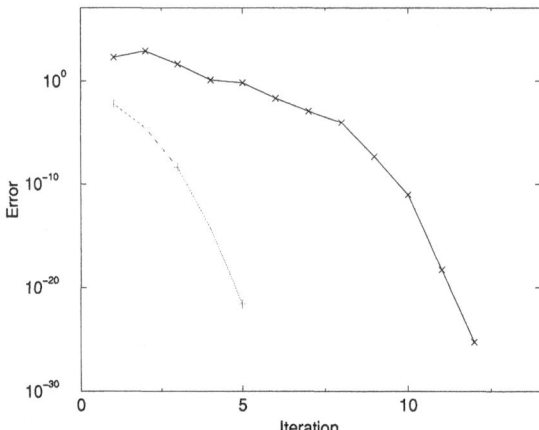

Fig. 1. Two typical examples of the Newton method convergence towards the solution of equation (16) for the problem of a superflow past a cylinder with $\xi/D = 1/10$ and a field $\psi_{(j)}$ discretized into $n = 128 \times 64 = 8190$ collocation points. The error measure is given by $\sum_{j=1}^{n} f_{(j)}^2(\psi)/n$. The convergence is faster than exponential, as expected for a Newton method.

a- Relaxation method In what remains of this section, we will write the NLSE under the following generic form, which is valid for both the Bose-Einstein condensates and the superflow past an obstacle:

$$\frac{\partial \psi}{\partial t} = -i \frac{\delta \mathcal{F}}{\delta \bar{\psi}} = i \left(\alpha \boldsymbol{\nabla}^2 \psi + [\Omega - V(\boldsymbol{x})] \psi - \beta |\psi|^2 \psi \right) + \boldsymbol{U} \cdot \boldsymbol{\nabla} \psi. \qquad (12)$$

When the extremum of $\mathcal{F}$ is a local *minimum*, the stationary solution ψ_S of (12) can be reached by integrating to relaxation the associated real Ginzburg-Landau equation (RGLE)

$$\frac{\partial \psi}{\partial t} = -\frac{\delta \mathcal{F}}{\delta \bar{\psi}} = \alpha \boldsymbol{\nabla}^2 \psi + [\Omega - V(\boldsymbol{x})] \psi - \beta |\psi|^2 \psi - i \boldsymbol{U} \cdot \boldsymbol{\nabla} \psi. \qquad (13)$$

Indeed, (12) and (13) have the same stationary solutions.

In our numerical computations, equation (13) is integrated to convergence by using the Forward-Euler/Backwards-Euler time stepping scheme

$$\psi(t + \sigma) = \Theta^{-1} \left[(1 - i \sigma \boldsymbol{U} \cdot \boldsymbol{\nabla}) + \sigma \left([\Omega - V(\boldsymbol{x})] - \beta |\psi(t)|^2 \right) \right] \psi(t) \qquad (14)$$

with

$$\Theta = \left[1 - \sigma \alpha \boldsymbol{\nabla}^2 \right]. \qquad (15)$$

The advantage of this method is that it converges to the stationary solution of (12) independently of the time step σ.

b- Newton method We use Newton's method [12] to find unstable stationary solutions of the RGLE.

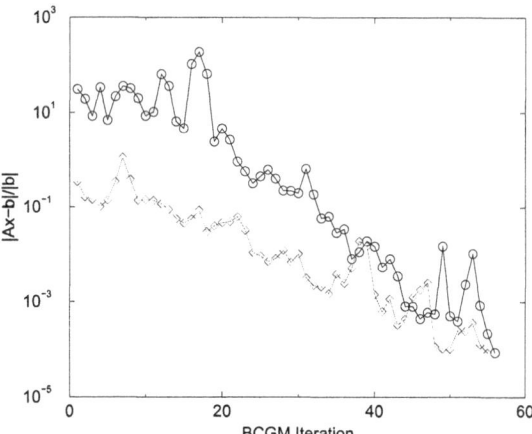

Fig. 2. Two typical examples of a bi-conjugate gradient method convergence corresponding to the case shown on figure 1. The convergence of the relative error achieved for the **x** solution of **Ax=b** is given by $|\mathbf{Ax\text{-}b}|/|\mathbf{b}|$, where $\mathbf{A}= [df_{(j)}/d\psi_{(k)}]$, $\mathbf{b}= -f_{(j)}(\psi)$ and $\mathbf{x}= \delta\psi_{(k)}$.

In order to work with a well-conditioned system [6], we search for the fixed points of (14). These can be found as the roots of

$$f(\psi) = \Theta^{-1} \left[(1 - i\,\sigma\,\boldsymbol{U}\cdot\boldsymbol{\nabla}) + \sigma\,\left([\Omega - V(\boldsymbol{x})] - \beta|\psi(t)|^2\right)\right]\psi(t) - \psi(t), \quad (16)$$

where Θ^{-1} was already introduced in equation (14). Calling $\psi_{(j)}$ the value of the field ψ over the j-th collocation point, finding the roots of $f(\psi)$ is equivalent iterating the Newton step

$$\psi_{(j)} = \psi_{(j)} + \delta\psi_{(j)} \tag{17}$$

up to convergence. Every Newton step (17) requires the solution for $\delta\psi_{(k)}$ of

$$\sum_k \left[\frac{df_{(j)}}{d\psi_{(k)}} \right] \delta\psi_{(k)} = -f_{(j)}(\psi). \tag{18}$$

This solution is obtained by an iterative bi-conjugate gradient method (BCGM) [13]. The BCGM uses the direct application of $[df_{(j)}/d\psi_{(k)}]$ over an arbitrary field φ to obtain an approximative solution of (18). Note that since the convergence of the time step (14) does not depend on σ, the roots found through this Newton iteration are also independent of σ. Therefore, σ becomes a free parameter that can be used to adjust the pre-conditioning of the system in order to optimise the convergence of the BCGM [6].

c- Implementation We use standard Fourier pseudo spectral methods [14]. Typical convergences of the Newton and bi-conjugate gradient iterations are shown in figures 1 and 2.

In the case of the radially symmetric Bose Condensate, $\psi(r, \tilde{t})$ is expanded as $\psi(r, \tilde{t}) = \sum_{n=0}^{N_R/2} \hat{\psi}_{2n}(\tilde{t}) T_{2n}(r/R)$, where T_n is the n-th order Chebyshev polynomial and $\hat{\psi}_{N_R}$ is fixed to satisfy the boundary condition $\psi(R, \tilde{t}) = 0$.

The time integration of the NLSE is done by using a fractional step (Operator-Splitting) method [15].

3 Attractive Bose-Einstein condensates

In this section, following reference [1], we study condensates with attractive interactions which are known to be metastable in spatially localized systems, provided that the number of condensed particles is below a critical value $\mathcal{N}_c$ [16]. Various physical processes compete to determine the lifetime of attractive condensates. Among them one can distinguish macroscopic quantum tunnelling (MQT) [17,18], inelastic two and three body collisions (ICO) [19,20] and thermally induced collapse (TIC) [18,21]. We compute the life-times, using both a variational Gaussian approximation and the exact numerical solution for the condensate wavefunction.

3.1 Computations of stationary states

a- Gaussian approximation A Gaussian approximation for the condensate density can be obtained analytically through the following procedure.

Inserting

$$\psi(r, \tilde{t}) = A(\tilde{t}) \exp\left(-r^2/2r_{\mathrm{G}}^2(\tilde{t}) + ib(\tilde{t})r^2\right) \tag{19}$$

into the action (1), where $\mathcal{F}$ is given by (4), yields a set of Euler-Lagrange equations for $r_{\mathrm{G}}(\tilde{t})$, $b(\tilde{t})$ and the (complex) amplitude $A(\tilde{t})$. The stationary solutions of the Euler-Lagrange equations produce the following values [5]:

$$\mathcal{N}(\mu) = \frac{4\sqrt{2\pi^3}\left(-8\mu + 3\sqrt{7 + 4\mu^2}\right)}{7|\tilde{a}|\left(-2\mu + \sqrt{7 + 4\mu^2}\right)^{3/2}}, \tag{20}$$

$$\mathcal{E} = \mathcal{N}(\mu)\left(-\mu + 3\sqrt{7 + 4\mu^2}\right)/7. \tag{21}$$

$\mathcal{N}$ is found to be maximal at $\mathcal{N}_c^G = 8\sqrt{2\pi^3}/|5^{5/4}\tilde{a}|$. The corresponding value of the chemical potential is $\mu = \mu_c^G = 1/2\sqrt{5}$.

Linearizing the Euler-Lagrange equations around the stationary solutions, yields the following expression for the eigenvalues [5]:

$$\lambda^2(\mu) = 8\mu^2 - 4\mu\sqrt{7 + 4\mu^2} + 2 \tag{22}$$

This qualitative behaviour is the generic signature of a Hamiltonian Saddle Node (HSN) bifurcation defined, at lowest order, by the normal form [22]

$$m_{eff}\ddot{Q} = \delta - \beta Q^2, \tag{23}$$

where $\delta = (1 - \mathcal{N}/\mathcal{N}_c)$ is the bifurcation parameter. The critical amplitude Q is related to the radius of the condensate [5]. We can relate the parameters β and m_{eff} to critical scaling laws, by defining the appropriate energy

$$\mathcal{E} = \mathcal{E}_0 + m_{eff}\dot{Q}^2/2 - \delta\, Q + \beta Q^3/3 - \gamma\delta. \tag{24}$$

From (23) it is straightforward to derive, close to the critical point $\delta = 0$, the universal scaling laws

$$\mathcal{E}_\pm = \mathcal{E}_c - \mathcal{E}_l\delta \pm \mathcal{E}_\Delta\delta^{3/2}, \tag{25}$$
$$\lambda_\pm^2 = \pm\lambda_\Delta^2\delta^{1/2}, \tag{26}$$

where $\mathcal{E}_c = \mathcal{E}_0$, $\mathcal{E}_l = \gamma$, $\mathcal{E}_\Delta = 2/3\sqrt{\beta}$ and $\lambda_\Delta^2 = 2\sqrt{\beta}/m_{eff}$.

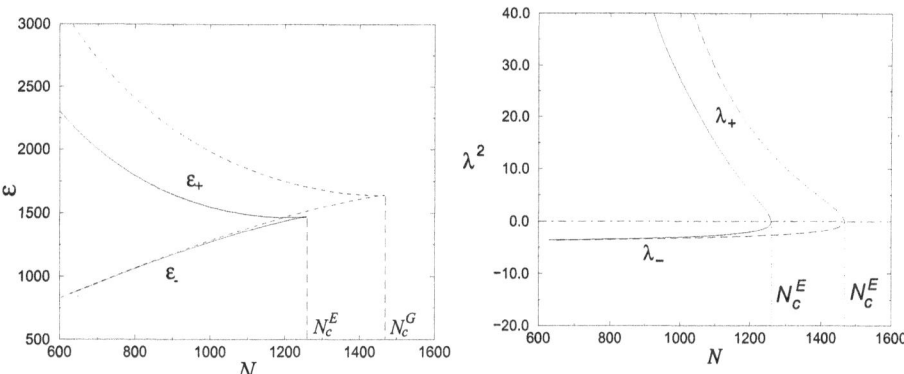

Fig. 3. Stationary solutions of the NLSE equation versus particle number $\mathcal{N}$. Left: value of the energy functional $\mathcal{E}_+$ on the stable (elliptic) branch and $\mathcal{E}_-$ on the unstable (hyperbolic) branch. Right: square of the bifurcating eigenvalue ($\lambda_\pm^2$); $|\lambda_-|$ is the energy of small excitations around the stable branch. Solid lines: exact solution of the NLSE equation. Dashed lines: Gaussian approximation.

b- Numerical branch following Using the branch-following method described in section 2, we have computed the exact stationary solutions of the NLSE. We use the following value $\tilde{a} = -5.74 \times 10^{-3}$, that corresponds to experiments with ^{7}Li atoms in a radial trap [23,16].

As apparent on Fig. 3, the exact critical $\mathcal{N}_c^E = 1258.5$ is smaller than the Gaussian one $\mathcal{N}_c^G = 1467.7$ [24,17]. The critical amplitudes corresponding to the Gaussian approximation can be computed from (20) and (21). One finds $\mathcal{E}_c = 4\sqrt{2\pi^3}/|5^{3/4}\tilde{a}|$, $\mathcal{E}_\Delta = 64\sqrt{\pi^3}/|5^{9/4}\tilde{a}|$ and $\lambda_\Delta^2 = 4\sqrt{10}$. For the exact solutions, we obtain the critical amplitudes by performing fits on the data. One finds $\mathcal{E}_\Delta = 1340$ and $\lambda_\Delta^2 = 14.68$. Thus, the Gaussian approximation captures the bifurcation qualitatively, but with quantitative 17% error on N_c [24], 24% error

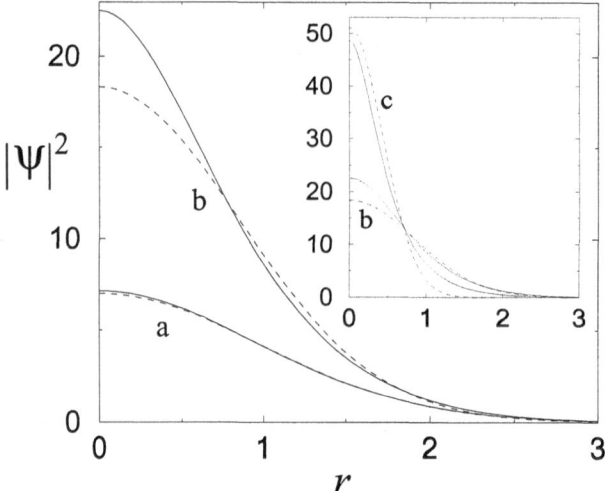

Fig. 4. Condensate density $|\psi|^2$ versus radius r, in reduced units (see text). Solid lines: exact solution of the NLSE equation. Dashed lines: Gaussian approximation. Stable (elliptic) solutions are shown for particle number $\mathcal{N} = 252$ (a) and $\mathcal{N} = 1132$ (b). (c) is the unstable (hyperbolic) solution for $\mathcal{N} = 1132$ (see insert).

on $\mathcal{E}_\Delta$ and 14% error on λ_Δ^2. Fig. 4 shows the physical origin of the quantitative errors in the Gaussian approximation. By inspection it is apparent that the exact solution is well approximated by a Gaussian only for small $\mathcal{N}$ on the stable (elliptic) branch.

3.2 Estimation of the condensate life-time

In this section, we estimate the decay rates due to thermally induced collapse, macroscopic quantum tunnelling and inelastic collisions.

a- Thermally induced collapse The thermally induced collapse (TIC) rate Γ_T is estimated using the formula [25]

$$\frac{\Gamma_T}{\omega} = \frac{|\lambda_+|}{2\pi} \exp\left[\frac{-\hbar\omega\left(\mathcal{E}_+ - \mathcal{E}_-\right)}{k_B T}\right] \tag{27}$$

where $\hbar\omega(\mathcal{E}_+ - \mathcal{E}_-)$ is the (dimensionalised) height of the nucleation energy barrier, T is the temperature of the condensate and k_B is the Boltzmann constant. Note that the prefactor characterises the typical decay time which is controlled by the slowest part of the nucleation dynamics: the top-of-the-barrier saddle point eigenvalue λ_+. The behaviour of Γ_T can be obtained directly from the universal saddle-node scaling laws (25) and (26). Thus the exponential factor and the prefactor vanish respectively as $\delta^{3/2}$ and $\delta^{1/4}$.

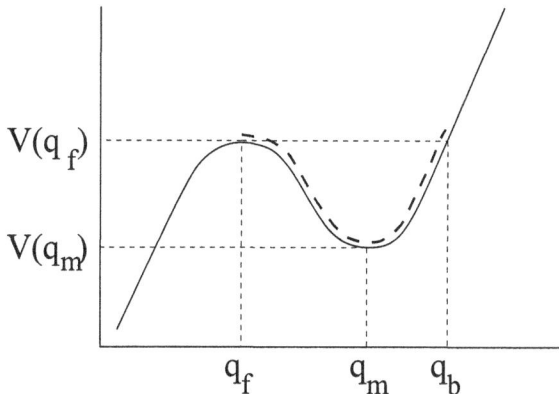

Fig. 5. The bounce trajectory is shown as dashes, above the potential $V(q)$.

b- Macroscopic quantum tunnelling We estimate the MQT decay rate using an instanton technique that takes into account the semi classical trajectory giving the dominant contribution to the quantum action path integral [18,17]. This trajectory is approximated as the solution of

$$\frac{d^2 q(t)}{dt^2} = -\frac{dV(q)}{dq},\tag{28}$$

$V(q)$ is a polynomial such that $-V(q)$ reconstructs the Hamiltonian dynamics. $V(q)$ is determined by the relations

$$V(q_m) \quad = -\mathcal{E}_+ \tag{29}$$
$$V(q_f) \quad = -\mathcal{E}_- \tag{30}$$
$$\partial_q^2 V(q_m) = |\lambda_+(\mathcal{N})| \tag{31}$$
$$\partial_q^2 V(q_f) = -|\lambda_-(\mathcal{N})|. \tag{32}$$

The bounce trajectory is displayed on Fig. 5 (dashed line) above the potential $V(q)$. The MQT rate is estimated as

$$\frac{\Gamma_Q}{\omega} = \sqrt{\frac{|\lambda_-|v_0^2}{4\pi}} \exp\left[\frac{-4}{\sqrt{2}} \int_{q_f}^{q_b} \sqrt{V(q) - V(q_f)}dq\right],\tag{33}$$

where v_0 is defined by the asymptotic form of the bounce trajectory $q(t)$ [18]: $q(\tau) \sim q_f + (v_0/|\lambda_-|)\exp[-|\lambda_-\tau|]$. Universal scaling laws can be derived close to criticality from (23), (25) and (26). The exponential factor in (33) follows the same scaling than $\sqrt{|\mathcal{E}_+ - \mathcal{E}_-|}dq$. It therefore vanishes as $\delta^{5/4}$. From the asymptotic form of $q(t)$, dq follows the same law as $v_0/|\lambda_-|$. Thus $v_0 \sim \delta^{3/4}$ and the prefactor vanishes as $\delta^{7/8}$.

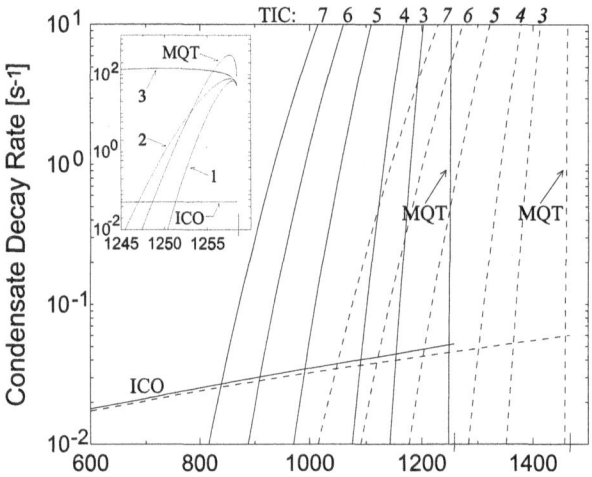

Fig. 6. Condensate decay rates versus particle number. ICO: inelastic collisions. MQT: macroscopic quantum tunnelling TIC: thermally induced collapse at temperatures $1nK$ (1),$2nK$(2), $50nK$ (3), $100nK$ (4), $200nK$ (5), $300nK$ (6) and $400nK$ (7). The insert shows the details of the cross-over region between quantum tunnelling and thermal decay rate. Solid lines: exact solution of the NLSE equation, dashed lines: Gaussian approximation.

c- Inelastic collision The inelastic collision rate (ICO) is estimated using the relation

$$\frac{d\mathcal{N}}{dt} = f_C(\mathcal{N}) \tag{34}$$

with

$$f_C(\mathcal{N}) = K \int |\psi|^4 d^3\tilde{\mathbf{x}} + L \int |\psi|^6 d^3\tilde{\mathbf{x}}, \tag{35}$$

where $K = 3.8 \times 10^{-4}\,\mathrm{s}^{-1}$ and $L = 2.6 \times 10^{-7}\,\mathrm{s}^{-1}$. The ICO rate can be evaluated from the stable branch alone. In order to compare the particle decay rate $f_C(\mathcal{N})$ to the condensate collective decay rates obtained for TIC and MQT, we compute the condensate ICO half-life as:

$$\tau_{1/2}(\mathcal{N}) = \int_{\mathcal{N}/2}^{\mathcal{N}} dn/f_C(n) \tag{36}$$

and plot $\tau_{1/2}^{-1}$

d- Discussion It is apparent by inspection of Fig. 6 that for a given value of $\mathcal{N}$ the exact and Gaussian approximate rates are dramatically different. We now compare the relative importance of the different exact decay rates. At $T \leq 1\,\mathrm{nK}$

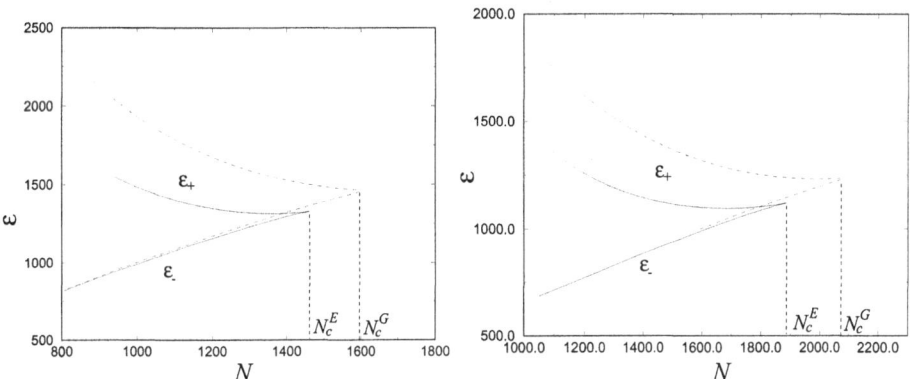

Fig. 7. Bifurcation diagram for two anisotropic traps. Left: $\omega_x = \omega_y = \omega_{\mathrm{iso}}$ and $\omega_z = \omega_{\mathrm{iso}}/5$. Right: $\omega_x = \omega_y = \omega_{\mathrm{iso}}/5$ and $\omega_z = \omega_{\mathrm{iso}}$. The continuous line corresponds to the exact solutions to the NLSE displayed on figures 9 et 10. The dashed line corresponds to the approximate Gaussian solutions of equation (38). The elliptic and hyperbolic branches are respectively labelled $\mathcal{E}_-$ and $\mathcal{E}_+$. The critical particle number are given by $N_c^E = 1460$ and $N_c^G = 1598$ (left), and by $N_c^E = 1886$. $N_c^G = 2080$ (right). The energy is given in units of $\hbar\omega_{\mathrm{iso}}$.

the MQT effect becomes important compared to the ICO decay in a region very close to $\mathcal{N}_c^E$ ($\delta \leq 8 \times 10^{-3}$) as it was shown in [17] using Gaussian computations but evaluating them with the exact maximal number of condensed particles $\mathcal{N}_c^E$. Considering thermal fluctuations for temperatures as low as $2\,\mathrm{nK}$, it is apparent on Fig. 6 (see insert) that the MQT will be the dominant decay mechanism only in a region extremely close to $\mathcal{N}_c$ ($\delta < 5 \times 10^{-3}$) where the condensates will live less than 10^{-1} s. Thus, in the experimental case of $^7\mathrm{Li}$ atoms, the relevant effects are ICO and TIC, with cross-over determined in Fig. 6.

3.3 Effect of anisotropy

In this section, the Gaussian approximation is compared to the exact numerical solution of a condensate in an anisotropic potential well. The effect of the anisotropy over the Gaussian approximation has been discussed in the literature [17,26].

The Gaussian solution for an axisymmetric potential with radial frequency ω_r and axial frequency ω_z is obtained by using the variational solution [17]

$$\psi(r,z) = \sqrt{\frac{N}{\pi^{3/2} d_r^2 d_z}} \exp\left(-\frac{r^2}{2d_r^2} - \frac{z^2}{2d_z^2}\right). \tag{37}$$

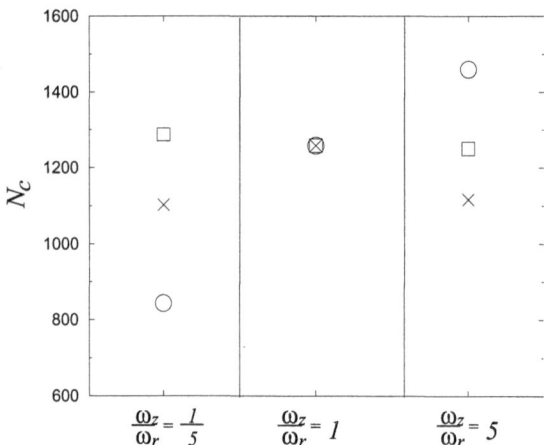

Fig. 8. Maximum number of particles condensed in the trap, for three different ratios ω_z/ω_r of the trap's characteristic frequencies. The points labelled with circles have the same radial frequency $\omega_r = \omega_{\mathrm{iso}}$. The points labelled as squares have the same arithmetic mean $(\omega_x + \omega_y + \omega_z)/3 = \omega_{\mathrm{iso}}$. Finally, the points labelled with x have the same geometric mean $(\omega_x \omega_y \omega_z)^{1/3} = \omega_{\mathrm{iso}}$.

Here, d_r and d_z are respectively the radial and axial length scales of the Gaussian distribution. By adimentionalizing the NLSE we obtain

$$\frac{\partial \psi}{\partial \tilde{t}} = i \left[\frac{1}{2} \nabla^2 \psi - \frac{1}{2} \left(\tilde{z}^2 + \frac{\omega_r^2}{\omega_z^2} \tilde{r}^2 \right) \psi - \left(\tilde{a} |\psi|^2 - \mu \right) \psi \right]. \qquad (38)$$

Here, $\tilde{r}$ and $\tilde{z}$ are adimensional length variables r/L_z and z/L_z, with $L_z = \sqrt{\hbar/m\omega_z}$, the axial characteristic length. The scattering length $\tilde{a}$ is also adimentionalize with respect to L_z. Just like in the isotropic case, the extremum of $\mathcal{E}$ associated to (37), gives the Gaussian solution.

We use the numerical method described in section 2 to obtain the exact numerical solution. The results shown below required ~ 50 hours of CPU on a Cray 90-8.

Figure 7 shows the bifurcation diagram of $\mathcal{E}$ for two different geometries, compared to the Gaussian approximations obtained from (37).

On figure 8 the maximum number of particles is plotted as a function of the anisotropy. We observe that, lowering ω_z at constant ω_r increases the maximum number of particles. It seems that the algebraic mean of the frequencies is the good parameter to determine the maximum number.

On figure 9 the anisotropic stationary solutions are plotted with $\omega_x = \omega_y = \omega_{\mathrm{iso}}$ and $\omega_z = \omega_{\mathrm{iso}}/5$. Here, $\omega_{\mathrm{iso}} = 908.41$ is the frequency used in the previous isotropic case. Both cases show that at the saddle-node bifurcation, the condensate is quasi-isotropic. Following the unstable branch, the degree of anisotropy increases.

Fig. 9. Surfaces of constant density for an attractive Bose-Einstein condensate, trapped in an anisotropic potential with characteristic frequencies $\omega_x = \omega_y = \omega_{\mathrm{iso}}$ and $\omega_z = \omega_{\mathrm{iso}}/5$. The first image shows a stable stationary solution with $N = 1137$ obtained for a value of the chemical potential $\mu = 4$. The second picture corresponds to the critical solution $N = 1460$, obtained for $\mu = 1.9$. The last picture is the unstable solution corresponding to $N = 1002$ and $\mu = -2$.

Fig. 10. Surfaces of constant density for an attractive Bose-Einstein condensate, trapped in an anisotropic potential with characteristic frequencies $\omega_x = \omega_y = \omega_{\mathrm{iso}}/5$ and $\omega_z = \omega_{\mathrm{iso}}$. The first image shows a stable stationary solution with $N = 1387$ obtained for a value of the chemical potential $\mu = 0.55$. The second picture corresponds to the critical solution $N = 1886$, obtained for $\mu = 0.31$. The last picture is the unstable solution corresponding to $N = 1382$ and $\mu = -0.4$.

4 Vortical nucleation solutions in a model of superflow

In this section, following references [2–4], we investigate the stationary stable and unstable (nucleation) solutions of the NLSE describing the superflow around a cylinder, using the numerical methods developed in section 2. We study a disc of diameter D, moving at speed U in a two-dimensional (2D) superfluid at rest. The NLSE (11) can be mapped into two hydrodynamical equations by applying Madelung's transformation [27,28]:

$$\psi = \sqrt{\rho} \exp\left(\frac{i\phi}{\sqrt{2}c\xi}\right). \tag{39}$$

The real and imaginary parts of the NLSE produce for a fluid of density ρ and velocity

$$v = \nabla\phi - U, \tag{40}$$

the following equations of motion

$$\frac{\partial \rho}{\partial t} + \nabla(\rho v) = 0 \tag{41}$$

$$\left[\frac{\partial \phi}{\partial t} - U \cdot \nabla\phi\right] + \frac{1}{2}(\nabla\phi)^2 + c^2[\rho - \Omega(x)] - c^2\xi^2 \frac{\nabla^2\sqrt{\rho}}{\sqrt{\rho}} = 0. \tag{42}$$

In the coordinate system x that follows the obstacle, these equations correspond to the continuity equation and to the Bernoulli equation [29] (with a supplementary *quantum pressure* term $c^2\xi^2\nabla^2\sqrt{\rho}/\sqrt{\rho}$) for an isentropic, compressible and irrotational flow. Note that, in the limit where $\xi/D \to 0$, the quantum pressure term vanishes and we recover the system of equations describing an Eulerian flow.

4.1 Bifurcation diagram and scaling in 2D

In this section, varying the ratio of the coherence length ξ to the cylinder diameter D, we obtain scaling laws in the $\xi/D \to 0$ limit.

a- Bifurcation diagram We present results for $\xi/D = 1/10$ which are representative of all ratios we computed. The functional $\mathcal{E}$ and energy $\mathcal{F}$ of the stationary solutions are shown in Fig. 11 as a function of the Mach number ($M = |U|/c$). The stable branch (a) disappears with the unstable solution (c) at a saddle-node bifurcation when $M = M^c \approx 0.4286$. The energy $\mathcal{F}$ has a cusp at the bifurcation point, which is the generic behaviour for a saddle-node. There are no stationary solutions beyond this point. When $M \approx 0.4282$, the unstable symmetric branch (c) bifurcates at a pitchfork to a pair of asymmetric branches (b). These branches have not yet been considered in the literature. Their nucleation energy barrier is given by $(\mathcal{F}_{b'} - \mathcal{F}_{a'})$ which is roughly half of the barrier for the symmetric branch $(\mathcal{F}_{c'} - \mathcal{F}_{a'})$.

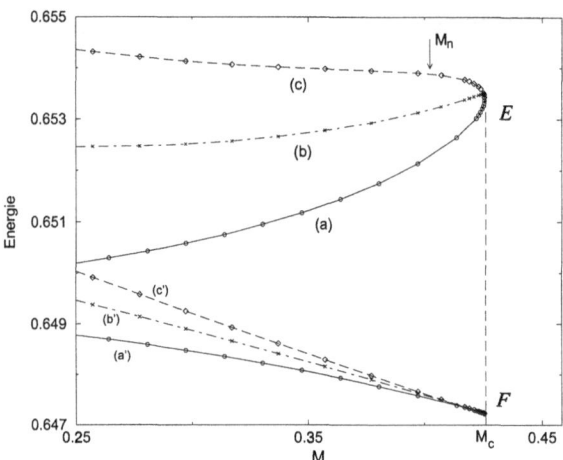

Fig. 11. Plot of the energy ($\mathcal{F}$), and functional ($\mathcal{E}$) versus Mach number ($M = |\boldsymbol{U}|/c$), with $D = 10\xi$. Stable state (a). Nucleation solutions: asymmetric branch (b) and symmetric branch (c). The diagram shows a saddle-node and a pitchfork bifurcation. The point where vortices cross the surface of the disc (see *fig.* 12) is labelled by M_n. The total fluid momentum is given by $-d\mathcal{F}/dU$ (see text).

We can relate branches in Fig. 11 to the presence vortices in the solution. When $M^\mathrm{pf} \leq M \leq M^c$, solutions are irrotational ($M^\mathrm{pf} \sim 0.405$ as indicated in Fig. 11). For $M \leq M^\mathrm{pf}$ the stable branch (a) remains irrotational (Fig. 12A) while the unstable branch (b) corresponds to a one vortex solution (Fig. 12B) and the unstable branch (c), to a two vortex solution (Fig. 12C). The distance between the vortices and the obstacle in branches (b) and (c) increases when M is decreased. Branch (c) is precisely the situation described in [30]. Furthermore, the value $M^c \approx 0.4286$ is close to the predicted value $\sqrt{2/11}$. However, the existence of a vortex-less interval $M^\mathrm{pf} \leq M \leq M^c$ was not predicted in [30]. Figure 12D shows the result of integrating the NLSE forward in time with, as initial condition, a slightly perturbed unstable symmetric stationary state (Fig. 12C). The perturbation drives the system over the nucleation barrier and cycles it, after the emission of two vortices, back to a stationary stable solution. This shows that the branch (c) corresponds to hyperbolic fixed points of NLSE.

Figures 12E,F show the phase of the field at the surface of the disc ($r = D/2$ and $\theta \in [0, 2\pi]$) for four different flow speeds. In both unstable branches, 2π-discontinuities, a diagnostic of vortex crossing, appear between $M = 0.40$ and $M = 0.41$.

b- Scaling laws We now characterize the dependence on ξ/D of the main features of the bifurcation diagram. When ξ/D is decreased, M^c and M^pf become

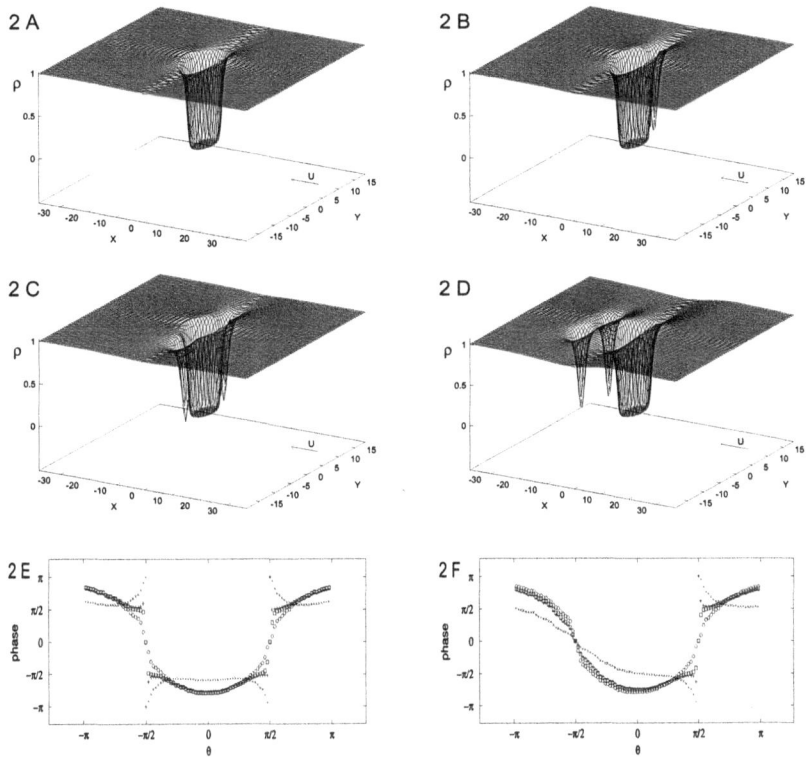

Fig. 12. Stationary states : stable (A), one vortex unstable (B), two vortices unstable (C). The surface indicates the fluid density around the cylinder $(M = 0.24, \xi/d = 0.1)$. (D) Shows the result of the NLSE integration, starting from a slightly perturbed stationary (C) state. Figures (E) and (F) display the phase of the complex field ψ at the surface of the cylinder versus the polar angle θ. Asymmetric branch (A), symmetric branch (B). $M = 0.4286$ (o), $M = 0.41$ ($\square$), $M = 0.40$ (+), $M = 0.30$ (×). The crossing out of the vortex produces a phase discontinuity at $M^{\mathrm{pf}} \sim 0.405$.

indistinguishable. In the limit where $\xi/D = 0$, the critical Mach number M^c will be that of an Eulerian flow M^c_{Euler}.

Figure 13 shows the convergence of M^c to the Eulerian critical velocity. This convergence can be characterized by fitting the polynomial law $M^c = K_1(\xi/D)^{K_2} + M^c_{\mathrm{Euler}}$ to $M^c(\xi/D)$. This fit is shown on Fig. 13 as a dotted line, yielding $K_1 = 0.322$, $K_2 = 0.615$ and $M^c_{\mathrm{Euler}} = 0.35$. This result is compatible with the analytical approximate results presented in [32,33], that predicts a square root ($K_2 = 0.5$) polynomial dependence on ξ/D.

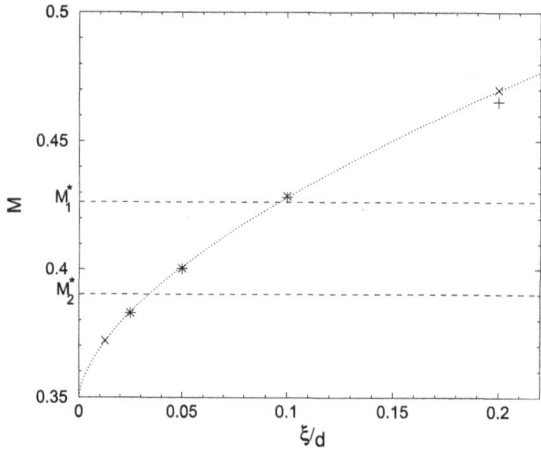

Fig. 13. Saddle-node bifurcation Mach number M^c (+) and pitchfork bifurcation Mach number M^{pf} (×), as a function of ξ/D. The dotted curve corresponds to a fit to the polynomial law $M^c = K_1(\xi/D)^{K_2} + M^c_{\text{Euler}}$ with $K_1 = 0.322$, $K_2 = 0.615$ and $M^c_{\text{Euler}} = 0.35$. The dashed lines $M^*_1 \approx 0.4264$ and $M^*_2 \approx 0.3903$ correspond respectively to first and second order compressible corrections to the $M^c = 0.5$ critical velocity computed using a local sonic criterion for an incompressible flow (see text). Note that the agreement, that is claimed to exist in [30,31,2], between the $\xi/D = 1/10$ numerical result and M^*_1, is just a coincidence.

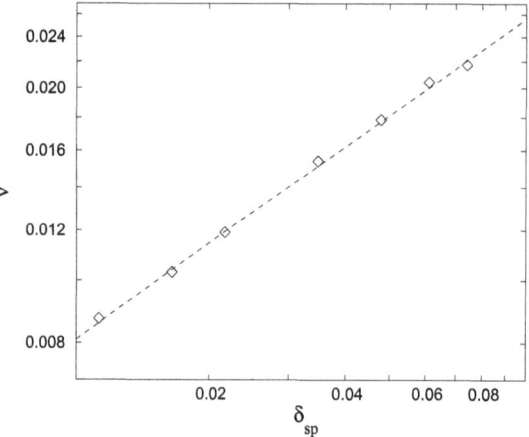

Fig. 14. Vortex emission frequency as a function of $\delta_{\text{sp}} = (M - M^c)/M^c \ll 1$ (with $M^c = 0.3817$), for a symmetric wake and $\xi/D = 1/20$. The dashed line shows a fit of a polynomial $\nu = K_1 \delta_{\text{sp}}^{1/2}$ with $K_1 = 0.081$. The obtained $\delta_{\text{sp}}^{1/2}$ law for the frequency is equivalent to the one expected for a dissipative system.

The dashed line M_1^* on Fig. 13 corresponds to the critical velocity $M^c = \sqrt{2/11}$ obtained in [30] by applying a local sonic criterion. Note that the agreement, that is obtained in [30,31,2], between the $\xi/D = 1/10$ numerical result and M_1^*, is clearly a coincidence. It is shown in [3] that an iterative scheme can be used to improve the $M^c = \sqrt{2/11}$ result. The next order [34] to the critical Mach number is $M_2^* = \sqrt{\sqrt{233} - 11}/(2\sqrt{7}) \approx 0.3903$.

d- Dynamical solutions The stationary solutions obtained in the above subsection provide adequate initial data for the study of dynamical solutions. Indeed, after a small perturbation, their integration in time will generate a dynamical evolution with very small acoustic emission. Therefore, this procedure corresponds to an efficient way to start vortical dynamics in a controlled manner.

Starting from a two-vortex unstable stationary solution at a supercritical Mach number $M^c = 0.9$, the evolution of the NLSE time integration shows a clearly periodical emission of vortex pairs (see Fig. 12). This emission conserves total circulation.

We have studied the behaviour of the frequency of vortex emission close to the bifurcation for such symmetrical wakes with different supercritical velocities (characterized by $\delta_{\mathrm{sp}} = (M - M^c)/M^c = -\delta > 0$). Our results, plotted on Fig. 14, are consistent with a $\delta_{\mathrm{sp}}^{1/2}$ scaling. Note that this scaling is unexpected in a Hamiltonian system and is related to dissipative scaling. Our system generates vortices that are carried away by the flow. It thus behaves in a dissipative way because the incident kinetic energy of the flow is irreversibly transferred to the vortical wake [3].

4.2 Subcriticality and vortex-stretching in 3D

In this section, using a 3D version of our code to integrate the NLSE, we study 3D instabilities of the basic 2D superflow.

a- Preparation method We used the 2D laminar stationary solution $\psi_{0V}(x,y)$ (corresponding to branch (a) of precedent section) and the one-vortex unstable stationary solution $\psi_{1V}(x,y)$ (branch (b)) to construct the 3D initial condition

$$\psi_{3\mathrm{D}}(x,y,z) = f_{\mathrm{I}}(z)\psi_{1V}(x,y) + [1 - f_{\mathrm{I}}(z)]\psi_{0V}(x,y). \tag{43}$$

The function $f_{\mathrm{I}}(z)$, defined by

$$f_{\mathrm{I}}(z) = (\tanh[(z - z_1)/\Delta_z] - \tanh[(z - z_2)/\Delta_z])/2,$$

takes the value 1 for $z_1 \leq z \leq z_2$ and 0 elsewhere, with Δ_z an adaptation length.

Figure 15 represents a 3D initial data prepared with this method for $\xi/D = 0.025$, $|U|/c = 0.26$ and $\Delta_z = 2\sqrt{2}\xi$ in the $[L_x \times L_y \times L_z]$ periodicity box ($L_x/D = 2.4\sqrt{2}\pi$, $L_y/D = 1.2\sqrt{2}\pi$ and $L_z/D = 0.4\sqrt{2}\pi$). The surface $|\psi_{3\mathrm{D}}| = 0.5$ draws the cylinder surface and the initial condition vortex line, with both ends pinned to the right side of the cylinder.

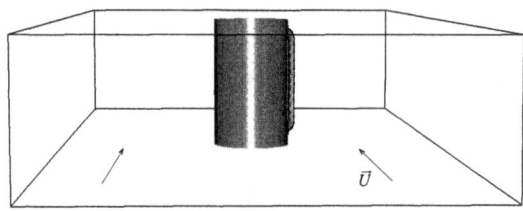

Fig. 15. Initial condition of a vortex pinned to the cylinder generated by eq.(43). The surface $|\psi_{3D}| = 0.5$ is shown for $\xi/D = 0.025$, $|U|/c = 0.26$ and $\Delta_z = 2\sqrt{2}\xi$ in the $[L_x \times L_y \times L_z]$ periodicity box ($L_x/D = 2.4\sqrt{2}\pi$, $L_y/D = 1.2\sqrt{2}\pi$ and $L_z/D = 0.4\sqrt{2}\pi$).

b- Short time dynamics Starting from the initial condition (43), the evolution of the NLSE time integration shows a short-time and a long-time dynamics.

During the short-time dynamics, the initial pinned vortex line rapidly contracts, evolving through a decreasing number of half-ring-like loops, down to a single quasi-stationary half-ring (see Figs. 16a, 16b, 16c). This evolution happens mainly on the plane perpendicular to the flow, provided that the initial vortex is long enough to contract to a quasi-stationary half-ring as shown on Fig. 16c. Otherwise, the vortex line collapses against the cylinder while moving upstream.

Note that this quasi-stationary half-ring has been used by Varoquaux [35,36] to estimate the nucleation barrier in a 3D experiment.

The dynamics of the half-ring situation (Fig. 16c) is very slow and can be shown to be close to a stationary field. Indeed, the local flow velocity v in an Eulerian flow around a cylindrical obstacle is known to vary from $v = |U|$ at infinity to $v = 2|U|$ at both sides of its surface. Moreover, the diameter d of a stationary vortex ring in an infinite Eulerian flow with no obstacle is given by [27]:

$$|U|/c = (\sqrt{2}\xi/d) \left[\ln\left(4d/\xi\right) - K\right], \tag{44}$$

where $|U|$ is the flow velocity at infinity and the vortex core model constant $K \sim 1$ is obtained by fitting the numerical results in [37]. Therefore, for the values used on Figs. 16, we expect that local velocities range from $v = 0.25$ to $v = 2 \times 0.25$. Equation (44) thus implies that the diameter of an hypothetical stationary half-ring should be bounded by $d(v = |U| = 0.25) = 18.8\xi$ and $d(v = 2|U| = 0.5) = 6.3\xi$. The diameter $d \approx 9\xi$ measured on the half-ring observed on Fig. 16c is consistent with its quasi-stationary behaviour. Similarly the diameter of the half-ring shown on Fig. 18 $d \approx 7.6\xi$ is also found to be between the corresponding bounds $d(0.35) = 11.4\xi$ and $d(2 \times 0.35) = 3\xi$.

b- Vortex stretching as a subcritical drag mechanism A small perturbation over the half-ring solution can drive the system into two opposite situations where the half-ring either starts moving upstream or downstream.

When driven upstream, the half-ring eventually collapses against the cylinder, dissipating its energy as sound waves. Otherwise, the vortex loop is stretched

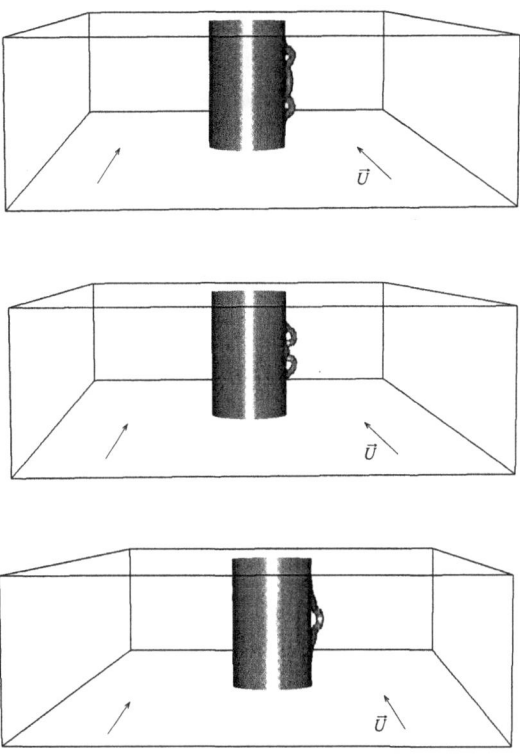

Fig. 16. Short-time dynamics for $\xi/D = 1/40$ and $|U|/c = 0.25$ starting from Fig. 15: A ($t = 5\xi/c$), B($t = 10\xi/c$) and C($t = 15\xi/c$). The contraction of the initial vortex line occurs in the plane perpendicular to the flow. The half-rings have a diameter compatible with that of a quasi-stationary half-ring (see text).

while the pinning points move towards the back of the cylinder. Figures 17 show the long-time dynamics for a stretching case with $\xi/D = 1/40$ and $|U|/c = 0.25$ starting from Fig. 16c. Figure 19 shows a later situation for $\xi/D = 1/20$ and $|U|/c = 0.35$ starting from Fig. 18. As the vortex loop grows, its backmost part remains oblique to the flow. The described vortex stretching mechanism consumes energy, thus generating drag. It can be responsible for the appearance of drag in experimental superflows if fluctuations are strong enough to nucleate the initial vortex loop (which is imposed extrinsically in our numerical system). Note that it takes place for 2D subcritical velocities.

Figure 20 displays several numerical and experimental [38] critical Mach numbers (V_c/C) with respect to D/ξ, which seem to follow a (-1) slope in a log-log plot. The squares stand for our numerical stretching cases while the crosses correspond to non-stretching cases. There is a frontier between the 3D numerical dissipative and non-dissipative cases [4]. For $1/30 < \xi/D < 1/20$, the frontier

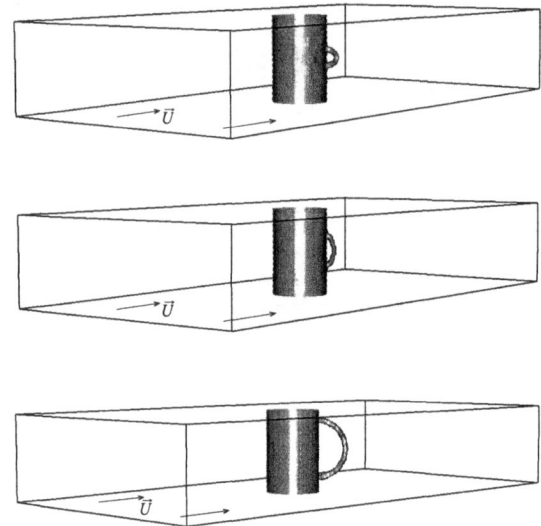

Fig. 17. Long-time dynamics for $\xi/D = 1/40$ and $|U|/c = 0.25$ starting from Fig. 16c. The half-ring moves downstream while growing.

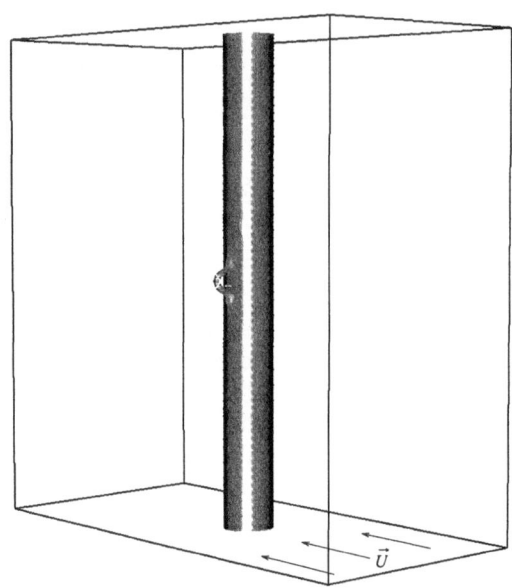

Fig. 18. Quasi-nucleation solution for $|U|/c = 0.35$ and $\xi/D = 1/20$ at time $t = 15\xi/c$.

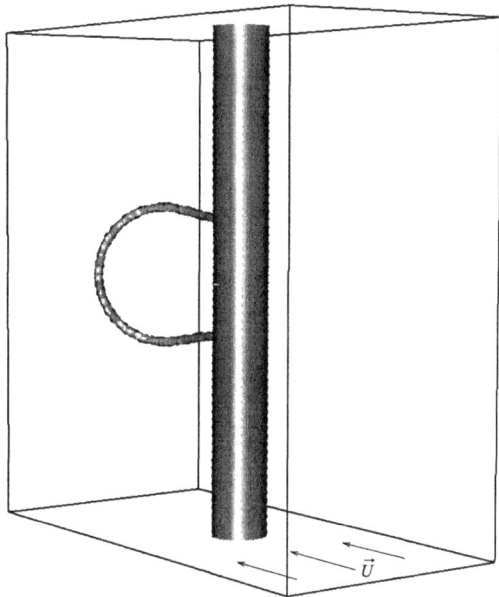

Fig. 19. Vortex stretching at $t = 150\xi/c$ with $|U|/c = 0.35$ and $\xi/D = 1/20$. The vortex line is oblique to the flow.

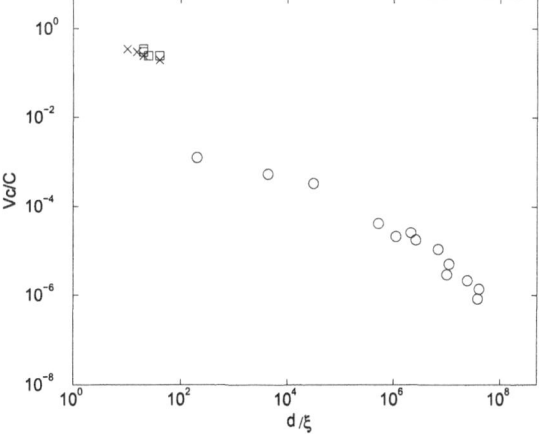

Fig. 20. Critical Mach number V_c/C versus scale ratio of numerical and experimental data D/ξ. Circles correspond to several experiments from [38]. Squares stand for our numerical stretching cases while crosses correspond to non-stretching cases [4].

corresponds to the expression $R_s = 5.5$ with

$$R_s \equiv |U|D/c\xi = MD/\xi. \tag{45}$$

This superfluid 'Reynolds' number is defined in the same way as the standard (viscous) Reynolds number $Re \equiv |U|D/\nu$ (with ν the kinematic viscosity). It has been shown, in the superfluid turbulent ($R_s \gg 1$) regime, that R_s is equivalent to the standard (viscous) Reynolds number Re [8–10]. Note that, for a Bose condensate of particles of mass m, the quantum of velocity circulation around a vortex, $\Gamma = 2\pi\sqrt{2}c\xi$, has the Onsager-Feynman value $\Gamma = h/m$ (h is Planck's constant) and the same physical dimensions L^2T^{-1} as ν.

The value of R_s divides the space of parameters into a laminar flow zone and a recirculating flow zone, very much like in the problem of a circular disc in a viscous fluid in which this frontier is also found to be around $Re \sim 5$. It therefore seems to exist some degree of universality between viscous normal fluids and superfluids modelled by NLSE as discussed in [8–10]. In the context of superfluid ^{4}He flow, the experimental critical velocity is known to depend strongly on the system's characteristic size D. It is often found to be well below the Landau value (based on the velocity of roton excitation) except for experiments where ions are dragged in liquid helium. Feynman's alternative critical velocity criterion $Rs \sim \log(D/\xi)$ is based on the energy needed to form vortex lines. It produces better estimates for various experimental settings, but does not describe the vortex nucleation mechanism [27].

In a recent experiment, Raman *et al.* have studied dissipation in a Bose-Einstein condensed gas by moving a blue detuned laser beam through the condensate at different velocities [11]. In their inhomogeneous condensate, they observed a critical Mach number for the onset of dissipation $M_{2D}^c/1.6$.

Our computations were performed for values of ξ/D comparable to those in Bose-Einstein condensed gas experiments. They demonstrate the possibility of a subcritical drag mechanism, based on 3D vortex stretching. It would be very interesting to determine experimentally the dependence of the critical Mach number on the parameter ξ/D and the nature (2D or 3D) of the excitations [4].

5 Conclusion

We have seen that the numerical tools developed in section 2 can be used in practice to obtain the stationary solutions of the NLSE. These methods have allowed us to find the full bifurcation diagrams of Bose-Einstein condensates with attractive interactions and superflows past a cylinder. Furthermore, the stationary solutions have given us efficient way to start vortical dynamics (in 2D and 3D) in a controlled manner.

6 Acknowledgements

We acknowledge useful scientific discussions with V. Hakim, L. Tuckerman and E. Varoquaux. This work was supported by the ECOS-CONICYT program no.

C96E01. Computations were performed at the Institut du Développement et des Ressources en Informatique Scientifique.

References

1. C. Huepe, S. Métens, G. Dewel, P. Borckmans, and M.-E. Brachet. Decay rates in attractive bose-einstein condensates. *Phys. Rev. Lett.*, 82(2):1616, 1999.

2. C. Huepe and M.-E. Brachet. Solutions de nucléation tourbillonnaires dans un modèle d'écoulement superfluide. *C. R. Acad. Sci. Paris*, 325(II):195–202, 1997.

3. C. Huepe and M. E. Brachet. Scaling laws for vortical nucleation solutions in a model of superflow. *to appear in Physica D*, 1999.

4. C. Nore, C. Huepe, and M. E. Brachet. Subcritical dissipation in three-dimensional superflows. *submitted to Phys. Rev. Lett.*, 1999.

5. C. Huepe. *Bifurcations et instabilités dans les condensats de Bose-Einstein et les écoulements superfluides*. PhD thesis, Ecole Normale Supérieure, 1999.

6. C. Mamun and L. Tuckerman. Asymmetry and hopf bifurcation in spherical couette flow. *Phys. Fluids*, 7(1):80, 1995.

7. Franco Dalfovo, Stefano Giorgini, Lev P. Pitaevskii, and Sandro Stringari. Theory of bose-einstein condensation in trapped gases. *Reviews of Modern Physics*, 71(3), 1999.

8. C. Nore, M. Abid, and M. Brachet. Kolmogorov turbulence in low-temperature superflow. *Phys. Rev. Lett.*, 78:3896, 1997.

9. C. Nore, M. Abid, and M. Brachet. Decaying kolmogorov turbulence in a model of superflow. *Phys. Fluids*, 9(9):2644, 1997.

10. M. Abid, M. E. Brachet, J. Maurer, C. Nore, and P. Tabeling. Experimental and numerical investigations of low-temperature superfluid turbulence. *Eur. J. Mech. B/Fluids*, 17(4):665, 1998.

11. C. Raman, M. Köhl, R. Onofrio, D.S. Durfee a nd C.E. Kuklewicz, Z. Hadzibabic, and W. Ketterle. Evidence for a critical velocity in a bose-einstein con densed gas. *Phys. Rev. Lett.*, 83(13):2502, 1999.

12. R. Seydel. *From Equilibrium to Chaos: Practical Bifurcation and Stability Analysis*. Elsevier, New York, 1988.

13. W. Press, S. Teukolsky, W. Vetterling, and B.Flannery. *Numerical Recipres in Fortran*. Cambridge Univ. Press, Cambridge, 1994.

14. D. Gottlieb and S. A. Orszag. *Numerical Analysis of Spectral Methods*. SIAM, Philadelphia, 1977.

15. R. Klein and A. J. Majda. Self-stretching of perturbed vortex filaments. *Physica D*, 53:267, 1991.

16. C. C. Bradley, C. A. Sackett, and R. G. Hulet. Bose-einstein condensation of lithium: Observation of limited condensate number. *Phys. Rev. Lett.*, 78(6):985, 1997.

17. M. Ueda and A. J. Leggett. Macroscopic quantum tunneling of a bose-einstein condensate with attractive interaction. *Phys. Rev. Lett.*, 80(8):1576, 1998.

18. H. T. C. Stoof. Macroscopic quantum tunneling of a bose condensate. *J. Stat. Phys.*, 87:1353, 1997.

19. H. Shi and W.-M. Zheng. Bose-einstein condensation in an atomic gas with attractive interactions. *Phys. Rev. A*, 55(4):2930, 1997.

20. R. J. Dodd, Mark Edwards, C. J. Williams, C. W. Clark, M. J. Holland, P. A. Ruprecht, and K. Burnett. Role of attractive interactions on bose-einstein condensation. *Phys. Rev. A*, 54(1):661, 1996.

21. C. A. Sackett, H. T. C. Stoof, and R. G. Hulet. Growth and collapse of a bose-einstein condensate with attractive interactions. *Phys. Rev. Lett.*, 80(10):2031, 1998.

22. J. Guckenheimer and P. Holmes. *Nonlinear Oscillations, Dynamical Systems and Bifurcations of Vector Fields*. Springer-Verlag, Berlin, 1983.

23. C. C. Bradley, C. A. Sackett, J. J. Tollett, and R. G. Hulet. Evidence of bose-einstein condensation in an atomic gas with attractive interactions. *Phys. Rev. Lett.*, 75(9):1687, 1995.

24. P. A. Ruprecht, M. J. Holland, K. Burnett, and Mark. Edwards. Time-dependent solution of the nonlinear schrödinger equation for bose-condensed trapped neutral atoms. *Phys. Rev. A*, 51(6):4704, 1995.

25. C.W. Gardiner. *Handbook of Stochastic Methods*. Springer-Verlag, Berlin, 1985.

26. V. M. Pérez-García, H. Michinel, and H. Herrero. Bose-einstein solitons in highly asymmetric traps. *Phys. Rev. A*, 57(5):3837, 1998.

27. R. J. Donnelly. *Quantized Vortices in Helium II*. Cambridge Univ. Press, Cambridge, 1991.

28. E. A. Spiegel. Fluid dynamical form of the linear and nonlinear schrödinger equations. *Physica D*, 1:236, 1980.

29. L. Landau and E. Lifschitz. *Mécanique des fluides*, volume 6. Editions Mir, 1989.

30. T. Frisch, Y. Pomeau, and S. Rica. Transition to dissipation in a model of super-flow. *Phys.Rev.Lett.*, 69:1644, 1992.

31. Y. Pomeau and S. Rica. Vitesse limite et nucléation dans un modèle de superfluide. *Comptes Rendus Acad. Sc. (Paris). Série II*, 316:1523, 1993.

32. C. Josserand. *Dynamique des Superfluides: Nucléation de vortex et T ransition de phase du permier ordre*. PhD thesis, Ecole Normale Supeérieure, 1997.

33. C. Josserand, Y. Pomeau, and S. Rica. Vortex shedding in a model of superflow. *Physica D*, 134(1):111–125, 1999.

34. V. Hakim. private communication. 1999.

35. O. Avenel, G.G. Ihas, and E. Varoquaux. The nucleation of vortices in superfluid ^{4}he: Answers and questions. *J. Low Temp.Phys.*, 93:1031–1057, 1993.

36. G.G. Ihas, O. Avenel, R. Aarts, R Salmelin, and E. Varoquaux. *Phys. Rev. Lett.*, 69(2):327, 1992.

37. C. A. Jones and P. H. Roberts. *J. Phys. A*, 15:2599, 1982.

38. J. Wilks. *The properties of liquid and solid helium*. Clarendon Press, Oxford, 1967.

Phase boundary dynamics in a one-dimensional non-equilibrium lattice gas

G.Oshanin[1], J.De Coninck[2], M.Moreau[1] and S.F.Burlatsky[3†]

[1] *Laboratoire de Physique Théorique des Liquides, Université Pierre et Marie Curie, 4, Place Jussieu, 75252 Paris, France*

[2] *Centre de Recherche en Modélisation Moléculaire, Université de Mons-Hainaut, 20, Place du Parc, 7000 Mons, Belgium*

[3] *Department of Chemistry, Massachusetts Institute of Technology, Cambridge, MA 02139 USA*

We study dynamics of a phase boundary in a one-dimensional lattice gas, which is initially put into a non-equilibrium configuration and then evolves in time by particles performing nearest-neighbor random walks constrained by hard-core interactions. Initial non-equilibrium configuration is characterized by an S-shape density profile, such that particles density from one side of the origin (sites $X \leq 0$) is larger (high density phase, HDP) than that from the other side (low-density phase, LDP). We suppose that all lattice gas particles, except for the rightmost particle of the HDP, have symmetric hopping probabilities. The rightmost particle of the HDP, which determines the position of the phase separating boundary, is subject to a constant force F, oriented towards the HDP; in our model this force mimics an effective tension of the phase separating boundary. We find that, in the general case, the mean displacement $\overline{X(t)}$ of the phase boundary grows with time as $\overline{X(t)} = \alpha(F)t^{1/2}$, where the prefactor $\alpha(F)$ depends on F and on the initial densities in the HDP and LDP. We show that $\alpha(F)$ can be positive or negative, which means that depending on the physical conditions the HDP may expand or get compressed. In the particular case when $\alpha(F) = 0$, i.e. when the HDP and LDP coexist with each other, the second moment of the phase boundary displacement is shown to grow with time sublinearly, $\overline{X^2(t)} = \gamma t^{1/2}$, where the prefactor γ is also calculated explicitly. Our analytical predictions are shown to be in a very good agreement with the results of Monte Carlo simulations.

---

† Present address: LSR Technologies, Inc., 898 Main St, Acton, MA 01720-5808 USA

I. INTRODUCTION.

A fundamental question concerning the behavior of systems out of equilibrium can be formulated as follows. Suppose that two different phases, composed of the same or of two different substances, are initially prepared in different regions of space and have a common interface. What is the future evolution for the system and for the phase-separating interface? This problem appears in the analysis of such diverse phenomena as expansion of the poisoned state in catalytic reactions, or, more generally, propagation of chemical fronts, dielectric breakdown, growth of dendrites and clusters in Ising magnets, spatial intermittency

O. Descalzi et al. (eds.), Instabilities and Nonequilibrium Structures VII & VIII, 69–108.
© 2004 *Kluwer Academic Publishers. Printed in the Netherlands.*

in hydrodynamics, wetting, rise of liquids in capillaries and many others (see [1–12] and references therein).

Theoretical analysis of the problem follows basically two distinct avenues. One type of approach is to describe the system evolution in terms of some appropriate set of starting equations, a standard list of which includes such non-linear differential field equations of different complexity as, e.g. Newell-Whitehead, viscous Burgers, Swift-Hohenberg, Cahn-Hilliard equations and etc [1–12]. These equations are usually referred to as "microscopic", with the understanding, however, that they don't involve atomic degrees of freedom but rather serve as elementary building blocks from which the analysis starts. Another approach consists in the direct study of models involving particles with microscopically defined dynamics [13–16]. The dynamic rules can be, for instance, chosen to construct a suitable cellular automaton, which converges asymptotically to the field equation in question, e.g. the Navier-Stokes equation [15–18], allowing then for much more efficient numerical analysis than simulations of the continuous-space counterpart. Alternatively, they can be deduced from realistic microscopic interactions with the intention to derive equations describing the time evolution of some macroscopic properties, such as, e.g. local particle densities [13,19–21]. Considerable progress has been made recently in this direction [13,19–21], which revealed, however, the fact that macroscopic equations derived on the basis of realistic microscopic interactions may have a different structure compared to the generally accepted field equations and can be reduced to them only under certain assumptions.

In the present paper* we study dynamics of the phase boundary in a two-phase microscopic model system consisting of identical hard-core particles which are placed initially on a one-dimensional, infinite in both directions lattice in a non-equilibrium, "shock"-like configuration. That is, particles mean densities from the left and from the right of the origin of the lattice (Fig.1), which we denote as ρ_- and ρ_+ respectively, are generally not equal to each other. We suppose, without lack of generality that $\rho_- \geq \rho_+ \geq 0$, and will call in what follows the phase which initially occupied the left half-line as the high-density phase (HDP), while the phase initially occupying the right half-line will be referred to as the low-density

---

*This paper is based partly on the talk given at the conference on Inhomogeneous Random Systems, Palaiseau, France, January 1997, and at the workshop on Instabilities and Non-Equilibrium Structures, Santiago, Chili, December 1997

phase or the LDP.

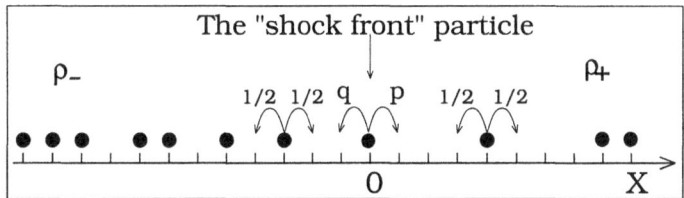

Fig. 1. Initial "shock" configuration of the lattice gas particles: ρ_- is the mean density of particles at the half-line $-\infty < X < 0$ and ρ_+, $(\rho_+ \leq \rho_-)$, is the mean density of particles at the half-line $0 < X < \infty$. All gas particles, excluding the "phase boundary" particle, have equal $(= 1/2)$ probabilities for jumps to the right and to the left. The PBP has asymmetric hopping probabilities.

Particles are then allowed to perform *symmetric*, (i.e. with equal probabilities for going to the left or to the right), hopping motion between the nearest lattice sites under the constraint that neither two particles can simultaneously occupy the same lattice site and can not pass through each other. Further on, we single out the rightmost particle of the HDP, which determines position of the phase boundary and which we will call in the following as the PBP - the "phase boundary" particle (Fig.1). We suppose that this only particle is subject to a constant force F which favors the PBP to jump in a preferential direction. Thus for the PBP the probabilities of going to the right (p) and to the left (q) will be different from each other. In most situations we will suppose that F is directed towards the HDP; we adopt the convention that in this case F is positive definite, $F \geq 0$; the PBP hopping probabilities p and q are related to the force and the reciprocal temperature β through $p/q = exp(-\beta F)$ and $p + q = 1$. From the physical point of view, such a *constant* force can be understood as an effective boundary tension derived from the solid-on-solid-model Hamiltonian of the phase-separating boundary [25–27] and mimic, in a mean-field fashion, the presence of attractive interactions between the lattice-gas particles which are not explicitly included into the model.

We hasten to remark that the system evolution in the case when long-range attractive interactions between the gas particles are present can be fairly more complex. First, in this case the hopping probabilities of any given particle are coupled to the instantaneous positions of all other gas particles and, consequently, evolution of the local densities is described by non-linear and non-local integro-differential equations (see [19,20] and references therein). Furthermore, the "boundary-tension" force F is generally time-dependent and reaches a constant value only at sufficiently large times. Moreover, this value will be explicitly dependent on the density distribution around the PBP resulting in a non-linear coupling between the dynamics of the PBP and of the lattice-gas particles. A mean-field-type analysis of this situation, which has interesting applications within the context of spreading of liquid monolayers on solid supports [22,23], has been presented recently in [28]; here, we assume that F is a fixed given parameter, which is independent of the particle density. We note that such an assumption is justified when particle-particle interactions are sufficiently weak and time t is sufficiently large [28].

We note now that several particular cases of the general model under study have been already discussed in the literature. Consequently, considering different limiting with respect to ρ_+, ρ_- and F situations we will be able to check our predictions against already known results. We mention some of these results:

(a) Dynamics of the PBP in the symmetric case $\rho_- = \rho_+ = \rho$ and in the absence of an external force F has been studied as early as 1965 by Harris [24], who has shown rigorously that the mean-square displacement of the PBP grows sublinearly with time,

$$\overline{X_r^2(t)} \; = \; \frac{1-\rho}{\rho} \, \sqrt{\frac{2t}{\pi}} \quad (1)$$

Here and henceforth the overline denotes the averaging with respect to different realizations of the PBP trajectories $X_r(t)$. Moreover, it was shown in [29] that $t^{-1/4}X_r(t)$ converges in distribution to a Gaussian variable with variance $(1-\rho)\sqrt{2/\pi}/\rho$.

(b) Further on, a rigorous probabilistic description of the situation with $\rho_- = 1$, $\rho_+ = 0$ and zero "boundary tension" force has been developed in [29]. It was proven that in this case the mean displacement of the PBP grows in time in proportion to $\sqrt{t \, log(t)}$ as $t \to \infty$.

(c) An opposite case when $F = \infty$, (such that the PBP performs totally directed random walk), while $\rho_+ = 0$ and $\rho_- = \rho$ has been considered in [30]. It was demonstrated that the mean displacement of the PBP obeys $\overline{X_r(t)} = -\alpha_{lim}\sqrt{t}$, in which law the prefactor α_{lim} is defined implicitly by

$$\sqrt{\frac{\pi}{2}} \, \alpha_{lim} \, \exp(\alpha_{lim}^2/2) \, [1 \, - \, \Phi(\alpha_{lim}/\sqrt{2})] \; = \; 1 - \rho, \quad (2)$$

where $\Phi(x)$ denotes the error function.

(d) A more general situation has been considered in [31] and subsequently, in [32], which works deal with the behavior of the driven PBP in the symmetric case $\rho_- = \rho_+ = \rho$. Both works have shown that at arbitrary negative values of the boundary tension force, $F \leq 0$, the mean-square displacement of the SFB follows

$$\overline{X_r(t)} = \alpha(F) \sqrt{t} \quad (3)$$

In [31] it was found analytically, in terms of a mean-field-type approach, and also confirmed by numerical Monte Carlo simulations that for arbitrary ρ and $p \geq q$ the parameter $\alpha(F)$ is determined by the following transcendental equation:

$$(\sqrt{\frac{\pi}{2}} \alpha(F) \exp(\alpha^2(F)/2) [1 + \Phi(\alpha(F)/\sqrt{2})] +$$

$$+ \frac{p - q(1 - \rho)}{p - q})(\sqrt{\frac{\pi}{2}} \alpha(F) \exp(\alpha^2(F)/2) \times$$

$$\times [1 - \Phi(\alpha(F)/\sqrt{2})] + \frac{q - p(1 - \rho)}{p - q}) = \frac{pq\rho^2}{(p - q)^2} \quad (4)$$

In [32], which has also established an interesting relation between the time evolution of a symmetric lattice gas with a single driven tracer and the evolution of the interfaces in a two-dimensional Potts model with Glauber dynamics, the PBP dynamics was analysed in terms of a rigorous probabilistic approach and the result in Eq.(4) has been rigorously proven.

An interesting observation made in [31] and subsequently, in [32], concerned the validity of the Einstein relation for the tracer diffusion in a one-dimensional hard-core lattice gas. An illuminating discussion of this issue and a considerable amount of new results establishing the Einstein relation for different interacting particle systems was presented recently in [33]. Now, Eq.(4) shows that in the case of a weak asymmetry, i.e. when $p - q \to 0$, the parameter $\alpha(F)$ is given exactly by

$$\alpha(F) = (p - q) \frac{1 - \rho}{\rho} \sqrt{\frac{2}{\pi}} \quad (5)$$

If one defines then the time-dependent mobility $\mu(t)$ of the driven PBP as $\mu(t) = lim_{F \to 0} X(t)/Ft$, it would yield, by virtue of Eq.(3), that $\mu(t) = t^{-1/2} lim_{F \to 0} \alpha(F)/F$, where $\alpha(F)$ is given by Eq.(5). On the other hand, the time-dependent diffusivity $D(t)$ of the PBP in absence of external force obeys $D(t) = (1 - \rho)/\rho(2\pi t)^{1/2}$ [24]. As it was shown in [31] and [32], the asymptotic form in Eq.(5) implies that $\mu(t) = \beta D(t)$, i.e. the Einstein relation holds exactly in the non-stationary regime[†].

Here we develop an analytical dynamical description of the lattice gas model with initial "shock"-like configuration of particles, which is based on the mean-field-type assumption that the average of the product of local realization-dependent variables, describing occupation of lattice sites, factorizes into the product of their average values, i.e. the local densities. Such an assumption permits us to derive the closed-form system of equations for the time evolution of the PBP mean displacement and of the density profiles around it. These equations allow for the analytical solution and explicit computation of the time evolution of the PBP mean displacement. Our main results are the following:

We find that at sufficiently large times the mean displacement of the PBP obeys Eq.(3), in which the prefactor $\alpha(F)$ is determined implicitly as the solution of the transcedental equation

$$q\,\rho_- \left\{1 \,+\, \sqrt{\pi/2}\,\alpha(F)\,\exp(\alpha^2(F)/2)\,[1 \,+\, sgn(\alpha(F))\,\Phi(|\alpha(F)|/\sqrt{2})]\right\}^{-1} -$$

$$-\,p\,\rho_+ \left\{1 \,-\, \sqrt{\pi/2}\,\alpha(F)\,\exp(\alpha^2(F)/2)\,[1 \,-\, sgn(\alpha(F))\,\Phi(|\alpha(F)|/\sqrt{2})]\right\}^{-1} \,=\, q\,-\,p, \quad (6)$$

where $sgn(\alpha) = 1$ for $\alpha > 0$ and $sgn(\alpha) = -1$ for $\alpha \leq 0$. Equation (6) holds for any relation between p, q and $\rho_\pm$ and reduces to the previously obtained Eq.(2) and Eq.(4) in the appropriate limits. This result has been found subsequently in [32].

Equation (6) predicts that three different regimes can take place depending on the relation between p/q and $\rho_\pm$:

(1) When $p(1-\rho_+) > q(1-\rho_-)$ the parameter $\alpha(F)$ is finite and positive definite, which means that the HDP expands compressing the LDP. In the particular case $p/q = 1$ and $\rho_+ = 0$ the parameter $\alpha(F)$ appears to be a positive, logarithmically growing with time function, which behavior agrees with the results of [29].

(2) When $p(1 - \rho_+) = q(1 - \rho_-)$, the parameter $\alpha(F) = 0$. This relation between the system parameters when the HDP and the LDP are in equilibrium with each other and the PBP mean displacement is zero, was found also in [33] from the analysis of the stationary

---

[†]Two of us, G.O. and S.F.B., wish to thank Professor J.L.Lebowitz who has suggested us to examine the question of validity of the Einstein relation in the non-stationary regime.

behavior in a finite one-dimensional lattice gas (see also Section VI.A). Despite the fact that the PBP mean displacement is zero, the fluctuations in the PBP position grow with time. We show here that in this case the mean-square displacement of the PBP obeys

$$\overline{X_r^2(t)} = \frac{(1 - \rho_-)(1 - \rho_+)}{(\rho_- + \rho_+ - 2\,\rho_-\,\rho_+)} \sqrt{\frac{8t}{\pi}}, \quad (7)$$

which reduces to the classical result in Eq.(1) in the limit $p = q$ and $\rho_- = \rho_+$. Eq.(7) is derived here using heuristic arguments based on the Einstein relation between the mobility and diffusivity of a test particle and is confirmed by Monte Carlo simulations.

(3) When $p(1 - \rho_+) < q(1 - \rho_-)$ the parameter $\alpha(F)$ is less than zero - the expanding LDP and the applied force effectively compress the HDP.

Further on, we show that the particles density profile as seen from the moving PBP, stabilizes around its position, approaching a constant value. We find that in the regime when the HDP expands (i.e. $\alpha(F) > 0$) the density profile around the PBP is given by

$$\rho(X < X(t); t) \approx \frac{\rho_-}{1 + I_+(|\alpha(F)|)} [1 + \alpha(F)\, t^{-1/2}(X(t) - X) + ...\,], \quad (8.a)$$

$$\rho(X > X(t); t) \approx \frac{\rho_+}{1 - I_-(|\alpha(F)|)} [1 + \alpha(F)\, t^{-1/2}(X(t) - X) + ...\,], \quad (8.b)$$

for $X(t) - X \ll \sqrt{t}/A$. Within the opposite limit ($\alpha(F) < 0$) when the HDP gets compressed, $\rho(X; t)$ follows

$$\rho(X < X(t); t) \approx \frac{\rho_-}{1 - I_-(|\alpha(F)|)} [1 + \alpha(F)\, t^{-1/2}(X(t) - X) + ...\,], \quad (9.a)$$

$$\rho(X > X(t); t) \approx \frac{\rho_+}{1 + I_+(|\alpha(F)|)} [1 + \alpha(F)\, t^{-1/2}(X(t) - X) + ...\,], \quad (9.b)$$

where

$$I_\pm(|\alpha(F)|) = \alpha^2(F) \int_0^\infty dz \, \exp(-\frac{\alpha^2(F)}{2}\,(z^2 \mp 2z) =$$

$$= \sqrt{\frac{\pi}{2}}\,|\alpha(F)|\, \exp(\frac{\alpha^2(F)}{2})\,[1 \pm \Phi(|\alpha(F)|/\sqrt{2})] \quad (10)$$

This paper is outlined as follows: in Section 2 we formulate our model and discuss the approximations involved. In Section 3 we write down basic equations, describing the time evolution of the PBP and of the lattice-gas particles in the particular case when the LDP is absent. Section 4 presents solutions of the dynamic equations in the case $\rho_+ = 0$ and discussion of the PBP dynamics at different values of the initial mean density ρ_- and of the "boundary tension" force F. Further on, in Section 5 we consider the general case when the LDP is present and evaluate the dynamical equations describing the time behavior of the system under study. Section 6 is devoted to the analysis of these equations and evaluation of their solutions. Next, Section 7 contains the description of the Monte Carlo simulations algorithm. Finally, in Section 8 we conclude with a brief summary of results and discussion.

II. THE MODEL.

Consider a one-dimensional, infinite in both directions regular lattice of unit spacing, the sites $\{X\}$ of which are partly occupied by identical particles. Suppose next that the initial configuration of particles is as depicted in Fig.1, i.e. all particles are placed on the lattice with the single occupancy condition, at random positions and in such a way that the mean particle density (ρ_-) at sites $X \leq 0$ is different from the mean particle density (ρ_+) at the sites with $X > 0$. As we have already mentioned, we supposed that $\rho_- \geq \rho_+ \geq 0$ and call the gas phase from the left of the PBP as the HDP and the gas phase from the right of the PBP as the LDP.

After deposition onto the lattice, the particles are allowed to move by attempting jumps to neighboring sites. The motion is constrained by the hard-core exclusion between the particles; that is, neither two particles can simultaneously occupy the same lattice site nor can pass through each other. For a given realization of the process the instantaneous particle configuration is described by an infinite set of time-dependent occupation variables $\{\tau_X(t)\}$, where each $\tau_X(t)$ assumes two possible values; namely, $\tau_X(t) = 1$ if the site X is occupied at time t and $\tau_X(t) = 0$ if this site is vacant. Position of the PBP at time t is denoted as $X_r(t)$, which is also a random, realization-dependent function.

More specifically, we define particles dynamics as follows: each particle waits a random, exponentially distributed time with mean 1 and then selects, with given probabilities, the direction of jump - to the right or to the left. When the direction of the jump is chosen, the particle attempts to jump onto the nearest site. If the target site is unoccupied by any other particle at this moment of time, the jump is instantaneously fulfilled. If the site is occupied, the particle remains at its position and waits till the next attempt. The process is memory-less and the choice of jump directions for different attempts is uncorrelated.

We will distinguish between the jump probabilities of the PBP and the jump probabilities of all other particles of the gas. The jump probabilities of the gas particles are *symmetric*, i.e. for them an attempt to jump to the right and an attempt to jump to the left occur with probability 1/2, while the jump probabilities of the PBP are *asymmetric*; it attempts to jump to the right with probability p and to the left with probability q, $p + q = 1$. These probabilities are related to the "boundary tension" force and the temperature $T = 1/\beta$ through the relation $p/q = exp(-\beta F)$.

Here we will develop a mean-field-type description of the time evolution of the system under study using an approximate picture, based on two assumptions:

We assume first that the average of the product of the occupation variables $\tau_X(t)$ of different sites factorizes into the product of their average values, which corresponds to the local equilibrium assumption[‡]. Under such an assumption we can describe the system evolution directly in terms of the realization-averaged values $\rho(X;t) =< \tau_X(t) >$, which define the local density of the site X at time t or, in other words, the probability that the site X is occupied by a lattice gas particle at time t. The resulting equations will then be closed with respect to $\rho(X;t)$, i.e. will not include higher-order correlation function. A non-trivial aspect of these equations, which is common for diverse front propagation problems [6], is that one of the boundary conditions is imposed in the moving frame.

Secondly, the evolution of the spatial position of the phase boundary will be described in terms of $P(X;t)$, which defines the probability of having the PBP at site X at time t. Anticipating that in the limit $t \to \infty$ the ratio

$$\frac{\overline{X_r(t)}}{\sqrt{\overline{X_r^2(t)}}} \to \infty, \quad (11)$$

i.e. that fluctuations in the PBP trajectory grow at essentially slower rate than its mean displacement, we will neglect fluctuations in the PBP trajectories, supposing that the position of the PBP at time t is a well-defined function of time, which is the same for all realizations

---

[‡]Note, that the density profiles around the PBP, as shown by Eqs.(8) and (9), tend to constant values on progressively larger and larger scales as the PBP advances; that is, there is no structure in the density profiles and they are merely the product measures. This "propagation of local equilibrium" [34] insures that the decoupling procedure involved is correct in a large time scale.

of the process, $X_r(t) = X(t)$. We note that such an assumption is quite consistent with rigorous results presented in [32].

These two simplifying assumptions will allow us to determine explicitly the dynamics of the PBP and to calculate the density profiles as seen from the PBP. Our analytical predictions will be checked against the results of Monte Carlo simulations of the stochastic process, described in the beginning of this Section, which allows a direct control of the validity of the local equilibrium assumption.

III. DYNAMICAL EQUATIONS IN THE ABSENCE OF THE LOW-DENSITY PHASE.

We consider first dynamics of the PBP in the particular case when the LDP is absent (Fig.2), i.e. $\rho_+ = 0$, and thus the jumps of the PBP away from the HDP are unconstrained.

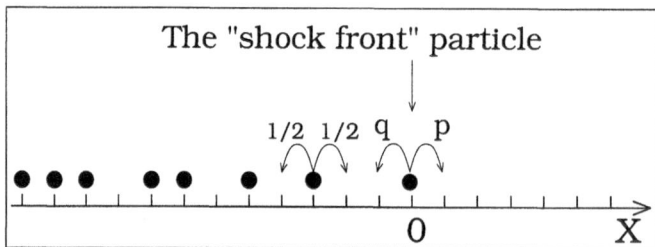

Fig. 2. Initial "shock" configuration of the lattice gas particles. All particles are placed at random positions and at a fixed mean density ρ_- from the left to the origin of the lattice, i.e. at sites $X < 0$. All sites $X > 0$ are vacant at $t = 0$.

We start with the derivation of the dynamical equations, which govern the time evolution of $\rho(X;t)$. In the continuous-time limit and under the assumption of the factorization of the occupation variables at different sites, dynamics of $\rho(X;t)$ is guided by the following balance equation

$$\dot{\rho}(X;t) = -\frac{1}{2}\,\rho(X;t)\,\{1\,-\,\rho(X+1;t)\,+\,1\,-\,\rho(X-1;t)\}\,+$$

$$+\frac{1}{2}\,(1\,-\,\rho(X;t))\,\{\rho(X+1;t)\,+\,\rho(X-1;t)\}, \quad (12)$$

where the terms in the first two lines describe the contribution due to the jumps from the occupied site X onto the neighboring unoccupied sites, while the terms in the third line account for possible arrivals of particles to unoccupied site X from the occupied adjacent sites. One may readily notice that in Eq.(12) non-linear terms cancel each other and it reduces to the discrete-space diffusion equation

$$\dot{\rho}(X;t) = \frac{1}{2}\{\rho(X+1;t) + \rho(X-1;t) - 2\,\rho(X;t)\} \quad (13)$$

We note now that in deriving Eq.(13) we have implicitly supposed that the jump probabilities of particles arriving from the site $X+1$ are symmetric. This means that Eq.(13) holds only for the sites X, which are inaccessible for the PBP, whose jumping probabilities are asymmetric by definition. Consequently, Eq.(13) is valid only for the sites X such that $X < X(t) - 1$. For $\rho(X(t)-1;t)$ we have instead of Eq.(12)

$$\dot{\rho}(X(t)-1;t) = \frac{1}{2}\{\rho(X(t)-2;t)\,(1 - \rho(X(t)-1;t)) - $$

$$- \rho(X(t)-1;t)\,(1 - \rho(X(t)-2;t))\}+$$

$$+ q\,(1 - \rho(X(t)-1;t))\,\rho(X(t)-2;t) - p\,\rho(X(t)-1;t) \quad (14)$$

The terms in the first two lines of Eq.(14) describe exchanges of particles between the sites $X(t) - 1$ and $X(t) - 2$. The particles which may be involved in these exchanges are the lattice gas particles which have symmetric jumping probabilities and here, again, the non-linear terms cancel each other. Two terms in the third line of Eq.(14) account for the effective change in the occupation of the $(X(t)-1)$-site due to jumps of the PBP and are defined in the frame of reference moving with the phase boundary. The first term describes creation of a vacancy at a previously occupied site $(X(t)-1)$ due to the unconstrained jump of the PBP away of the gas phase. The second one accounts for the effective creation of a particle at $(X(t)-1)$ in the event when the PBP jumps onto unoccupied site $(X(t)-1)$ and the site $(X(t)-2)$ is occupied prior to the jump.

Similar reasonings yield the following equation for the time evolution of the probability distribution $P(X;t)$, which defines the PBP dynamics,

$$\dot{P}(X;t) = -P(X;t)\{p + q\,(1 - \rho(X(t)-1;t))\} + $$

$$+ p\,P(X-1;t) + q\,(1 - \rho(X(t);t))\,P(X+1;t) \quad (15)$$

Multiplying both sides of Eq.(15) by X and summing over all lattice sites we find that the displacement of the PBP obeys

$$\dot{X}(t) \; = \; p \; - \; q \; + \; q \, f(1;t), \quad (16)$$

where we took into account the normalization condition $\sum_X P(X;t) = 1$ and denoted as $f(\lambda;t)$ the pair-wise correlation function

$$f(\lambda;t) \; = \; \sum_X P(X;t) \, \rho(X - \lambda;t) \quad (17)$$

Equation (17) defines the probability of having at time moment t a particle at distance λ from the PBP, or, in other words, can be interpreted as the density profile as seen from the moving PBP.

We now turn to the time evolution of $f(\lambda;t)$. Differentiating the pair-wise correlation function in Eq.(17) with respect to time, we have

$$\dot{f}(\lambda;t) \; = \; \sum_X \{\dot{\rho}(X - \lambda;t) \, P(X;t) \; + \; \dot{P}(X;t) \, \rho(X - \lambda;t)\} \quad (18)$$

We notice now that again the behavior for $\lambda = 1$ and $\lambda > 1$ has to be considered separately. In the domain $\lambda > 1$ we find, taking advantage of Eqs.(13) and (15), that $f(\lambda;t)$ obeys

$$\dot{f}(\lambda;t) \; = \; \frac{1}{2} \, \{f(\lambda - 1;t) \; + \; f(\lambda + 1;t) \; - \; 2 \, f(\lambda;t)\} \; -$$

$$- \, (p - q) \, f(\lambda;t) \; + \; p \, f(\lambda - 1;t) \; + \; q \, f(\lambda + 1;t) \; -$$

$$- \, q \, \sum_X P(X;t) \, \rho(X - 1;t) \, \rho(X - \lambda - 1;t) \; +$$

$$+ \, q \, \sum_X P(X;t) \, \rho(X - 1;t) \, \rho(X - \lambda;t) \quad (19)$$

We proceed further on making the same simplifying assumption, which underlies the derivation of Eqs.(12) to (15), i.e. assuming that the average of the product of the occupation variables decouples into the product of the average values. This means that the two last terms in Eq.(19) can be rewritten as

$$\sum_X P(X;t) \, \rho(X - 1;t) \, \rho(X - \lambda - 1;t) \; =$$

$$= \{\sum_X P(X;t)\, \rho(X-1;t)\}\, \{\sum_{X'} P(X';t)\, \rho(X'-\lambda-1;t)\}, \quad (20.a)$$

and

$$\sum_X P(X;t)\, \rho(X-1;t)\, \rho(X-\lambda;t) \;=\;$$

$$=\; \{\sum_X P(X;t)\, \rho(X-1;t)\}\, \{\sum_{X'} P(X';t)\, \rho(X'-\lambda;t)\} \quad (20.b)$$

Decoupling of the third-order correlation functions as in Eqs.(20) permits us to cast Eq.(19) into the following form

$$\dot{f}(\lambda;t) \;=\; \frac{1}{2}\, \{f(\lambda-1;t) \;+\; f(\lambda+1;t) \;-\; 2\, f(\lambda;t)\} \;-$$

$$-\, f(\lambda;t) \;+\; p\, f(\lambda-1;t) \;+\; q\, f(\lambda+1;t) \;-$$

$$-\, q\, f(1;t)\, (f(\lambda+1;t) \;-\; f(\lambda;t)) \quad (21)$$

Equation (21) does not include now the third-order correlation functions and thus is closed with respect to $f(\lambda;t)$.

Next, using Eqs.(18), (15) and (14) and decomposing the third-order correlation functions into the product of pair-wise correlations, we obtain for the time evolution of $f(1;t)$:

$$\dot{f}(1;t) \;=\; \frac{1}{2}\, \{f(2;t) \;-\; f(1;t)\} \;-$$

$$-\, f(1;t) \;+\; 2\, q\, f(2;t)\, (1 \;-\; f(1;t)) \;+$$

$$+\, p\, (f(0;t) \;-\; f(1;t)) \;+\; q\, f^2(1;t) \quad (22)$$

From Eq.(22) we can now deduce the boundary condition for Eq.(21). Setting in Eq.(21) the correlation parameter λ equal to 1 and comparing the terms in the rhs of Eq.(21) against the terms in the rhs of Eq.(12) we can infer that $f(1;t)$ obeys

$$\frac{1}{2}\, (f(1;t) \;-\; f(0;t)) \;=\; p\, f(1;t) \;-\; q\, f(2;t)\, (1 \;-\; f(1;t)) \quad (23.a)$$

Another pair of boundary conditions will be

$$f(\lambda;0) \;=\; \left(\sum_X P(X;t)\,\rho(X-\lambda;t)\right)\Bigg|_{t=0} \;=\; \rho_-, \quad (23.b)$$

and

$$f(\lambda \to \infty;t) \;\to\; \rho_-, \quad (23.c)$$

which mean that initially the lattice gas particles are uniformly distributed, with mean density ρ_-, on the half-line $X < 0$, and that the density of the lattice gas at large separations from the phase boundary is equal to its unperturbed value.

Equations (16) and (21) to (23) constitute a closed system of equations which allows a complete determination of $X(t)$. Solution of these equations will be discussed in the next section.

IV. SOLUTION OF DYNAMICAL EQUATIONS IN THE CASE $\rho_+ = 0$.

We now turn to the continuous-space limit and rewrite our equations expanding $f(\lambda\pm1;t)$ into the Taylor series and retaining terms up to the second order in powers of the lattice spacing. We then find that $f(\lambda;t)$ obeys

$$\dot{f}(\lambda;t) \;=\; \frac{1}{2}\frac{\partial^2}{\partial\lambda^2}f(\lambda;t) \;-\; \dot{X}(t)\frac{\partial}{\partial\lambda}f(\lambda;t), \quad (24)$$

while Eq.(23.a) transforms to

$$\frac{1}{2}\frac{\partial}{\partial\lambda}f(\lambda;t)\Bigg|_{\lambda=1} \;=\; \dot{X}(t)\,f(1;t), \quad (25)$$

where, by virtue of Eq.(16), we have replaced the multiplier $(p - q + qf(1;t))$ by $\dot{X}(t)$. We note that Eqs.(24) and (25) hold for any relation between p and q (any orientation of the force F), but in the absence of the LDP the analysis of the case $p > q$ does not make any sense. Consequently, in this section we will consider only the case when $p \le q$.

A. Expansion of the gas phase.

Let us first consider the solution of Eqs.(24) and (25) supposing that $X(t) > 0$, $(X(0) = 0)$. Conditions when such a behavior takes place will be defined below. We notice that the structure of Eqs.(24) and (25) calls for the scaling solution in terms of variable $\omega = (\lambda - 1)/X(t)$; $0 \le \omega \le \infty$. In terms of this variable Eqs.(24) and (25) can be rewritten as

$$\frac{\partial^2}{\partial \omega^2} f(\omega) \; + \; (\frac{d}{dt} X^2(t)) \; (\omega - 1) \; \frac{\partial}{\partial \omega} f(\omega) \; = \; 0, \quad (26)$$

and

$$\left. \frac{\partial}{\partial \omega} f(\omega) \right|_{\omega=0} \; = \; (\frac{d}{dt} X^2(t)) \; f(\omega = 0), \quad (27)$$

while Eqs.(23.b) and (23.c) collapse into a single equation

$$f(\omega = \infty) \; = \; \rho_- \quad (28)$$

Solution of Eqs.(26) to (28) can be readily obtained in an explicit form if we assume that $dX^2(t)/dt = A^2$, where A is a time-independent constant, $0 \leq A < \infty$. Such an assumption actually makes sense if we recollect results of [29] and [30–32], which demonstrated that in two extreme situations, i.e. when $p/q = 0$ (totally directed walk of the PBP) and when $p/q = 1$ (no force exerted on the PBP), the PBP displacement shows the same generic behavior $X(t) \sim \sqrt{t}$. Hence, one can expect that for arbitrary p/q, $0 \leq p/q \leq 1$, the PBP displacement should also grow in proportion to $\sqrt{t}$.

Time-independent A ($p/q > 1$). The general solution of Eq.(26) has the form

$$f(\omega) \; = \; C_1 \int_0^\omega dz \; \exp(-\frac{A^2}{2} (z^2 - 2z)) \; + \; C_2, \quad (29)$$

where C_1 and C_2 are adjustable constants. Substitution of Eq.(29) into Eq.(27) gives

$$C_1 \; = \; A^2 \, C_2, \quad (30)$$

while Eq.(28) yields the second relation

$$C_1 \int_0^\infty dz \; \exp(-\frac{A^2}{2} (z^2 - 2z)) \; + \; C_2 \; = \; \rho_- \quad (31)$$

Consequently, we have for the density profile

$$f(\omega) \; = \; \frac{\rho_-}{1 + I_+(A)} \; \{1 \; + \; A^2 \int_0^\omega dz \; \exp(-\frac{A^2}{2} (z^2 - 2z))\}, \quad (32)$$

where the function $I_+(A)$ has been made explicit in Eq.(10).

The density $f(\omega)$ in Eq.(32) is a function of A, which still remains undetermined. To define A we notice that $X(t) \sim \sqrt{t}$ behavior and Eq.(16) imply that

$$f(\omega = 0) \; \rightarrow \; \frac{q - p}{q}, \; \text{as } t \; \rightarrow \; \infty, \quad (33)$$

84

and consequently, we have that in the limit $t \to \infty$ the parameter A approaches a constant, time-independent value which obeys

$$I_+(A) = \frac{p - q(1 - \rho_-)}{q - p} \quad (34)$$

Equation (34) implicitly determines A as a function of p/q and ρ_-. Numerical solution of this equation is presented in Fig.3.

Now, a simple analysis shows that Eq.(34) has a unique positive solution for any p and q which satisfy $p > q(1 - \rho_-)$, (or, since $p = 1 - q$, such q which are less than $1/(2 - \rho_-)$). When $p/q \to 1 - \rho_-$, the parameter $A \to 0$ as

$$A \approx \sqrt{\frac{2}{\pi}} \frac{p - q(1 - \rho_-)}{q\rho_-}, \quad (35)$$

and is exactly equal to zero for $p/q = 1 - \rho_-$. It means that in the domain of parameters such that $p > q(1 - \rho_-)$, the gas phase expands and the phase boundary moves as $X(t) = A\sqrt{t}$, $A > 0$. Before we turn to the analysis of the behavior of the PBP in the regime $p < q(1 - \rho_-)$, let us mention some other interesting aspects of Eqs.(34) and (32).

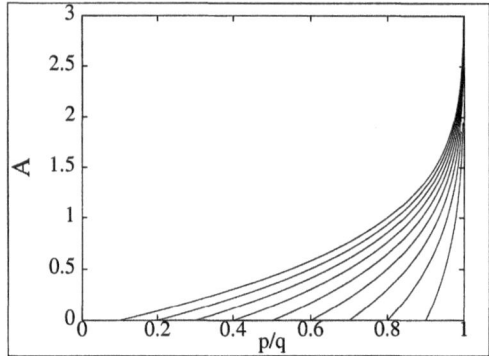

Fig. 3. Numerical solution of Eq.(34). The parameter A is plotted versus the ratio p/q for different values of the mean density ρ_-. The curves from the left to the right correspond to $\rho_- = 0.9, 0.8, 0.7, \ldots, 0.1$.

Time-dependent A ($p/q = 1$). We note that Eq.(34) predicts that A diverges logarithmically when $p/q \to 1$ (Fig.3). Namely,

$$A \approx \sqrt{-2 \, ln(1 - \frac{p}{q})}, \quad (36)$$

which means apparently that when $p = q$ the parameter A is some increasing function of time. This is, of course, consistent with the result of [29] which states that the mean displacement of the PBP obeys $X(t) \sim \sqrt{t\, ln(t)}$ for the lattice gas with $\rho_- = 1$ and $p = q = 1/2$. Let us now estimate, in terms of our approach, the behavior of A for arbitrary ρ_- and $p = q = 1/2$. For $p = q$ our Eq.(16) reduces to

$$\dot{X}(t) \; = \; f(\omega = 0) \quad (37)$$

Next, supposing that $X(t)$ still follows the law $X(t) = A\sqrt{t}$, in which the prefactor A may be a slowly varying function of time, such that $A/\sqrt{t} \to 0$ when $t \to \infty$, we find that the representation of $f(\lambda; t)$ in terms of a single scaled variable ω is still appropriate; weak time-dependence of parameter A actually results in the appearence of vanishing in time correction terms. We have then that the boundary condition in Eq.(25) reads

$$\left.\frac{\partial}{\partial \omega} f(\omega)\right|_{\omega=0} \; \approx \; \frac{A^3}{2\sqrt{t}}, \; \text{when } t \; \to \; \infty \quad (38)$$

On the other hand, we can calculate the derivative of $f(\omega)$ directly, using Eq.(32). This gives

$$\left.\frac{\partial}{\partial \omega} f(\omega)\right|_{\omega=0} \; \approx \; \frac{\rho_- A}{\sqrt{2\pi}} \; \exp(-A^2/2) \quad (39)$$

Comparing next the rhs of Eqs.(38) and (39), we infer that the parameter A obeys

$$A^2 \; \exp(A^2/2) \; \approx \; \sqrt{\frac{2\rho_-^2 t}{\pi}}, \quad (40)$$

which yields

$$A \; \approx \; \sqrt{ln(\frac{2\rho_-^2 t}{\pi})} \quad (41)$$

Equation (41) thus generalizes the result of [29] for arbitrary initial mean density ρ_-.

Wandering of the PBP in the critical case $p/q = 1 - \rho_-$. Here we present some heuristic estimates of the time evolution of the second moment of the distribution $P(X; t)$ in the case when the gas phase does not "wet" the region $X > 0$, i.e. when $X(t) = 0$. To do this, let us recall the Einstein relation between the diffusion coefficient D of a particle, which performs an unconstrained symmetric random walk in absence of external forces, and the mobility μ of the same particle in the case when an external constant force is present. The

Einstein relation states that $\mu = \beta D$. Of course, it is not clear *apriori* whether the Einstein relation between the diffusion coefficient and the mobility should hold also for the tracer particle diffusing in a one-dimensional lattice gas; indeed, it may be invalidated because of the hard-core interactions. This question has been addressed for the first time in [33], in which work several important advancements have been made. To illustrate some of the results obtained in [33], which are relevant to the model under study, let us first define the mobility of the tracer particle:

$$\mu \;=\; lim_{t \to \infty} \mu(t), \quad (42)$$

where $\mu(t)$ denotes

$$\mu(t) \;=\; lim_{F \to 0} \frac{X(t)}{Ft}, \quad (43)$$

i.e. $\mu(t)$ is the ratio of the mean displacement $X(t)$ of the tracer particle, diffusing in the presence of constant external force F, and $F\,t$; the ratio being taken in the limit when the external force tends to the critical value (zero) at which the mean displacement vanishes. Next, the diffusion coefficient of the tracer particle is defined by

$$D \;=\; lim_{t \to \infty} D(t) \;=\; lim_{t \to \infty} \{\frac{\overline{X_r^2(F=0,t)}}{2t}\}, \quad (44)$$

where $\overline{X_r^2(F=0,t)}$ denotes the mean-square displacement of the tracer particle in the case when the external force is equal to its critical value (zero). Now, for the tracer diffusion in a one-dimensional hard-core lattice gas one has that $\overline{X_r^2(F=0,t)}$ obeys Eq.(1), while $X(t)$ is determined by Eqs.(3) and (4).

One readily notices now that the Einstein relation holds trivially for infinitely large systems, since here both μ and D are equal to zero [33]. A more striking result obtained in [33] concerned the case when the one-dimensional lattice is a closed ring of length L. It was shown that here both μ and D are finite, both vanish with the length of the ring as $1/L$ and obey the Einstein relation $\mu(L) = \beta D(L)$ exactly! Next, [31] and subsequently, [32], focused on the non-stationary behavior in infinite systems and showed that the Einstein relation holds in an even more general sense: namely, the time-dependent mobility $\mu(t)$ and the diffusivity $D(t)$ obey

$$\mu(t) \;=\; \beta\, D(t), \quad (45)$$

at times t sufficiently large, such that the asymptotical regimes described by Eqs.(1) and (3) are established.

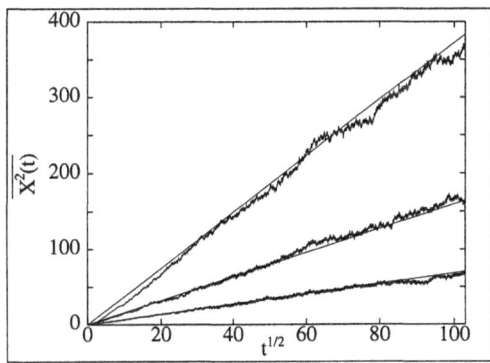

Fig. 4. Mean-square displacement of the PBP in the critical case $p/q = 1 - \rho_-$. The solid lines depict our analytical prediction in Eq.(48). The noisy lines give the corresponding results of Monte Carlo simulations. The curves from the top to the bottom correspond to $\rho_- = 0.3$, 0.5 and 0.7 respectively.

Now, in the situation under study we have non-zero critical force ($A = 0$ when $p = q(1 - \rho_-)$ or, in other words, when $F = F_c = -\beta^{-1}ln(1 - \rho_-)$). We thus define the time-dependent mobility $\mu(t)$ as

$$\mu(t) \;=\; lim_{F \to F_c} \frac{X(t)}{(F - F_c)\,t}, \quad (46)$$

which yields, by virtue of Eq.(35), the following result

$$\mu(t) \;=\; \beta\,\frac{1 - \rho_-}{\rho_-}\,\sqrt{\frac{2}{\pi t}} \quad (47)$$

Assuming next that the generalized Einstein relation in Eq.(45) holds in this case, we find that in the critical case $p = q(1 - \rho_-)$ the mean-square displacement of the PBP obeys:

$$\overline{X_r^2(F = F_c, t)} \;=\; \frac{1 - \rho_-}{\rho_-}\,\sqrt{\frac{8t}{\pi}}, \quad (48)$$

which is surprisingly similar to the classic result in Eq.(1).

In Fig.4 we compare our analytical prediction in Eq.(48) against the results of Monte Carlo simulations, performed at three different values of the gas phase densities ρ_-. It shows that our Eq.(48) is in a good agreement with the numerical results. This means that the Einstein relation in Eq.(45) holds even in such a "pathological" situation, in which the critical value of the external force is not equal to zero and the particle density is different from both sides of the test particle.

Density profiles. Let us now analyse the form of the density profiles as seen from the PBP. In Figs.5 and 6 we plot the result in Eq.(32) versus the scaled variable ω for different initial mean densities and different values of the ratio p/q.

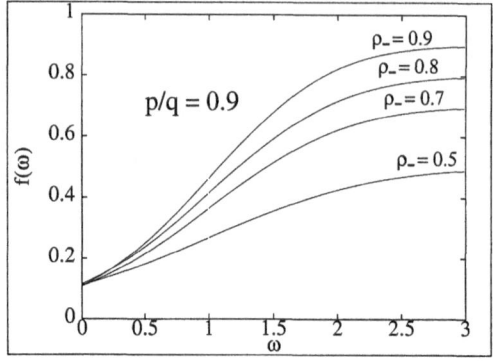

Fig. 5. Expansion of the gas phase. Plot of $f(\omega)$, Eq.(32), versus $\omega = (\lambda - 1)/X(t)$ at fixed p/q and ρ_-. The corresponding values of the parameter A for the curves from the top $(\rho_- = 0.9)$ to the bottom $(\rho_- = 0.5)$ are $A = 1.37, 1.3, 1.25$ and 1.

In Fig.5 we depict $f(\omega)$ for fixed $p/q = 0.9$, which corresponds to fixed "boundary tension" force F, and different initial mean densities ρ_-. In the range of used parameters, all corresponding values of A are of the same order $(A \approx 1)$ and the density curves look quite similar; starting from the same value at $\omega = 0$, $f(\omega = 0) = 1 - p/q = 0.1$, they quite rapidly, within a few units of ω, approach their unperturbed initial values.

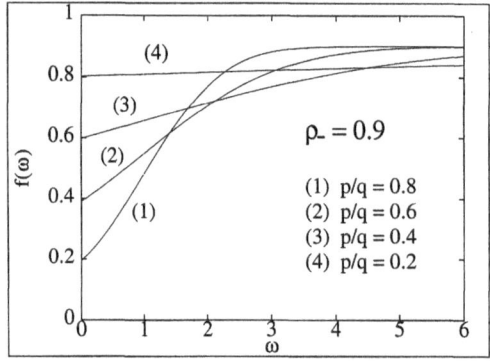

Fig. 6. Expansion of the gas phase. Plot of $f(\omega)$, Eq.(32), versus ω at fixed density $\rho_- = 0.9$ and variable ratio p/q. The corresponding values of the parameter A are 0.09 (4), 0.31 (3), 0.58 (2) and 1 (1).

We note, however, that on the X-scale it does not mean that the density past the rightmost particle rapidly reaches the unperturbed value ρ_-. Instead, at sufficiently large times the density stays almost constant and equal to $1 - p/q$ within a macroscopically large region $\sim X(t)$.

Now, in Fig.6 we plot $f(\omega)$ versus ω in the opposite case when ρ_- is fixed and the "boundary tension" force is varied. Here the density profiles display rather strong dependence on the parameter A. When A is small $f(\omega)$ shows almost linear dependence on ω (curves (3) and (4)). The reason for such a behavior is that here the phase boundary moves essentially slower $(A < 1)$, compared to the typical displacements of the gas particles, which then have sufficient time to equilibrate the density profile past the PBP. In the opposite case of relatively large values of A, $(A > 1)$, such an equilibration does not take place and the dependence of $f(\omega)$ on ω is progressively more pronounced the larger A is.

It may also be worth-while to discuss the shapes of the density profiles in terms of the variables λ and t. First, from Eq.(32) we have that in the limit of small ω, i.e. $\lambda \ll X(t)$, the density obeys

$$f(\lambda; t) \approx (1 - \frac{p}{q}) [1 + \frac{A(\lambda - 1)}{\sqrt{t}} + ...], \quad (49)$$

which means that past the PBP the density is almost constant in the region whose size grows in proportion to $X(t)$. Next, within the opposite limit, i.e. at distances λ which exceed considerably $X(t)$, we obtain from Eq.(32) the following result

$$f(\lambda, t) \approx \rho_- - (1 - \frac{p}{q}) \frac{A\sqrt{2t}}{\lambda} \exp(-\frac{\lambda^2}{2t}) + ... \quad (50)$$

Equation (50) shows that at large separations from the phase boundary the density approaches the unperturbed value ρ_- exponentially fast. The approach is from below and is weakly (only through the prefactors) dependent on the parameters p/q and A.

Mass of particles and mean density. We close this subsection with a brief analysis of the time evolution of the integral characteristic of the propagating gas phase; namely, of the "mass" $M(t)$ and the mean density $\rho_{mean} = M(t)/X(t)$ of lattice gas particles at sites $X > 0$ at time t.

The parameter $M(t)$, which measures the amount of the gas-phase particles which emerged up to time t in the previously empty half-line $X > 0$, is formally defined as

$$M(t) = \int_0^{X(t)} dX \, \rho(X; t) \quad (51)$$

Changing the variable of integration, we find that $M(t)$ can be rewritten as

$$M(t) \;=\; \int_0^{X(t)} d\lambda \; f(\lambda;t) \;=\;$$

$$=\; X(t) \int_0^1 d\omega \; f(\omega) \;=\; M \; t^{1/2}, \quad (52)$$

where M is given by

$$M \;=\; A \; (1 - \frac{p}{q}) \; exp(A^2/2) \quad (53)$$

Figure 7 displays the plot of the prefactor M versus p/q and shows that M is a monotonically increasing function of p/q. In contrast to the parameter A, M remains finite for $p = q$, which means that bulk contribution to the "mass", as it could be expected intuitively, comes from the lattice gas particles, whose motion is constrained by hard-core exclusions and whose mean displacement grows only as $\sqrt{t}$, without an additional logarithmic factor which is specific only to the PBP.

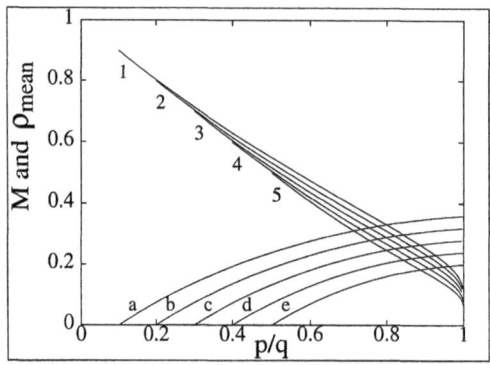

Fig. 7. Expansion of the gas phase. Plots of the parameter M and the mean density ρ_{mean} in the half-line $X > 0$ versus p/q for different initial densities ρ_-. The curves (a) to (e) show the value of the parameter M for $\rho_- = 0.9, 0.8, 0.7, 0.6$ and 0.5 respectively. The curves (1) to (5) display the corresponding mean densities ρ_{mean}.

Finally, we depict in Fig.7 the mean density on the interval $X \in [0, X(t)]$, defined as

$$\rho_{mean} \;=\; \frac{M(t)}{X(t)} \;=\; (1 - \frac{p}{q}) \; exp(A^2/2) \quad (54)$$

Figure 7 shows that despite of the exponential factor $exp(A^2/2)$ the mean density ρ_{mean} rapidly decreases with an increase of p/q and is a slowly increasing function of ρ_-.

B. Compression of the gas phase.

Let us next address the question of the PBP dynamics in the case $p < q(1 - \rho_-)$, when $X(t)$ is expected to be less than zero and thus the gas phase to be effectively compressed by the "boundary tension" force exerted on the PBP. Recollecting the results of [30–32] we suppose that here $X(t)$ obeys $X(t) = -B\sqrt{t}$, $B > 0$, and define the scaled variable as $\omega = (\lambda - 1)/B\sqrt{t}$, where ω is positive definite $0 \le \omega \le \infty$. In terms of this variable Eqs.(24) takes the form

$$\frac{\partial^2}{\partial \omega^2} f(\omega) + B^2 (\omega + 1) \frac{\partial}{\partial \omega} f(\omega) = 0, \quad (55)$$

while the boundary condition in Eq.(25) reads

$$\frac{\partial}{\partial \omega} f(\omega) \bigg|_{\omega=0} = -B^2 f(\omega = 0) \quad (56)$$

Again, the boundary and initial conditions in Eqs.(23.b) and (23.c) collapse into a single Eq.(28).

The general solution of Eq.(55) can be written down as

$$f(\omega) = C_1 \int_0^\omega dz \, \exp(-\frac{B^2}{2}(z^2 + 2z)) + C_2, \quad (57)$$

where C_1 and C_2 are to be chosen in such a way that Eqs.(28) and (56) are satisfied. Inserting Eq.(55) into Eqs.(28) and (56) we then obtain

$$C_1 = -B^2 C_2, \quad (58)$$

and

$$C_2 = \rho_- \{1 - B^2 \int_0^\infty dz \, \exp(-\frac{B^2}{2}(z^2 + 2z))\}^{-1} \quad (59)$$

Consequently, the density profile past the PBP can be expressed in terms of B and ω as

$$f(\omega) = \rho_- \{1 - B^2 \int_0^\omega dz \, \exp(-\frac{B^2}{2}(z^2 + 2z))\}/$$

$$/\{1 - B^2 \int_0^\infty dz \, \exp(-\frac{B^2}{2}(z^2 + 2z))\} \quad (60)$$

Next, Eq.(16) implies that also in this case $f(\omega = 0) \to (q - p)/q$ as $t \to \infty$, which yields eventually the following closed-form equation for the parameter B:

92

$$1 \; - \; B^2 \int_0^\infty dz \; \exp(-\frac{B^2}{2}(z^2 + 2z)) \; = \; \frac{q\rho_-}{q-p} \quad (61)$$

Equation (61) can be put into a more compact form if we express the integral over dz in terms of the probability integral. We then obtain

$$I_-(B) \; = \; \frac{q(1 - \rho_-) - p}{q - p}, \quad (62)$$

where $I_-(B)$ is defined in Eq.(10). In Fig.8 we present the numerical solution of Eq.(62), plotting the prefactor B as a function of the ratio p/q at different values of the density ρ_-.

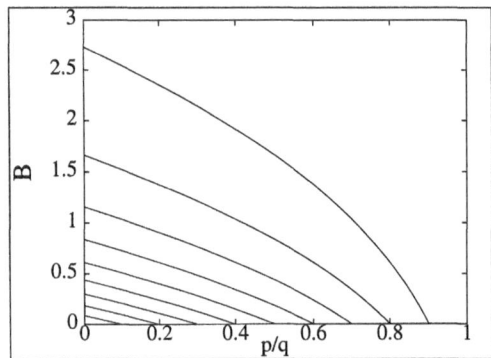

Fig. 8. Compression of the gas phase. Numerical solution of Eq.(62). The parameter B is plotted versus the ratio p/q for different values of the mean density ρ_-. The curves from top to bottom correspond to $\rho_- = 0.1, 0.2, 0.3, \ldots, 0.9$.

Equation (62) resembles the form of Eq.(34), which determines the parameter A, but differs from it in two aspects; first, the rhs of Eq.(62) is exactly the rhs of Eq.(34) but taken with the opposite sign, which insures that B is positive for $p < q(1 - \rho_-)$, and second, the sign before the probability integral in brackets is opposite to that in Eq.(34). The latter circumstance is responsible for the fact that B tends to the limiting value B_{lim} when $q \to 1$ ($p \to 0$). When $q \to p/(1 - \rho_-)$ the parameter B tends to zero exactly in the same way as the parameter A in Eq.(35) taken with the opposite sign, which means that the prefactor in $X(t)$ does not have a discontinuity at the "critical" point $p/q = 1 - \rho_-$ both for its value and for its slope.

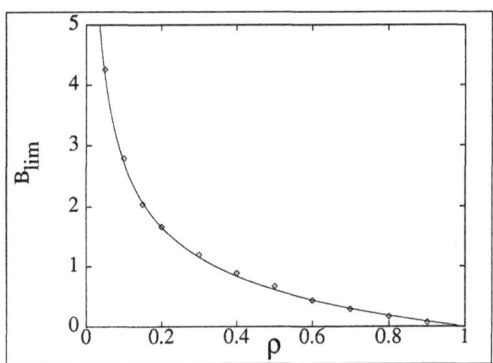

Fig. 9. Plot of the parameter B_{lim} versus the mean particle density ρ_-. The solid line shows the numerical solution of Eq.(63). The diamonds represent the associated Monte Carlo simulation results.

Now, in the limit $q = 1$ $(p = 0)$, Eq.(62) reduces to

$$\sqrt{\frac{\pi}{2}} \, B_{lim} \, \exp(B_{lim}^2/2) \, [1 \, - \, \Phi(B_{lim}/\sqrt{2})] \; = \; 1 - \rho_-, \quad (63)$$

which was obtained previously in [30,31] (see Eq.(2) in the present paper). Within the limit $\rho_- \to 1$ Eq.(63) yields

$$B_{lim} \; \approx \; \sqrt{\frac{2}{\pi}} \, (1 - \rho_-), \quad (64)$$

which shows that B_{lim}, as it could be expected intuitively, tends to zero when the density tends to 1.

When the gas is very dilute, i.e. $\rho_- \ll 1$, we may expect that B_{lim} is large. Expanding the probability integral as

$$\Phi(B_{lim}/\sqrt{2}) \; \approx \; 1 \, - \, \sqrt{\frac{2}{\pi}} \, B_{lim}^{-1} \, \exp(-B_{lim}^2/2) \, +$$

$$+ \sqrt{\frac{2}{\pi}} \, B_{lim}^{-3} \, \exp(-B_{lim}^2/2) \; - \; ... \, , \quad (65)$$

we find, upon substitution of Eq.(65) into the Eq.(63), the following result

$$B_{lim} \; \approx \; \frac{1}{\sqrt{\rho_-}}, \quad (66)$$

i.e. B_{lim} diverges when $\rho_- \to 0$ in proportion to the inverse of the square-root of the particle mean density. In Fig.9 we present the numerical solution of Eq.(63) together with the results of Monte Carlo simulations. Obviously, the agreement is very good.

Finally, in Fig.10 we combine the results of the subsections A and B and plot both analytical and Monte Carlo results obtained for the dependence of the prefactor $\alpha(F) = X(t)/\sqrt{t}$ on the ratio p/q and the density ρ_-. Again, we find very good agreement between our analytical predictions and numerical results, which support the validity of the approximations involved in our analysis.

Density profiles. Consider now the density profiles as seen from the PBP in the compression regime. In Fig.11 we plot $f(\omega)$ versus ω for different values of p/q at fixed ρ_-.

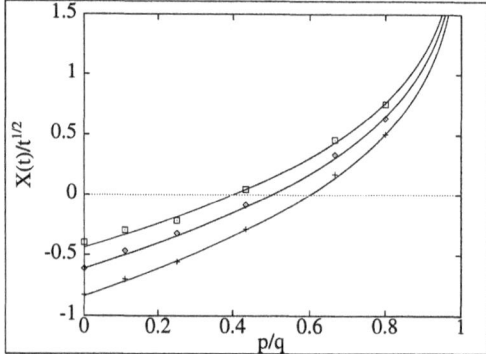

Fig. 10. Theoretical and experimental results for the prefactor in the dependence $X(t) = \alpha(F)\sqrt{t}$ as the function of the ratio p/q. The curves from top to bottom correspond to different values of the density ρ_-: the upper curve gives $\alpha(F)$ for $\rho_- = 0.6$, the lower curve corresponds to $\rho_- = 0.4$ and the curve in the middle - to $\rho_- = 0.5$. Symbols denote the results of Monte Carlo simulations.

Figure 11 shows that similarly to the behavior in the expansion regime, the density profiles are quite sensitive to the value of the parameter B. When B is smaller than unity, $f(\omega)$ shows almost linear dependence on ω, while in the case when $B > 1$ this dependence is non-linear and $f(\omega)$ rapidly drops from $f(\omega = 0) = 1 - p/q$ to ρ_-.

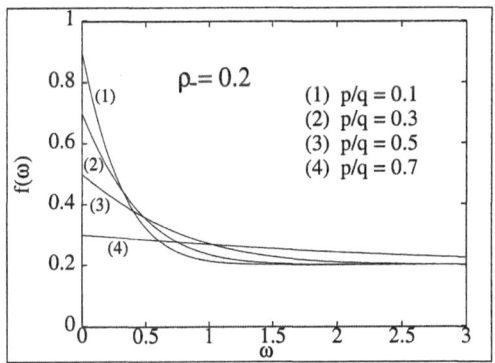

Fig. 11. Compression of the gas phase. Plot of $f(\omega)$, Eq.(60), versus the scaled variable ω at fixed density $\rho_- = 0.2$ and variable ratio p/q. The corresponding values of the parameter B are 1.52 (1), 1.21 (2), 0.83 (3) and 0.34 (4).

We finally present explicit results for $f(\lambda; t)$. In the limit of small λ, such that $\lambda \ll X(t)/B^2$, we have

$$f(\lambda; t) \approx (1 - \frac{p}{q}) [1 - \frac{B(\lambda - 1)}{\sqrt{t}} + ...], \quad (67)$$

which shows that the density is almost constant, (being only slightly less than $(1 - p/q)$), in the spatial region whose size is of the order of the PBP mean displacement.

For large λ we obtain from Eq.(60)

$$f(\lambda; t) \approx \rho_- + (1 - \frac{p}{q}) \frac{B\sqrt{2t}}{\lambda} exp(-\lambda^2/2t) + ... , \quad (68)$$

i.e. similarly to the behavior in the expansion regime, the density approaches the unperturbed value ρ_- exponentially fast and with a rate which is weakly (only through the pre-exponential factor) dependent on the parameter B and the ratio p/q.

V. DYNAMICAL EQUATIONS IN THE PRESENCE OF THE LOW-DENSITY PHASE.

Let us now consider the time evolution of the local density $\rho(X; t)$ and of the probability distribution $P(X; t)$ in the general case when the LDP is present and $\rho_- \geq \rho_+ \geq 0$.

One readily notices that also in this case Eq.(12) and, consequently, Eq.(13), describe the time evolution of the realization-averaged occupation variable $\rho(X; t)$ for all X excluding the sites $X = X(t) \pm 1$. Dynamical equation describing evolution of $\rho(X;t)$ at $X = X(t) - 1$ will be, however, somewhat modified as compared to Eq.(14). We have here

$$\dot{\rho}(X(t) - 1;t) = \frac{1}{2}\left(\rho(X(t) - 2;t) - \rho(X(t) - 1;t)\right) +$$

$$+ q\left(1 - \rho(X(t) - 1;t)\right)\rho(X(t) - 2;t) -$$

$$- p\left(1 - \rho(X(t) + 1;t)\right)\rho(X(t) - 1;t), \quad (69)$$

in which we account that the hops of the PBP in positive direction can be constrained by the LDP particles by introducing a factor $(1 - \rho(X(t) + 1;t))$. In a similar fashion, we find that at the site $X = X(t) + 1$ the local particle density obeys

$$\dot{\rho}(X(t) + 1;t) = \frac{1}{2}\left(\rho(X(t) + 2;t) - \rho(X(t) + 1;t)\right) -$$

$$- q\left(1 - \rho(X(t) - 1;t)\right)\rho(X(t) + 1;t) +$$

$$+ p\left(1 - \rho(X(t) + 1;t)\right)\rho(X(t) + 2;t) \quad (70)$$

Next, for the time evolution of the distribution function $P(X;t)$ we obtain the following equation

$$\dot{P}(X;t) = - P(X;t)\left[p\left(1 - \rho(X + 1;t)\right) + q\left(1 - \rho(X - 1;t)\right)\right] +$$

$$+ \left(1 - \rho(X;t)\right)\left[p\,P(X - 1;t) + q\,P(X + 1;t)\right], \quad (71)$$

which differs from the corresponding equation of the previous sections, Eq.(15), by the factors $(1 - \rho(X + 1;t))$ and $(1 - \rho(X;t))$ in the first and third terms respectively; these factors account, in a mean-field-type fashion, for the fact that hops of the PBP in the positive direction can take place only if the corresponding lattice sites are free of the LDP particles at this moment of time.

Further on, multiplying both sides of Eq.(71) by X and summing over all lattice sites we have that the mean displacement of the PBP obeys:

$$\dot{X}(t) = p - q - p\,f(\lambda = -1;t) + q\,f(\lambda = 1;t), \quad (72)$$

which thus generalizes Eq.(16) for the case of non-zero density of the LDP; the factor $f(-1;t)$ on the right-hand-side of Eq.(72) accounts for the hindering effects of the LDP particles on the PBP dynamics.

Consider now the time evolution of the correlation function $f(\lambda;t)$, defined in Eq.(17). By virtue of Eqs.(18), (13) and (71) we find that the evolution of this property is guided by:

$$\dot{f}(\lambda;t) = \frac{1}{2} [f(\lambda+1;t) + f(\lambda-1;t) - 2 f(\lambda;t)] -$$

$$- f(\lambda;t) [1 - p f(-1;t) - q f(1;t)] +$$

$$+ p f(\lambda-1;t) [1 - f(-1;t)] + q f(\lambda+1;t) [1 - f(1;t)], \quad (73)$$

which holds for all λ excluding $\lambda = \pm 1$. In the limit $\rho_- \to 0$, i.e when $f(-1;t) \to 0$, this equation reduces to Eq.(21). In the continuous-space limit Eq.(73) attains the form

$$\dot{f}(\lambda;t) = \frac{1}{2} \frac{\partial^2 f(\lambda;t)}{\partial \lambda^2} - [p - q - p f(-1;t) + q f(1;t)] \frac{\partial f(\lambda;t)}{\partial \lambda} \quad (74)$$

which is exactly Eq.(24) with $\dot{X}(t)$ defined by Eq.(72).

Further on, we find that the correlation function $f(\lambda;t)$ at the left-hand adjacent to the PBP site (for $\lambda = 1$) obeys

$$\dot{f}(1;t) = \frac{1}{2} (f(2;t) - f(1;t)) + p f(0;t) (1 - f(-1;t)) -$$

$$- q f(1;t) (1 - f(1;t)) - 2 p f(1;t) (1 - f(-1;t)) +$$

$$+ 2 q f(2;t) (1 - f(1;t)) \quad (75)$$

Comparing now Eq.(75) with Eq.(73) we have the following condition on $f(\lambda;t)$ at $\lambda = 1$:

$$\frac{1}{2} (f(0;t) - f(1;t)) = q f(2;t) (1 - f(1;t)) - p f(1;t) (1 - f(-1;t)) \quad (76)$$

Next, from Eqs.(18), (70) and (71) we can derive

$$\dot{f}(-1;t) = \frac{1}{2} (f(-2;t) - f(-1;t)) -$$

$$- p f(-1;t) (1 - f(-1;t)) - 2 q f(-1;t) (1 - f(1;t)) +$$

$$+ 2 p \, f(-2;t) \, (1 \, - \, f(-1;t)) \, + \, q \, f(0;t) \, (1 \, - \, f(1;t)), \quad (77)$$

which allows us to deduce the boundary condition on $f(\lambda;t)$ at the point $\lambda = -1$:

$$\frac{1}{2} \, (f(0;t) \, - \, f(-1;t)) \, = \, - \, q \, f(-1;t) \, (1 \, - \, f(1;t)) +$$

$$+ \, p \, f(-2;t) \, (1 \, - \, f(-1;t)) \quad (78)$$

In the continuous-space limit Eqs.(76) and (78) reduce to

$$\frac{1}{2} \, \left. \frac{\partial f(\lambda;t)}{\partial \lambda} \right|_{\lambda=\pm 1} \, = \, \dot{X}(t) \, f(\pm 1;t), \quad (79)$$

which represent two boundary conditions for the continuous-space Eq.(74). Equations (74) and (79), with the initial conditions

$$f(\lambda;t)|_{t=0} \, = \, \rho_+ \text{ for } \lambda < 0, \quad (80.a)$$

$$f(\lambda;t)|_{t=0} \, = \, \rho_- \text{ for } \lambda > 0, \quad (80.b)$$

and the boundary conditions

$$f(\lambda;t)|_{\lambda \to \infty} \, = \, \rho_-, \quad (81.a)$$

$$f(\lambda;t)|_{\lambda \to -\infty} \, = \, \rho_+, \quad (81.b)$$

constitute a closed system of equations which allows to compute $X(t)$ and the density profiles for arbitrary relation between p and q, as well as for arbitrary ρ_+ and ρ_-.

VI. SOLUTION OF DYNAMICAL EQUATIONS IN THE GENERAL CASE
$$\rho_- \geq \rho_+ \geq 0.$$

In this section we will derive explicit results for the dynamics of the mean displacement of the PBP and also for the density distribution around it. As it was done in the previous sections, we will discuss separately the behavior in the case when the HDP expands, compressing the LDP, and when, on the contrary, the LDP and the external force F compress the HDP.

A. Expansion of the high-density phase.

We again set $X(t) = A\sqrt{t}$ and suppose first that $A \geq 0$. Conditions at which such a behavior takes place will be specified below. For $\lambda \geq 1$ (past the PBP) we then have

$$f(\omega) = \frac{\rho_-}{1 + I_+(A)} \{1 + A^2 \int_0^\omega dz \, \exp(-\frac{A^2}{2}(z^2 - 2z)\}, \quad (82)$$

where $\omega = (\lambda - 1)/A\sqrt{t}$ and $I_+(A)$ is defined in Eq.(10). In front of the PBP, i.e. for $\lambda \leq -1$, the scaled density profile is given by

$$f(\theta) = \frac{\rho_+}{1 - I_-(A)} \{1 - A^2 \int_0^\theta dz \, \exp(-\frac{A^2}{2}(z^2 + 2z)\}, \quad (83)$$

in which we have denoted $\theta = -(\lambda + 1)/A\sqrt{t}$ and $I_-(A)$ is made explicit in Eq.(10). The density distributions $f(\omega)$ and $f(\theta)$ for different values of the parameters A, $\rho_\pm$ and p (q) are depicted in Figs.5,6 and 11 respectively.

Equations (82) and (83) contain the parameter A, which has not yet been specified. To determine A we take advantage of Eq.(72) which yields the following condition on the local densities at the sites adjacent to the PBP position:

$$q \left(1 - f(\lambda = 1; t)\right) = p \left(1 - f(\lambda = -1; t)\right) \quad (84)$$

Upon substitution of Eqs.(82) and (83) into the latter equation we find that A (in case when $A \geq 0$) obeys the following transcendental equation:

$$\frac{q \, \rho_-}{1 + I_+(A)} - \frac{p \, \rho_+}{1 - I_-(A)} = q - p, \quad (85)$$

which generalizes our Eq.(34) and also the result of [31] (Eq.(4) of the present paper) for the case when the particle densities from the left and from the right of the PBP are different and the density of the LDP is not zero. One directly verifies that Eq.(85) reduces to Eq.(34) when we set $\rho_+ = 0$, while setting $\rho_+ = \rho_-$ we recover Eq.(4).

Let us now find the conditions under which the parameter A is positive, i.e. the HDP expands. To do this, we simply notice that when $A = 0$ both $I_+(A)$ and $I_-(A)$ are equal to zero, which means that the "critical" relation between p, q and $\rho_\pm$ is:

$$q \left(1 - \rho_-\right) = p \left(1 - \rho_+\right) \quad (86)$$

Equation (86) implies that A vanishes, (i.e. the LDP and the HDP are in equilibrium with each other), when the probability of the PBP to go towards the HDP times the density of vacancies in this phase is exactly equal to the probability of going towards the LDP times the density of vacancies in this phase. When $p(1 - \rho_+) \geq q(1 - \rho_-)$ the HDP expands.

We note that Eq.(86) was previously obtained in [33] from the analysis of the stationary states in a one-dimensional lattice gas placed in a finite box of length L. By explicit calculation of the distribution function of the PBP position in the general case when the numbers of the lattice gas particles from the right and from the left of the PBP are not equal, it was found [33] that in the limit $L \to \infty$ the PBP is localized at point X_0, which divides the system in proportion given by Eq.(86).

Equation (86) can also be rewritten using the definition of the external force F. Upon some algebra, we find then that the critical force at which both phases are in equilibrium with each other is given by

$$F_c = \beta^{-1} \ln(\frac{1 - \rho_+}{1 - \rho_-}) \quad (87)$$

Now, let us discuss the behavior of the parameter A in the limit when A is small or large, and calculate the diffusivity of the PBP in the critical case $A = 0$. In the limit of small A, i.e. when p, q and $\rho_\pm$ are close to their "critical" values determined by Eqs.(86) and (87), both $I_\pm(A) \approx \sqrt{\pi/2A}$. Substituting these expressions into Eq.(85) we find

$$A \approx \sqrt{\frac{2}{\pi}} \frac{p(1 - \rho_+) - q(1 - \rho_-)}{q\rho_- + p\rho_+}, \quad (88)$$

which is valid when $A \ll 1$. Eq.(88) allows for the computation of the PBP mobility, which we determine following the arguments presented in Section IV as

$$\mu(t) = lim_{F \to F_c} \frac{X(t)}{(F - F_c)\, t} =$$

$$= t^{-1/2}\, lim_{F \to F_c} \frac{A}{(F - F_c)} \quad (89)$$

Substituting Eq.(88) into Eq.(89) and taking the limit $F \to F_c$, we find

$$\mu(t) = \beta \frac{(1 - \rho_-)(1 - \rho_+)}{(\rho_- + \rho_+ - 2\rho_-\rho_+)} \sqrt{\frac{2}{\pi t}}, \quad (90)$$

which yields, by virtue of Eq.(45), the result presented in Eq.(7).

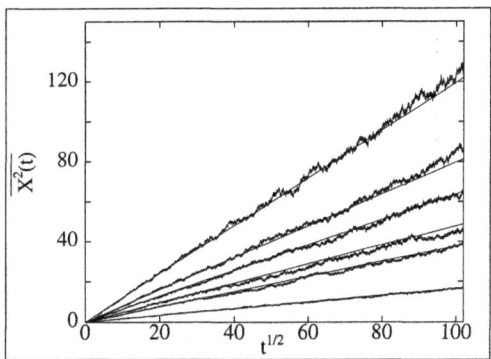

Fig. 12. Mean-square displacement of the PBP in the critical case $p/q = (1 - \rho_-)/(1-\rho_+)$, Eq.(86). The solid lines show our analytical prediction from Eq.(7) and the noisy lines give the results of Monte Carlo simulations. The curves from top to bottom correspond to the following values of the parameters (ρ_-, ρ_+): The first two curves are the analytical result in Eq.(1) and numerical data for the symmetric cases $(0.4, 0.4)$ and $(0.5, 0.5)$. The lower curves correspond to $(0.6, 0.5)$, $(0.9, 0.4)$, $(0.7, 0.5)$ and $(0.8, 0.2)$ respectively.

Equation (7) generalizes the classical result in Eq.(1) for the situation in which the mean particles densities for both sides of the tracer particle are different from each other. One can directly verify that Eq.(7) reduces to Eq.(1) when $\rho_- = \rho_+$, while setting $\rho_+ = 0$ we recover our previous result in Eq.(48). In Fig.12 we compare our analytical prediction in Eq.(90) against the results of Monte Carlo simulations, which shows that an approximate approach developed here represents a fair description of the PBP dynamics.

Next, in Section IV we have demonstrated that the prefactor A diverges when $p \to q$. Consequently, we can expect that even in the presence of the LDP the prefactor A can attain large values when $\rho_+ \ll 1$ and $p/q \to 1$. Setting in Eq.(85) $q = p$ and using the expansion in Eq.(65) we find from Eq.(85) that A is defined in the limit $\rho_+ \to 0$ by

$$A \approx \sqrt{2\,ln(\rho_-/\rho_+)}, \quad (91)$$

i.e. A grows as a square-root of the logarithm of ρ_+ when $\rho_+ \to 0$.

Finally, we estimate the behavior of the ratio δ of the particle densities immediately past and in front of the PBP. At zero moment of time this ratio is evidently $\delta = \delta_0 = \rho_-/\rho_+$. Our results in Eqs.(82) and (83) suggest that after some transient period of time the density profiles past and in front of the PBP attain stationary forms with respect to the variable

$\omega = (\lambda - 1)/X(t)$. Consequently, we have that, as the time evolves, the ratio of the particle densities immediately past and in front of the PBP tends to a constant value

$$\delta \;=\; \delta_0 \,\frac{1 - I_-(A)}{1 + I_+(A)} \quad (92)$$

Eq.(92) holds for arbitrary values of A. In the asymptotic limits when A is small or large, we find from Eq.(92) the following explicit asymptotic forms for δ:

$$\delta \;\approx\; \delta_0 \left(1 \;-\; \left(\frac{\pi}{2} - 1\right) A^2 \;+\; \ldots \right), \text{ when } A \ll 1, \quad (93)$$

and

$$\delta \;\approx\; \delta_0 \,\frac{exp(-A^2/2)}{\sqrt{2\pi}\,A^3} \text{ when } A \gg 1 \quad (94)$$

Therefore, the parameter δ, as it could be expected intuitively, is always less than δ_0. Complete dependence of δ on the parameter A is presented in Fig.14.

B. Compression of the high-density phase.

Consider now the behavior in the regime when the LDP and the applied force compress the HDP. Setting $X(t) = -B\sqrt{t}$, where B is supposed to be a positive constant, we find from Eqs.(74) and (79) that the particle density for $\lambda \geq 0$ (i.e. at sites $X < X(t)$) obeys Eq.(60), in which the variable ω is defined as $\omega = (\lambda - 1)/B\sqrt{t}$ and the parameter A is replaced by B. From the other side of the PBP, i.e for $\lambda \leq 0$, we have

$$f(\theta) \;=\; \frac{\rho_+}{1 + I_+(B)} \left\{1 \;+\; B^2 \int_0^\theta dz \, \exp(-\frac{B^2}{2}(z^2 - 2z))\right\}, \quad (95)$$

where the scaled variable $\theta = -(\lambda + 1)/B\sqrt{t}$. The function $f(\theta)$ is depicted in Figs.5 and 6. Substituting Eqs.(60) and (95) into Eq.(84) we arrive at the following transcendental equation for the parameter B:

$$\frac{q\,\rho_-}{1 - I_-(B)} \;-\; \frac{p\,\rho_+}{1 + I_+(B)} \;=\; q \;-\; p \quad (96)$$

Eq.(96) thus generalizes the result in Eq.(62) for the case of the non-zero particles density in the LDP.

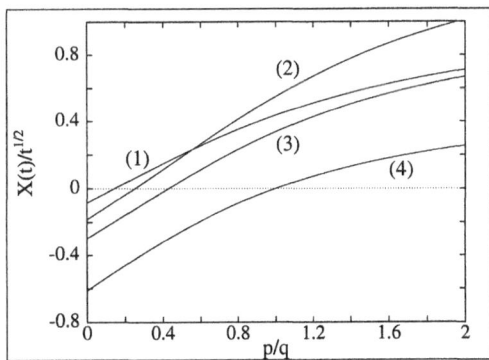

Fig. 13. Numerical solution of Eq.(6). Prefactor $\alpha(F)$, Eq.(3), is plotted versus the ratio p/q for different initial mean densities from the left and from the right of the origin. The curves from (1) to (4) correspond respectively to $(\rho_- = 0.9, \rho_+ = 0.3)$, $(0.8, 0.2)$, $(0.7, 0.3)$ and $(0.5, 0.5)$.

Now, noticing that Eqs.(96) and (85) can be cast into one another by the substitution $\pm I_\pm(A) \rightarrow \mp I_\mp(B)$, we can construct a general equation for the parameter $\alpha(F)$ in Eq.(3). This equation is presented in Eq.(6) and holds for arbitrary relation between $\rho_\pm$ and p/q, describing hence both the expansion and the compression regimes. In Fig.13 we present the numerical solution of Eq.(6), plotting $\alpha(F)$ as a function of the ratio p/q for different values of ρ_- and ρ_+.

Finally, we analyze the behavior of the parameter δ, which is defined as the ratio of the particle density immediately past the PBP and the particle density immediately in front of the PBP. Using Eqs.(60) and (95) we find

$$\delta \; = \; \delta_0 \; \frac{1 + I_+(B)}{1 - I_-(B)} \quad (97)$$

Numerical plot of $\delta(B)$ is presented in Fig.14.

Asymptotic behavior of the parameter δ in the limits when B is small or large readily follows from our Eqs.(93) and (94). Here we have

$$\delta \; \approx \; \delta_0 \left(1 \; + \; (\frac{\pi}{2} - 1) \, B^2 \; + \; ... \right), \text{ when } B \ll 1, \quad (98)$$

and

$$\delta \; \approx \; \delta_0 \; \sqrt{2\pi} \; B^3 \; exp(B^2/2), \text{ when } B \gg 1, \quad (99)$$

which means that in the compression regime the parameter δ is always greater than δ_0.

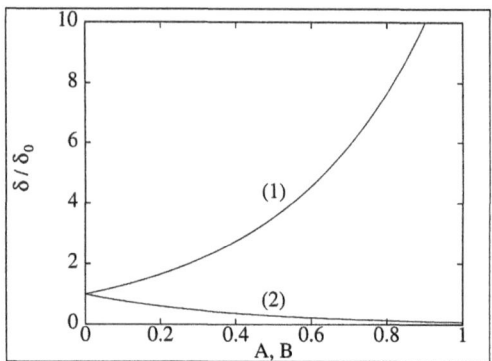

Fig. 14. Plot of the parameter δ, Eqs.(92) and (97), versus the parameters A and B. The curve (1) describes the behavior of $\delta(B)$ in the compression regime. The curve (2) defines $\delta(A)$ in the expansion regime.

VII. MONTE CARLO SIMULATIONS.

In order to check our analytical predictions, derived in terms of a mean-field approximation, we have performed Monte Carlo simulations of the process defined in the begining of Section II. The simulation algorithm was defined as follows:

We constructed first a one-dimensional regular lattice of unit spacing and length $2L+1$, sites of which were labelled by integers of the interval $[-L, L]$. In all simulations we took $L = 10^3$. At the zero moment of the MC time the particles were placed randomly on the lattice with the prescribed mean densities and the constraint that two particles can never simultaneously occupy the same site. To do this, we have called, for each lattice site from the interval $[-L+1, -1]$ independently, a random number from the interval $[0, 1]$. In case when the random number produced by the generator was less that ρ_- a particle was created on this lattice site. In case when the random number was greater than ρ_- the site was left empty. The same routine was performed for the sites with positive numbers $[1, L-1]$; here a particle was created at the corresponding site in case when the random number was less than ρ_+ and the site was left empty if the random number was greater than ρ_+. The phase boundary particle was placed at the origin. Additionally, we have prescribed that the sites $X = \pm L$ are occupied by particles. The particles at theses sites $X = \pm L$ are made immobile, blocking the lattice from both sides and preventing other particle to leave the system.

The subsequent particle dynamics employed in our simulations follows the definitions of Section 2 closely. We call for a random integer number X from the interval $[-L+1, L-1]$. Here three different events may take place:

(i) If the site X is vacant, a new site is considered.

(ii) If the site X is occupied by a particle, we first increase the MC time by unity and then let the particle choose, at random, a potential jump direction. This is done again by calling a random number from the interval $[0, 1]$. If the random number is less than 0.5, the particle attempts to jump to the site $X - 1$; otherwise, it attempts to jump to the site $X + 1$. The jump is fulfilled if at this moment of the MC time the adjacent site in the chosen direction is vacant (not occupied by any other particle or the PBP). Otherwise, the particle remains at X.

(iii) If the site X appears to be occupied by the PBP, we increase the MC time by unity and consider a random number from the interval $[0, 1]$; in case when this number is less than the prescribed value q, the PBP attempts to jump to the site $X - 1$. Otherwise, it attempts to jump to the site $X + 1$. The jump is fulfilled if at this moment of MC time the adjacent site in the chosen direction is vacant. Otherwise, the PBP remains at X.

In simulations we have followed the time evolution of several different properties: the PBP displacement, squared displacement (under the critical conditions) and the occupations of the sites $X = \pm L \mp 100$. Time behavior of these properties was plotted versus the "physical time" t, which is the time needed for each particle to move once, on average, or in other words, $t = \text{MCtime}/\text{number of particles}$. We have observed that for all values of the parameters $\rho_\pm$ and q, used in our simulations, the stationary regime in which the ratio $X_r(t)/\sqrt{t}$ approaches a constant value is established for displacements of order of 200 lattice units. To get the spatially resolved behavior, in computation of the PBP displacement each realization of the process was interrupted at the moment when the absolute value of the PBP displacement reaches the value of 500 lattice. For calculation of the mean displacement we used typically 10^2 realizations for each set of parameters $\rho_\pm$ and q. Results of these simulations are presented in Figs.9 and 10. Further on, computing the squared displacement we interrupted each realization of the process at the moment when the span of the PBP trajectory is equal to 10^2. Mean-square displacement was obtained by averaging over 2×10^3 realizations. Results for mean-square displacement of the PBP are presented in Figs.4 and 12. In all cases, we have obtained remarkably good agreement between our analytical predictions and simulation results. Finally, the measurements of the occupations of the sites $X = \pm L \mp 100$ were performed in order to be sure that the perturbances created by the PBP do not spread during the simulation time through the whole system and do not lead to artificial behaviors associated with the finite-size effects. We have observed that actually the mean densities of these sites don't vary with time and are equal to the unperturbed values $\rho_{\pm}$.

VIII. CONCLUSIONS.

To conclude, we have examined in terms of a mean-field-type approach the dynamics of the phase boundary propagation in a one-dimensional hard-core lattice gas which was initially put into a non-equlibrium, "shock"-like configuration and then allowed to evolve in time by particles attempting to hop to neighboring unoccupied sites. The "shock" configuration means that particle mean densities from the left and from the right of the origin are different. All particles of the lattice gas, except the particle separating the high- and the low-density phases, have symmetric hopping probabilities, while the phase boundary particle is subject to a constant force F and has asymmetric hopping probabilities. We have shown that the mean displacement of the PBP follows the generic law $X(t) = \alpha(F)\sqrt{t}$, in which the parameter $\alpha(F)$ can be both positive and negative, depending on the relation between the magnitude of the force and the initial mean densities. This prefactor is determined implicitly, in a form of the transcendental Eq.(6) for arbitrary magnitude of the force and arbitrary relation between the particle densities in the high- and low-density phases. In several asymptotic limits we find explicit formulae for the prefactor. Further on, we have shown that when F is equal to the critical value F_c, Eq.(87), the parameter $\alpha(F)$ is exactly equal to zero. In this case the high- and the low-density phases are in equilibrium with each other. We have found that here the mean-square displacement $\overline{X_r^2(t)}$ of the PBP follows $\overline{X_r^2(t)} \sim \gamma\sqrt{t}$, i.e. shows a sub-diffusive behavior. The form of the prefactor γ is determined explicitly, Eq.(7). Our analytical findings are in a very good agreement with the results of numerical simulations.

The authors wish to thank J.L.Lebowitz and R.Kotecky for helpful and encouraging discussions. Financial support from the FNRS and the COST Project D5/0003/95 is gratefully acknowledged.

[1] M.C.Cross and P.C.Hohenberg, Rev. Mod. Phys. **65**, 851 (1993)

[2] M.Ben-Amar, P.Pelcé and P.Tabeling, *Nonlinear Phenomena Related to Growth and Form* (Plenum, New York, 1991)

[3] Y.Pomeau and M.Ben Amar, in *Solids Far from Equilibrium*, ed.: C.Godreche (Cambridge University Press, Cambridge, 1992)

[4] A.M.Cazabat, F.Heslot, S.M.Troian and P.Carles, Nature **346**, 824 (1990)

[5] J.S.Langer, in *Chance and Matter*, eds.: J.Souletie, J.Vannimenus and R.Stora (North-Holland, Amsterdam, 1987)

[6] P.Collet and J.P.Eckmann, *Instabilities and Fronts in Extended Systems* (Princeton University Press, Princeton, 1990)

[7] P.Collet, in: *Wetting Phenomena*, eds.: J.De Coninck and F.Dunlop (Springer-Verlag, Berlin, 1990)

[8] P.Devillard and H.Spohn, Europhys. Lett. **17**, 113 (1992)

[9] M.Kardar and J.O.Indekeu, Europhys. Lett. **12**, 161 (1990)

[10] A.Malevanets, A.Careta and R.Kapral, Phys. Rev. E **52**, 4724 (1995)

[11] J.Armero, J.M.Sancho, J.Casademunt, A.M.Lacasta, L.Ramirez-Piscina and F.Sagués, Phys. Rev. Lett. **76**, 3045 (1996)

[12] U.Ebert, W.van Saarloos and C.Caroli, Phys. Rev. Lett. **77**, 4178 (1996)

[13] J.L.Lebowitz, E.Presutti and H.Spohn, J. Stat. Phys. **51**, 841 (1988)

[14] F.J.Alexander, C.A.Laberge, J.L.Lebowitz and R.K.P.Zia, J. Stat. Phys. **82**, 1133 (1996)

[15] J.Hardy, O.de Pazzis and Y.Pomeau, Phys. Rev. A **13**, 1949 (1976)

[16] U.Frisch, B.Hasslacher and Y.Pomeau, Phys. Rev. Lett. **56**, 1505 (1986)

[17] C.Appert and S.Zaleski, Phys. Rev. Lett. **64**, 1 (1990)

[18] D.H.Rothman and J.M.Keller, J. Stat. Phys. **52**, 1119 (1988)

[19] G.Giacomin and J.L.Lebowitz, Phys. Rev. Lett. **76**, 1094 (1996)

[20] G.Giacomin and J.L.Lebowitz, J. Stat. Phys. **87**, 37 (1997)

[21] M.Z.Guo, G.C.Papanicolaou and S.R.S.Varadhan, Commun. Math. Phys. **118**, 31 (1988)

[22] S.F.Burlatsky, G.Oshanin, A.M.Cazabat and M.Moreau, Phys. Rev. Lett. **76**, 86 (1996)

[23] S.F.Burlatsky, G.Oshanin, A.M.Cazabat, M.Moreau and W.P.Reinhardt, Phys. Rev. E **54**, 3832 (1996)

[24] T.E.Harris, J. Appl. Prob. **2**, 323 (1965)

[25] D.B.Abraham, P.Collet, J.De Coninck and F.Dunlop, Phys. Rev. Lett. **65**, 195 (1990)

[26] D.B.Abraham, P.Collet, J.De Coninck and F.Dunlop, J. Stat. Phys. **61**, 509 (1990)

[27] J.De Coninck and F.Dunlop, J. Stat. Phys. **47**, 827 (1987)

[28] G.Oshanin, J.De Coninck, A.M.Cazabat and M.Moreau, Phys. Rev. E **58**, R20 (1998); J. of Mol. Liquids **76**, 195 (1998)

[29] R.Arratia, Z. Ann. Probab. **11**, 362 (1983)

[30] S.F.Burlatsky, G.Oshanin, A.Mogutov amd M.Moreau, Phys. Lett. A **166**, 230 (1992)

[31] S.F.Burlatsky, G.Oshanin, M.Moreau and W.P.Reinhardt, Phys. Rev. E **54**, 3165 (1996)

[32] C.Landim, S.Olla and S.B.Volchan, Commun. Math. Phys. **192**, 287 (1998)

[33] P.Ferrari, S.Goldstein and J.L.Lebowitz, Diffusion, Mobility and the Einstein Relation, in: *Statistical Physics and Dynamical Systems*, eds.: J.Fritz, A.Jaffe and D.Szasz (Birkhäuser, Boston, 1985) p.405

[34] H.Spohn, *Large Scale Dynamics of Interacting Particles*, (Springer Verlag, New York, 1991), Ch.3

Conventional Bose-Einstein condensation

A. Verbeure

Instituut voor Theoretische Fysica, K.U.Leuven, Celestijnenlaan 200D,
B-3001 Leuven, Belgium
Andre.Verbeure@fys.kuleuven.ac.be

Abstract

These lecture notes are intended to be a pedestrian pedagogical introduction
to conventional Bose-Einstein condensation with an emphasis on first principles, sensitivities related to boundary conditions, the position of spontaneous
symmetry breakdown and the place of external fields (traps) in this matter.
The notes do not contain new results but can be considered as a chapter in a
textbook about undergraduate soft condensed matter topics.

O. Descalzi et al. (eds.), Instabilities and Nonequilibrium Structures VII & VIII, 109–130.

1 Introduction

In these lectures we discuss the conventional Bose-Einstein condensation (BEC). The readers, professional professors, are warned about the fact that we explain only simple facts of BEC which make the phenomenon understandable from the microscopic point of view. They should not expect high tech stands nor hype stories. The readers, students, graduate students and unexperienced individuals are warned as well, I keep it easy because also professors may try to read the text.

The theoretical prediction of the BEC phenomenon is due to Satyendra Nath Bose and Albert Einstein [1] and is dated 1924-1925. BEC is one of the basic quantum phenomena in nature because it is a property of the free or ideal gas of bosons. It is a property which is not necessarily the result of a special interaction between the particles. Einstein's arguments based on arguments of Bose statistics and the classical expression for the density of states lead to the phenomenon of a macroscopic accumulation of boson particles in the ground state of the ideal Bose gas (IBG). In particular difficulties in understanding the phase transition in finite volume argumentations was the basis for Uhlenbeck's criticism [2]. Years later (1937), due to the work of Kramers making clear that a transition is made sharp only in the thermodynamic limit, Uhlenbeck [3] accepted Einstein's work as a statement holding in the thermodynamic limit.

From the experimental point of view, Kapitza [4] and Allen and Misener [5] discovered in 1938 the λ-point in ^{4}He (bosons) as the transition point from the normal liquid (called He I) to the superfluid phase (called He II) at a temperature $T_\lambda = 2,17$ K, if ^{4}He is treated as an IBG, its BEC transition temperature T_c would be very close to T_λ, namely $T_c = 3,14$ K. BEC for the IBG corresponds to a macroscopic occupation of the ground state, a phenomenon which is related to a coherence property of the condensate mode. Inspired by these observations Tisza and London [6] gave an explanation to superfluidity and superconductivity on the basis of BEC, although the spectral properties of these phenomena showed completely different from that of the IBG. It was Bogoliubov who brought this spectral problem to a solution. For a recent extensive review on this question, see [7].

In the 1950's this idea was accepted as an important thesis, which was supported by neutron experiment.

Finally 1995 is an important year, by now considered as the year that BEC is discovered in the IBG. We refer of course to Bose condensation for trapped alkali metals (E. Cornell, C. Wieman, JILA, Boulder; R. Hulet, Rice Univ. Houston Texas; W. Kelterle, MIT, Cambridge Mass.).

We finish the introduction by referring to the book of A. Pais [8] with a title "Subtle is the lord", where on p. 413 the author guesses that BEC is Einstein's 6th most important contribution. It is an interesting idea to discover which ranking BEC will get in the 21st century.

2 Conventional BEC for IBG

2.1 Two-body interacting Bose gas

Consider a system of bosons of mass m enclosed in a cubic box $\Lambda = L \times L \times L \ldots \leq \mathbb{R}^d$. The boundary conditions play an important role. For periodic boundary conditions, consider the dual volume

$$\Lambda^* = \left\{ k \in \mathbb{R}^d \; ; \; k_\alpha = \frac{2\pi}{L} n_\alpha \; ; \; n_\alpha = 0, \pm 1, \ldots, \; \alpha = 1, \ldots \nu \right\}$$

and denote by V the volume of Λ.

Consider the interaction potential between bosons given by:

$$\varphi \; : \; x \in \mathbb{R}^d \to \varphi(x) \in \mathbb{R}$$

which we take nice i.e. $\varphi \in L^1(\mathbb{R}^\nu)$ and positive definite:

$$0 \leq v(q) = \int dx \; \varphi(x) e^{-iqx} \leq v(0) \,.$$

The Hamiltonian defined on Fock space $\mathcal{F}_\Lambda$ is given by:

$$H_\Lambda = \sum_{k \in \Lambda^*} \varepsilon_k \, a_k^* a_k + \frac{1}{2v} \sum_{k_1, k_2, q \in \Lambda^*} v(q) \, a_{k_1+q}^* a_{k_2-q}^* a_{k_1} a_{k_2} \tag{1}$$

where $\varepsilon_k = \dfrac{\hbar^2 k^2}{2m} - \mu$,

$$a_k^* = a^*(\psi_k) = \int_\Lambda dx \; \frac{e^{ikx}}{\sqrt{V}} a^*(x)$$

$$\psi_k(x) = \frac{1}{\sqrt{V}} e^{ikx} \quad \text{(the wave function)}$$

and the boson commutation relations

$$[a_k, a_{k'}^*] = \delta_{k,k'} \; ; \; [a_k, a_{k'}] = 0 \,,$$

μ is the chemical potential.

Remark that the wave functions ψ_k with $k \neq 0$ are periodic functions describing particular localizations of the boson particles in the space variables. On the other hand the wave function ψ_0 (i.e. $k = 0$) is a constant function, describing a completely delocalized state. This function plays a very particular role in the phenomenon of condensation.

The Hamiltonian (1) describes for non-trivial interactions φ, a general two-body interacting system of bosons. It is still a challenge in theoretical physics to show that these systems show Bose-Einstein condensation, except for a few trivial functions φ.

2.2 Ideal Bose gas

We consider the ideal Bose gas, i.e. we take $\varphi = 0$ in (1), and get

$$H_\Lambda = \sum_{k \in \Lambda^*} \varepsilon_k a_k^* a_k \tag{2}$$

Remark that thermodynamic stability for this system is guaranteed, i.e. H_Λ is bounded from below if and only if $\mu < 0$.

We consider the ground state of n bosons. Take

$$|n_1 k_1; n_2 k_2 \ldots\rangle = \frac{a_{k_1}^{*n_1}}{\sqrt{n_1!}} \cdots \frac{a_{k_r}^{*n_r}}{\sqrt{n_r!}} \cdots |0\rangle$$

where $|0\rangle$ is the Fock vacuum state satisfying:

$$a_k |0\rangle = 0 \ , \ \forall k \in \Lambda^*.$$

Remark that they are eigenfunctions of the Hamiltonian (2):

$$H_\Lambda |n_1 k_1; \ldots; n_r k_r; \ldots\rangle = \sum_i \varepsilon_{k_i} n_i |n_1 k_1; \ldots; n_r k_r; \ldots\rangle.$$

The minimal energy of the n bosons state is:

$$H_\Lambda |n, k = 0; 0, \ldots\rangle = n\,\varepsilon_0 |n; 0; 0 \ldots\rangle.$$

The energy density is

$$u_0 = \frac{n}{V}\varepsilon_0 = \rho\,\varepsilon_0$$

where $\rho = n/V > 0$ is the density of particles and $n = \rho V$ i.e. the number of particles is proportional to the volume. This indicates that for a system of bosons one can realize a macroscopic occupation of the ground state by particles. This is the basis of BEC. We stress here also that in the state of lowest energy, i.e. the ground state all particles are collected in the individual particle wave functions $\psi_0(x) = 1/\sqrt{V}$ spreading these particles uniformly over the whole volume. In other words there is a maximal overlap of the particle wave functions. Remark also that the minimal energy of the particles is compatible with the uncertainty relations coming from the momenta $|k| \simeq 1/L$. The corresponding energy density is given by: $u_0 \simeq \rho\,1/L^2$.

Let us now look at temperature states. We look at the occupation of the level for an arbitrary $k \in \Lambda^*$. The occupation of the number of particles is given by:

$$\langle N_k \rangle = \frac{tr \ e^{-\beta H_\Lambda} N_k}{tr \ e^{-\beta H_\Lambda}}$$

where $N_k = a_k^* a_k$, and easily computed to be:

$$\langle N_k \rangle = \frac{1}{e^{\beta \varepsilon_k} - 1}.$$

The total density in the volume Λ is:

$$\rho = \frac{\langle N_\Lambda \rangle}{V} = \frac{1}{V} \sum_{k \in \Lambda^*} \frac{1}{e^{\beta \varepsilon_k} - 1} . \tag{3}$$

All this is volume dependent and therefore boundary dependent. In order to make basic universal statements one takes the *thermodynamic limit* i.e. one let L or the volume tend to infinity while keeping the density $\rho = \langle N_\Lambda \rangle / V$ constant. For each limit volume Λ a fixed density ρ fixes a value of the chemical potential $\mu_L(\beta, \rho)$ and vice versa. Let us consider now the thermodynamic limit of the density equation (3), where we use $\mu(\beta, \rho) = \lim_{L \to \infty} \mu_L \leq 0$, then (3) becomes:

$$
\begin{aligned}
\rho &= \lim_{L \to \infty} \frac{\langle N_\Lambda \rangle}{V} \\[2mm]
&= \lim_{L \to \infty} \left(\frac{1}{V} \frac{1}{e^{-\beta \mu_L} - 1} + \frac{1}{V} \sum_{k \neq 0} \frac{1}{e^{\beta \varepsilon_k} - 1} \right) \\[2mm]
\rho &= \rho_0(\beta, \rho) + \rho(\beta, \mu(\beta, \rho))
\end{aligned}
\tag{4}
$$

with:

$$\rho(\beta, \mu) = \left(\frac{1}{2\pi} \right)^3 \int dk \frac{1}{e^{\beta \left(\frac{k^2}{2m} - \mu \right)} - 1} \; ; \; \hbar = 1$$

$$\rho_0(\beta, \rho) = \lim_{L \to \infty} \frac{1}{V} \frac{1}{e^{-\beta \mu_L} - 1} = \lim_{L \to \infty} \frac{\langle a_0^* a_0 \rangle}{V} .$$

Clearly $\rho_0(\beta, \rho)$ is the density of particles in the state of lowest energy ε_0 for a fixed total density ρ.

Also $\mu \to \rho(\beta, \mu)$ is monotonically increasing and

$$\rho(\beta, \mu) \leq \rho(\beta, 0) \equiv \rho_c(\beta) = \left(\frac{1}{2\pi} \right)^\nu \int dk \frac{1}{e^{\beta \frac{k^2}{2m}} - 1} . \tag{5}$$

One checks that $\rho_c(\beta)$, called the *critical density*, is finite for all dimensions $d > 2$. In the rest of the paper we limit ourself to the case of dimension $d = 3$.

If the density ρ is small, in particular if $\rho < \rho_c(\beta)$, then there exists a unique $\mu(\beta, \rho) < 0$, such that $\rho = \rho(\beta, \mu(\beta, \rho))$, and equation (4) is satisfied with $\rho_0(\beta, \rho < \rho_c) = 0$. If the density ρ is large enough, in particular if $\rho > \rho_c(\beta)$ then equation (4) yields

$$\rho - \rho_c(\beta) = \rho_0(\beta, \rho) > 0 \tag{6}$$

and one has a macroscopic occupation of the ground state, i.e. Bose-Einstein condensation has taken place in the $k = 0$ mode.

We analyze further the occurrence of condensation. Suppose we have $\rho_0 > 0$, i.e. we have condensation in the mode $k = 0$,

$$0 < \rho_0 = \lim_{L \to \infty} \frac{1}{V} \frac{1}{e^{-\beta \mu_L} - 1} = \lim_{L \to \infty} \frac{1}{V} \frac{1}{-\beta \mu_L}$$

or equivalently $\mu_L \simeq -1/V$, i.e. anyway $\lim_{L \to \infty} \mu_L = 0$.

Remark also that there is no condensation in the other modes $k \neq 0$, e.g. not in the first excited mode. Indeed

$$\inf_{k \neq 0} \frac{k^2}{2m} = \frac{1}{2m} \left(\frac{2\pi}{L} \right)^2 \inf_{n \neq 0} \left(n_1^2 + n_2^2 + n_3^2 \right) \simeq \frac{1}{V^{2/3}}$$

and

$$\varepsilon_k \simeq \frac{1}{V^{2/3}} + \frac{c}{V} .$$

Hence

$$\lim_{L \to \infty} \frac{1}{V} \frac{1}{e^{\beta \varepsilon_k} - 1} \bigg|_{k \neq 0} \simeq \lim_{L \to \infty} \frac{1}{V} \frac{1}{\left(\frac{1}{V^{2/3}} + \frac{c}{V} \right)} = \lim_{L \to \infty} \frac{1}{V^{1/3}} = 0 .$$

2.2.1 The critical temperature

We defined the critical density $\rho_c(\beta)$ in formula (6). Clearly $T \to \rho_c(\beta)$ is monotonically increasing to infinity. Hence for any given density ρ, there exists a critical β_c value such that $\rho = \rho_c(\beta_c)$ or a critical temperature $T_c = 1/k_B \beta_c$ depending on the mass m of the bosons and on the density ρ.

2.2.2 Exercise

Show that

$$\frac{\rho - \rho_c(\beta)}{\rho} = \begin{cases} (1 - T/T_c)^{3/2} & \text{if } T < T_c \\ 0 & \text{if } T \geq T_c \end{cases}$$

2.2.3 Generalized condensation [9-16]

Above we considered conventional BEC with a macroscopic occupation of the ground state $k = 0$, on the basis of saturation of the modes $k \neq 0$. The concept of generalized conventional BEC was discovered and worked out by many authors. condensation of modes $k \neq 0$ can appear accompanied or not with condensation in the mode $k = 0$. Without going into the details of this subtle phenomenon we illustrate it on the basis of taking the thermodynamic limit in a volume dependent way instead of in terms of increasing cubic boxes. We take increasing parallelepipeds: $\Lambda = L_1 \times L_2 \times L_3$ and hence the volumes as follows:

$$V = V^{\gamma_1} V^{\gamma_2} V^{\gamma_3} \quad \text{with} \quad \gamma_1 + \gamma_2 + \gamma_3 = 1 .$$

Taking cubic boxes means: $\gamma_1 = \gamma_2 = \gamma_3 = 1/3$, take now $\gamma_1 = 1/2$, $\gamma_2 = \gamma_3 = 1/4$, then

$$\frac{k^2}{2m} = \frac{1}{2m}(2\pi)^2 \left(\frac{n_1^2}{V} + \frac{n_2^2}{V^{1/2}} + \frac{n_3^2}{V^{1/2}} \right) \ ; \ n_\alpha = 0, \pm 1, \dots.$$

The density relation (4), becomes now:

$$\rho = \frac{1}{V} \sum_{\substack{n_1=0,\pm 1,\dots \\ n_2=n_3=0}} \frac{1}{e^{\beta \varepsilon_k} - 1} + \frac{1}{V} \sum_{n_\alpha, n_{\alpha \neq 1} \neq 0}^{''} \frac{1}{e^{\beta \varepsilon_k} - 1}.$$

In the limit $L \to \infty$ one gets (since μ_L is still of the order $-\frac{B}{V}$ for $\rho > \rho_c(\beta)$)

$$\rho - \rho_c(\beta) = \lim_L \frac{1}{V} \sum^{'} \frac{1}{e^{\beta \varepsilon_k} - 1} = \sum_{n_1=0,\pm 1,\dots} \frac{1}{A \, n_1^2 + B} = \rho_0 > 0$$

for A and B constants. We get BEC in infinitely many modes. Here we mention the results for two models in which we consider periodic boundary conditions yielding conventional and generalized conventional BEC.

1. *Imperfect or mean field Bose gas* [17,18]
   The model is described by the following local Hamiltonians:

   $$H_\Lambda^{MF} = \sum_{k \in \Lambda^*} \varepsilon_k a_k^* a_k + \frac{1}{V} N_\Lambda^2 \ ; \ \lambda > 0. \tag{7}$$

   The model is thermodynamically stable for all values of $\mu \in \mathbb{R}$. One gets the following result: if $\rho > \rho_c(\beta)$, the model shows BEC in the mode $k = 0$ and $\mu = \lim_L \mu_L = 2\lambda \rho > 0$.

2. *Weak interaction model* [19,16]
   The model is described by the local Hamiltonians:

   $$H_\Lambda^{WI} = H_\Lambda^{MF} + \frac{g}{2V} \sum_{k \in \Lambda^*} N_k^2 \ ; \ g > 0. \tag{8}$$

   If $\rho > \rho_c(\beta)$, then $\mu = \lim_L \mu_L = 2\lambda \rho$. and there is generalized condensation of the following type:

   (a) $\lim_L \frac{\langle N_k \rangle}{V} = 0$ for all $k \in \Lambda^*$, including $k = 0$, i.e. no condensation in the mode $k = 0$.

   (b) $\lim_{\delta > 0} \lim_{L \to \infty} \frac{1}{V} \sum_{|k| < \delta} \langle N_k \rangle = \rho_0 > 0$.

One remarks that the notion of type of condensation is a rather subtle affair. Thermodynamic quantities, energy particle, free energy densities are not sensitive to the type condensation. Generalized conventional condensation is a finite-size effect which as its effect e.g. on the level of the fluctuations, because these are governed by the way the energy gap behaves as a function of the size of the volume [20,21].

3 BEC and spontaneous symmetry breaking (SSB)

In this section we want to make clear that the occurence of BEC is always accompanied with the phenomenon of SSB. We want also to make clear that SSB about and its relation to off-diagonal long range order. We treat the IBG explicitly as the prototype example but we realize that there are many exact results in the literature concerning SSB also for interacting systems.

Heuristically, what is SSB ? In general a system is defined by a Hamiltonian or a set of local Hamiltonians. A system has a (discrete or continuous) symmetry if these Hamiltonians are left invariant under this symmetry. The system shows SSB if one or other state equation has a solution which is not invariant under the symmetry. Hence in order to detect SSB for a system one has to specify state equations. As we are talking here about equilibrium properties we specify here the equilibrium equations.

Here we consider the *algebra of observables* $\mathcal{A}$, generated by the boson creation and annihilation operators $a_k^{\#}$. A generic element of this algebra is of the form

$$a_{k_1}^* a_{k_2}^* \ldots a_{k_n}^* a_{p_1} \ldots a_{p_m}$$

for arbitrary $(k_1, k_2 \ldots k_n, p_1, \ldots p_m)$, e.g. the energy observable is given by $H_\Lambda = \sum_{k \in \Lambda^*} \varepsilon_k a_k^* a_k$, the momentum observable by $P_\Lambda = \sum_k k a_k^* a_k$.

A *state* or an expectation value functional on $\mathcal{A}$ is given by a linear functional ω satisfying:

$$\omega(\mathbb{1}) = 1 \ , \ \omega(A^*A) \geq 0 \ \text{ for all } \ A \in \mathcal{A}.$$

Hence a state is known on all its observables if all values

$$\omega\left(a_{k_1}^* a_{k_2}^* \ldots a_{k_n}^* a_{p_1} \ldots a_{p_m}\right)$$

are known.

Let us now consider a Gibbs state or equilibrium state $\omega_{\beta,\Lambda}$ at inverse temperature β:

$$\omega_{\beta,\Lambda}(A) = \frac{tr \, e^{-\beta H_\Lambda} A}{tr \, e^{-\beta H_\Lambda}} \ , \ A \in \mathcal{A} \tag{9}$$

if the system H_Λ is stable, i.e. if $Z = tr \, e^{-\beta H_\Lambda} < \infty$.

Let

$$\alpha_t^\Lambda(A) = e^{it H_\Lambda} \, A \, e^{-it H_\Lambda} \ , \ t \in \mathbb{R}$$

be the dynamics of the observables, then for all $A, B \in \mathcal{A}$:

$$\omega_{\beta,\Lambda}\left(A \, \alpha_{i\beta}^\Lambda(B)\right) = \frac{1}{Z_\Lambda} \, tr \, e^{-\beta H_\Lambda} \, A \, e^{-\beta H_\Lambda} \, B \, e^{\beta H_\Lambda}$$

$$= \frac{1}{Z_\Lambda} \, tr \, e^{-\beta H_\Lambda} \, B \, A = \omega_{\beta,\Lambda}(BA)$$

i.e. the Gibbs state satisfies the equations, called KMS-equations: $\forall A, B \in \mathcal{A}$

$$\omega_{\beta,\Lambda}\left(A\,\alpha_{i\beta}^{\Lambda}(B)\right) = \omega_{\beta,\Lambda}(BA)\,. \tag{10}$$

Conversely, one checks that any state ω satisfying these equations (10) for finite Λ equals the Gibbs state $\omega_{\beta,\Lambda}$.

This shows that the Gibbs states (9) are equivalently defined by the KMS-equations (10). This means as far as equilibrium states are concerned there is no SSB for finite volume systems. This means also that in order to detect SSB one has to go to the thermodynamic limit situation $L \to \infty$. Therefore, let the limit dynamics α_t be defined as: $\alpha_t = \lim\limits_{L\to\infty} \alpha_t^v$, then the limit *equilibrium or KMS-conditions* are defined as follows.

The state ω is an equilibrium state at inverse temperature $\beta = 1/k_B T$ for the system dynamics α_t, if $\forall A, B \in \mathcal{A} \cap Dom(\alpha_{i\beta})$ holds:

$$\omega(A\,\alpha_{i\beta}\,B) = \omega(BA)\,. \tag{11}$$

In particular, any solution ω of (11) is an equilibrium state, and the possibility of finding solutions breaking the symmetry is now open. Let τ be a map of the algebra of observables into itself representing a symmetry operation, i.e. such that $\tau\,\alpha_t = \alpha_t\,\tau$ (τ commutes with the dynamics), if there exists a state ω solution of (11) such that $\omega \cdot \tau \neq \omega$, then the system (α_t) shows SSB. This is the intrinsic definition of spontaneous symmetry breaking and the challenging difficulty of SSB is indeed to show for each system that the equations (11) do have or do not have a symmetry breaking solution. There are many results in the literature about the nonexistence of SSB, there are only a few results about the existence of SSB.

For the sake of illustration of the subtleties occuring in this matter we prove the appearance of SSB for the IBG.

Denote for all $f, g \in \mathcal{S}$ (Schwartz functions)

$$a^*(f) = \int dk\, a_k^*\, f(k)$$

then the canonical commutation relations become

$$[a(f), a^*(g)] = (f, g)$$

$$[a(f), a(g)] = 0$$

and the IBG-dynamics

$$\alpha_t\, A = \lim_L \alpha_t^\Lambda\, A = \lim_L e^{it[H_\Lambda,\cdot]}\, A$$

with

$$\lim_L [H_\Lambda, a^*(f)] = a^*(hf)$$

$$(hf)(k) = \varepsilon_k\, f(k) \ ; \ \varepsilon_k = \frac{k^2}{2m} - \mu$$

or

$$\alpha_t\, a^*(f) = a^*\left(e^{ith}f\right)$$

and the KMS-condition (11) can be formulated rigorously.

Consider first *the case $\mu < 0$*, i.e. $\rho(\beta, \mu) < \rho_c(\beta)$, then $h > 0$ as operator or $\varepsilon_k > 0$ for all $k \in \Lambda^*$.

Take then $A = \mathbb{1}$ and $B = a^*(f)$, (11) becomes

$$\omega\left(a^*(f)\right) = \omega\left(a^*(e^{-\beta h}f)\right).$$

By induction: $\forall n \in \mathbb{N}^+$,

$$\omega\left(a^*(e^{-\beta hn}f)\right) = \omega\left(a^*(e^{-\beta h(n-1)}f)\right) = \cdots = \omega\left(a^*(f)\right).$$

Taking the limit $n \to \infty$, one gets for all $f \in \mathcal{F}$,

$$\omega\left(a^*(f)\right) = 0.$$

Take now $A = a^*(f)$ and $B = a(g)$, then (11) becomes

$$\omega\left(a^*(f)\alpha_{i\beta}a(g)\right) = \omega\left(a(g)a^*(f)\right)$$

and using the commutation relations

$$\omega\left(a^*(f)a\left((e^{\beta h} - 1)g\right)\right) = (g, f).$$

Substituting $g \to 1/(e^{\beta h} - 1)g$, one gets

$$\omega\left(a^*(f)a(g)\right) = \left(g, \frac{1}{e^{\beta h} - 1}f\right)$$

$$= \int dk \frac{\bar{g}(k)\,f(k)}{e^{\beta\varepsilon_k} - 1}.$$

Analogously, one gets in general

$$\omega\left(a^*(f_1)\ldots a^*(f_n)a(g_1)\ldots a(g_m)\right)$$

$$= \delta_{n,m}\,\mathrm{perm}\left(\omega\left(a^*(f_i)a(g_j)\right)\right)_{i,j=1,\ldots,n}$$

this means that in the case $\mu < 0$ i.e. low density $\rho(\beta, \mu) < \rho_c(\beta)$ one gets a unique solution for the equilibrium equations (11).

Consider now *the case $\mu = 0$*, and $\rho > \rho_c(\beta)$ or $\rho - \rho_c(\beta) \equiv \rho_0 > 0$ and the operator $h \geq 0$ or $\varepsilon_{k=0} = 0$. Now we show that there exists a homogeneous solution ω_φ ($\varphi \in [0, 2\pi]$) of (11), such that:

$$\omega_\varphi\left(a^*(f)\right) \neq 0 \quad \text{for any} \quad f \in \mathcal{S} \quad \text{such that} \quad f(0) \neq 0.$$

Anyway, put

$$\omega_\varphi\left(a^*(f)\right) = \sqrt{\rho_0}\,e^{i\varphi}\,f(0)\,, \quad \varphi \in [0, 2\pi] \tag{12}$$

where ω_φ is a solution of (11) depending on the free complex phase φ and where ρ_0 is any positive constant which will turn out later indeed to be equal to the condensate density.

Introduce

$$b^*(f) = a^*(f) - \sqrt{\rho_0}\, e^{i\varphi} f(0)$$

then also

$$[b(f), b^*(g)] = (f, g)\mathbb{1} \; ; \; [b(f), b(g)] = 0$$

$$\alpha_t\, b^*(f) = b^*(e^{ith} f)\,.$$

By definition: $\forall f \in \mathcal{S}$

$$\omega_\varphi\left(b^*(f)\right) = 0\,.$$

As $\mathcal{R}(e^{\beta h} - 1)$ is dense in $\mathcal{S}$, one derives as above: $\forall f, g \in \mathcal{S}$

$$\omega_\varphi\left(b^*(f_1)\dots b^*(f_n)b(g_1)\dots b(g_m)\right)$$

$$= \delta_{n,m} \operatorname{perm}\left(\omega_\varphi(b^*(f_i)b(g_j))\right)_{i,j=1,\dots n}$$

solving equations (11) in the $b^\#$-variables, i.e. we have infinitely many solutions $\omega_\varphi(\varphi \in [0, 2\pi])$ of the equilibrium equations (11).

Let us now work a bit further on this solution ω_φ, in particular remark that (12) holds. Take the L^2-normalized characteristic function χ_Λ of the volume Λ:

$$\chi_\Lambda(x) = \begin{cases} \frac{1}{\sqrt{V}} & \text{if } x \in \Lambda \\ 0 & \text{if } x \notin \Lambda \end{cases}$$

and since we have Λ with periodic boundary conditions as in Sect. 2.1 the Fourier transform gives

$$f(k) = \frac{\hat{\chi}_\Lambda}{\sqrt{V}}(k) = \delta_{k,0}$$

then

$$\lim_{L \to \infty} \omega_\varphi\left(\frac{a^*(\hat{\chi}_\Lambda)}{\sqrt{V}}\right) = \lim_L \frac{\omega_\varphi(a_0^*)}{\sqrt{V}} = \sqrt{\rho_0}\, e^{i\varphi}\,. \qquad (*)$$

Also one has the two-point function:

$$\omega_\varphi\left(a^*(f)a(g)\right) = \omega_\varphi\left(b^*(f)b(g)\right) + \rho_0\, f(0)\, \bar{g}(0)$$

$$= \int dk \frac{f(k)\bar{g}(k)}{e^{\beta \frac{k^2}{2m}} - 1} + \rho_0\, f(0)\bar{g}(0)$$

and compute as a special case

$$\lim_{L \to \infty} \frac{\omega_\varphi\left(a^*(\hat{\chi}_\Lambda)a(\hat{\chi}_\Lambda)\right)}{V} = \lim_{L \to \infty} \frac{\omega_\varphi(a_0^* a_0)}{V} = \rho_0$$

or equivalently, together with formula (12) we proved

$$\rho_0 = \lim_L \frac{\omega_\varphi(a_0^* a_0)}{V} = \left|\lim_L \frac{\omega_\varphi(a_0^*)}{\sqrt{V}}\right|^2$$

the appearance of *off-diagonal long range order* in the case of condensation ($\rho_0 > 0$).

Consider now the one-parameter group of gauge transformations of the first kind, i.e. the group $\{\tau_\varphi | \varphi \in [0, 2\pi]\}$ of transformations of $\mathcal{A}$

$$\tau_\varphi \; : \; a^*(f) \to e^{i\varphi} a^*(f) \; ; \; \forall f \in \mathcal{S} \, .$$

The dynamics $(dt)_t$ of the system is invariant under the group $(\tau_\varphi)_\varphi$ because

$$\tau_\varphi H_\Lambda = H_\Lambda \, .$$

Hence: $\tau_\varphi \, \alpha_t = \alpha_t \, \tau_\varphi$ for all $t \in \mathbb{R}$, and $\varphi \in [0, 2\pi]$. This means that the system has the gauge symmetry.

On the other hand, consider the solution ω_φ of (11) and compute using (12):

$$\omega_\varphi \left(\tau_{\varphi'} a^*(f) \right) = e^{i\varphi'} \omega_\varphi \left(a^*(f) \right)$$

$$= e^{i(\varphi + \varphi')} \sqrt{\rho_0} \, f(0) = \omega_{\varphi + \varphi'} \left(a^*(f) \right)$$

yielding that for $\varphi' \neq 0$ and $\rho_0 > 0$:

$$\omega_\varphi \cdot \tau_{\varphi'} \neq \omega_\varphi$$

i.e. the solution ω_φ not invariant under the gauge transformations. We proved rigorously for the IBG that if there is condensation ($\rho_0 > 0$), there is *spontaneous symmetry breakdown* of the gauge symmetry.

All these results are explicitly proved for the IBG, but we have a general theorem about this (see [22]). For any arbitrary interacting bose system, if there is condensation then there is spontaneous symmetry breaking of the gauge symmetry group of the type as described above.

The next topic which we want to address is concerned with the by now standard knowledge of the fact that SSB is related to the Goldstone theorem [23], talking about the appearance of the Goldstone boson. On might ask the question: can one construct explicitly this Goldstone boson or field. This question is explicitly solved for the case of SSB of the gauge transformation symmetry as above in [24]. The general theorem can be found in [25]. Here we report on [24].

It is immediately checked that the gauge transformations τ_φ are implemented as follows

$$\tau_\varphi \, A = e^{i\varphi N} \, A \, e^{-i\varphi N}$$

where N is the number operator

$$N = \int dx \, n(x) \;\; \text{and} \;\; n(x) = a^*(x) a(x)$$

the generator of the one-parameter group of gauge transformations. From above, it is clear that SSB implies the existence of a state ω_0 satisfying

$$\omega_0 \left(\tau_\varphi \, a^*(y) \right) \neq \omega_0 \left(a^*(y) \right) \;\; \text{for} \;\; \varphi \neq 0$$

or equivalently (by translation invariance of the state and (∗))

$$\omega_0 \left(\left[\int dx\, a^*(x) a(x) \ ,\ a^*(y) \right] \right) = \omega_0 \left(a^*(y) \right) = \sqrt{\rho_0} \,. \tag{13}$$

Using the *homogeneity* of the state ω_0, (13) is equivalent to

$$\frac{1}{V} \omega_0 \left(\left[\int dx\, (n(x) - \omega_0(n(x))) \ ,\ \int_V dy\, (a^*(y) - \omega_0(a^*(y))) \right] \right) = \sqrt{\rho_0} > 0 \,.$$

Using locality: $[n(x), a^*(y)] = 0$ if $x \neq y$, (13) is equivalent to

$$\omega_0 \left([F_\Lambda(n) \ ,\ F_\Lambda(a^*)] \right) = \sqrt{\rho_0} \tag{13a}$$

where

$$F_\Lambda(n) = \frac{1}{\sqrt{V}} \int_V dx\, (n(x) - \omega_0(n(x)))$$

$$F_\Lambda(a^*) = \frac{1}{\sqrt{V}} \int_V dy\, (a^*(y) - \omega_0(a^*(y))) \,.$$

Remark that for general observable $A \in \mathcal{A}$:

$$F_{k,\Lambda}(A) = \frac{1}{\sqrt{V}} \int_V dx\, e^{ikx} \left(A(x) - \omega_0(A(x)) \right)$$

is the k-mode *fluctuation operator* of the observable A in the state ω_0. The $\lim_{L \to \infty} F_{k,\Lambda}(A) \equiv F_k(A)$ can be given a sense on the basis of noncommutative central limit theorems (see [26]).

Let

$$H = \int \varepsilon_k \, dP(k)$$

be a spectral resolution of the thermodynamic limit Hamiltonian in the state ω_0, then one can compute for $k \neq 0$

$$\omega_0 \left(F_k(n)^2 \right) = \varepsilon_k \coth \beta \varepsilon_k / 2$$

$$\omega_0 \left(F_k(a^*)^2 \right) = \frac{1}{\varepsilon_k} \coth \beta \varepsilon_k / 2 \,.$$

The Goldstone theorem learns that for long range interactions ε_k tends to an $\varepsilon_0 \neq 0$; ε_0 is then also called a plasmon frequency. That means that the limit $k \to 0$ of this result is meaningful and the operators $F(n)$ and $F(a^*)$ are well defined operators. This implies also that the limit $\Lambda \to \infty$ of the relation (13a) exists leading to the commutation relation

$$[F(n), F(a^*)] = \sqrt{\rho_0}$$

i.e. $(F(n), F(a^*))$ forms a *canonical pair of coordinates of the Goldstone boson*. One remarks that this commutator vanishes if the condensate i.e. if SSB disappears.

The Goldstone theorem learns also that for short range interactions there is no energy gap, i.e. $\lim_{k \to 0} \varepsilon_k = 0$. Then for $\beta < \infty$ (finite temperature) and small k:

$$\omega_0 \left(F_k(n)^2 \right) \simeq \frac{1}{\beta} \quad \text{or} \quad F_k(n) \simeq \frac{1}{\sqrt{\beta}}$$

$$\omega_0 \left(F_k(a^*)^2 \right) \simeq \frac{1}{\beta \varepsilon_k^2} \quad \text{or} \quad F_k(a^*) \simeq \frac{1}{\sqrt{\beta} \varepsilon_k}$$

the fluctuation of the creation operators have abnormal fluctuations and the degree of abnormality is determined by the behaviour of the spectrum ε_k in the neighbourhood of $k = 0$. We have also for the commutator the behaviour:

$$[F_k(n), F_k(a^*)] \simeq \frac{\sqrt{\rho_0}}{\beta \varepsilon_k}.$$

A particular interesting situation is the ground state case $T = 0$ or $\beta = \infty$. Then we have

$$\omega_0 \left(F_k(n)^2 \right) \simeq \varepsilon_k \quad \text{or} \quad F_k(n) \simeq \sqrt{\varepsilon_k}$$

$$\omega_0 \left(F_k(a^*)^2 \right) \simeq \frac{1}{\varepsilon_k} \quad \text{or} \quad F_k(a^*) \simeq \frac{1}{\sqrt{\varepsilon_k}}.$$

If one defines the abnormal fluctuations

$$\tilde{F}(n) = \lim_{k \to \infty} F_k(n)/\sqrt{\varepsilon_k}$$

$$\tilde{F}(a^*) = \lim_{k \to 0} \sqrt{\varepsilon_k} \, F_k(a^*)$$

on the basis of above they are well defined and nontrivial and in combination with (13a) they satisfy the commutation rule:

$$\left[\tilde{F}(n), \tilde{F}(a^*) \right] = \sqrt{\rho_0}.$$

The pair $(\tilde{F}(n), \tilde{F}(a^*))$ is again the canonical pair of the ground state Goldstone boson connected with the SSB. One can also study the dynamics of this pair and find out that it separates off from the other variables of the system and behaves as an harmonic oscillator pair of variables living on a time scale $\tilde{t} = \lim_{k \to 0} t/\varepsilon_k$ and an equivalently re-scaled energy frequency equal to 2.

4   BEC in traps viewed from conventional BEC

Since 1995 and the fine experiments with trapped Bose gases there is a lot of work [27-29] done on these boson systems. In this section we try to understand in which sense this type of condensation can be understood as a conventional Bose-Einstein condensation. Which is the main problem in order to fit these systems with conventional systems ?

One considers bosons in traps, these are external confining potentials making the system inhomogeneous in the space translations. A basic ingredient for systems with conventional BEC is that they are homogeneous. It is clear that the type of condensation for trapped boson systems can only be considered as a conventional condensation in some limits making the system effectively homogeneous. Trapped boson condensation is only detected for extreme diluted gases in the trap. Is the limit of extreme dilution creating the link with conventional condensation ? This is not simply understood. Moreover in spite of the extreme dilution, it seems also necessary to take into account interactions between the atoms in order to explain the experimental data. Also this is not simply understood.

Because of all these reasons we confine ourselves to a pedestrian description of the situation.

4.1 IBG in an harmonic potential

Suppose that we have an IBG in an external 3d-potential:

$$V_e(x) = \frac{1}{2}(\omega_1^2 x_1^2 + \omega_2^2 x_2^2 + \omega_3^2 x_3^2) \, .$$

It is an easy text book exercise to find out that the Hamiltonian for N bosons is given by:

$$H_N = \sum_{i=1}^{N} \sum_{j=1}^{3} \omega_j \left(a_{j,i}^* a_{j,i} + \frac{1}{2} \right) \, ; \quad \text{for } \bar{h} = 1 \, ; \, m = 1$$

whose spectrum is given by

$$E_n = \sum_{j=1}^{3} \left(n_j + \frac{1}{2} \right) \omega_j \, ; \, n = (n_1, n_2, n_3) \in \mathbb{N}^3 \, .$$

Clearly, the ground state $(T = 0)$ corresponds to $n = 0$ with a wave function:

$$\varphi(x) = \prod_{i=1}^{N} \varphi_0(x^i) \, , \, x^i = (x_{1,i}, x_{2,i}, x_{3,i}) \in \mathbb{R}^3$$

with

$$\varphi_0(x) = \left(\frac{\omega_0}{\pi} \right)^{3/2} \exp \left[-\frac{1}{2}(\omega_1 x_1^2 + \omega_2 x_2^2 + \omega_3 x_3^2) \right]$$

ω_0 is the geometric average of the oscillator frequencies given by: $\omega_0 = (\omega_1 \omega_2 \omega_3)^{1/3}$; hence the density distribution

$$n(x) = |\varphi_0(x)|^2$$

is a gaussian with the size of the cloud fixed by the oscillator length scale $a_0 = \omega_0^{-1/2}$ and hence the velocity distribution $(\hat{\varphi}_0(k))$ is also a gaussian centered around zero momentum with size approximately $a_0^{-1} = \omega_0$. Remark that interactions change drastically the form of the gaussian peak visible for the noninteracting case.

Let us now turn to the finite temperature situation, and consider the grand canonical ensemble. The density is given by

$$\left\langle \frac{N_N}{N} \right\rangle = \frac{1}{N} \sum_{k \in \mathbb{N}^3} \frac{1}{e^{\beta(E_n - \mu)} - 1}$$

where N_N is the number operator for the N boson particles, and the energy density per particle by

$$\left\langle \frac{H_N}{N} \right\rangle = \frac{1}{N} \sum_{n \in \mathbb{N}^3} \frac{E_n}{e^{\beta(E_n - \mu)} - 1}.$$

As for the uniformly homogeneous boson gas, separate off the lowest energy E_0-term, and

$$\left\langle \frac{N_0}{N} \right\rangle = \frac{1}{N} \frac{1}{e^{\beta(E_0 - \mu)} - 1}.$$

If the chemical potential μ tends to $E_0 = (\omega_1 + \omega_2 + \omega_3)/2$ then the density of lowest energy particles $\langle N_0 \rangle/N$ becomes finite and the lowest energy state gets a macroscopic occupation leading to BEC.

Then the critical density is defined as:

$$\rho_c(\beta) = \frac{\langle N_N \rangle - \langle N_0 \rangle}{N} = \frac{1}{N} \sum_{n \neq 0} \frac{1}{e^{\beta(\omega_1 n_1 + \omega_2 n_2 + \omega_3 n_3)} - 1} < \infty.$$

But it is good to remember that all this is volume dependent and has no singularities nor a thermodynamic limit playing a central role in the concept of BEC for homogeneous systems. Therefore BEC in traps is and remains so far a finite particle problem which strictly speaking no phase transition takes place, so no BEC in trapped gases.

For these reasons, theoreticians consider a semiclassical approximation for the excited states. This is a limit procedure by which the level spacing becomes smaller and smaller. This can be realized by taking the limit N tending to infinity and ω_0 tending to zero such that $N\omega_0^3 < \infty$ remains constant. The constant ρ plays the role of density of particles. One gets

$$\rho - \rho_0 = \int_0^\infty \frac{\rho(\varepsilon) d\varepsilon}{e^{\beta \varepsilon} - 1}$$

where $\rho(\varepsilon)$ is the density of states calculated from E_n:

$$\rho(\varepsilon) = \frac{1}{2} \varepsilon^2.$$

Remark that for the $d = 3$ IBG one has

$$\rho(\varepsilon) = \left(\frac{1}{2\pi} \right)^2 2^{3/2} \sqrt{\varepsilon}.$$

It is interesting to compare this procedure of limit with the exact results for the free boson gas in scaled external fields [30,31].

Consider the one-particle Hamiltonian in d-dimensional boxes Λ of sides L and with periodic boundary conditions:

$$h_L = -\frac{\Delta}{2m} + V_e\left(\frac{x}{L}\right) - \mu_L$$

with: $V_e(x) = c|x|^\alpha$; $c, \alpha > 0$.
The dynamics α_t^Λ is given by:

$$\alpha_t^\Lambda a^*(f) = a^*(e^{ith_L} f) \ , \ f \in \mathcal{F} \ .$$

For finite L the system is clearly not space translation invariant or homogeneous. But the system becomes clearly again homogeneous in the limit of L tending to infinity. Also in this limit the position x and momentum p observables satisfy:

$$[x, p] = i \, \mathbb{1} \ .$$

But the variable $y = x/L \in [-1, 1]$ becomes a relevant parameter such that

$$\lim_{L \to \infty} [y, p] = [y, x] = 0$$

i.e. y becomes a classical parameter always with the physical meaning of a position. There is indeed a formal similarity of this limit with the N tending to infinity but $N\omega_0^3 = \rho$ constant for the oscillator model above. It means that $\omega_0 \simeq 1/N^{1/3}$ or alternatively introducing the variable $x/N^{1/3}$ i.e. L is replaced by $N^{1/3}$. Anyway the limit L tending to infinity is a sense full thermodynamic limit yielding the following exact results. The density formula (4) in the thermodynamic limit becomes:

$$\rho = \rho_0 + \tilde{\rho}(\beta, \mu)$$

where

$$\tilde{\rho}(\beta, \mu) = \left(\frac{1}{2\pi}\right)^d \int dk \int_{-1}^1 dy \frac{1}{e^{\beta\left(\frac{k^2}{2m} + V_e(y) - \mu\right)} - 1} \ .$$

The critical density is now:

$$\tilde{\rho}_c(\beta) = \tilde{\rho}(\beta, 0) \ .$$

We have also that $\tilde{\rho}_c(\beta)$ is finite in the following cases:

(i)   for $\nu = 1$ if $\alpha \le 1$

(ii)   for $\nu = 2$ if $\alpha \le 2$

(iii)   for $\nu = 3$

Hence if $\tilde{\rho}_c(\beta)$ and for all densities $\rho > \tilde{\rho}_c(\beta)$ there is Bose-Einstein condensation i.e. $\rho_0 > 0$, ρ_0 is the density of the condensate. One checks that $\mu = \lim_L \mu_L < 0$ if $\rho < \tilde{\rho}_c(\beta)$ and $\mu = 0$ if $\rho \ge \tilde{\rho}_c(\beta)$.
Remark that the condensation is concentrated in the points (k, y) such that

$$\frac{k^2}{2m} + V_e(y) = 0 \ .$$

If $V_e(y) = c|y|^\alpha$, then in the points $k = y = 0$. In general the condensation takes place in the zeros of the external potential.

4.2 Interacting bosons in external potential

We limit ourself here with a formal description of what is usually done in this case without any claim on originality nor in content nor in the formulation. Moreover we constrain the material to the ground state $(T = 0)$.

We consider a two-body interacting system of bosons via a potential V put in an external field V_e:

$$H = \int dx\, a^*(x) \left(-\frac{\Delta}{2m} + V_e(x) \right) a(x)$$

$$+ \frac{1}{2} \int dx\, dx'\, a^*(x)\, a^*(x')\, V(x - x')\, a(x')\, a(x)$$

with

$$[a(x), a^*(x')] = \delta(x - x') \; ; \; [a(x), a(x')] = 0 \,.$$

Let ω_0 be a ground state of this system and introduce

$$a(x, t) = \varphi(x, t) + b(x, t)$$

where

$$\varphi(x, t) = \omega_0\left(a(x, t)\right) \in \mathbb{C}$$

is called the order parameter or the (classical) wave function of the condensate. Remark that if $\varphi(x, t) \neq 0$ there one has a broken gauge symmetry for the state ω_0. There is already a serious problem hidden here, namely the question whether such a ground state exists for which interactions V and for which external potentials V_e. The problem is even more serious considering that the system is not homogeneous. Worthless to say that this problem is far from being solved.

Following the spirit and interpretation of formula (12) one calls

$$\rho_0(x, t) = |\varphi(x, t)|^2$$

the density of the condensate. Remark that $b(x, t)$ is again a boson annihilation operator

$$[b(x, t), b^*(x', t)] = \delta(x - x') \; , \; [b(x, t), b(x', t)] = 0 \,.$$

The dynamical equation of the annihilation operator (Heisenberg equation of motion) is readily derived:

$$i\frac{\partial}{\partial t} a(x, t) = [H, a(x, t)]$$

$$= \left(-\frac{\Delta}{2m} + V_e(x) + \int dx'\, a^*(x', t)\, V(x - x')\, a(x', t) \right) a(x, t) \,.$$

Take now the effective potential

$$V(x - x') = g \, \delta(x - x')$$

$$g = \frac{4\pi}{m} a$$

$$a = \text{scattering length}$$

then one gets:

$$i\frac{\partial}{\partial t} a(x, t) = \left(-\frac{\Delta}{2m} + V_e(x) + \frac{g}{2} a^*(x, t) \, a(x, t) \right) a(x, t) .$$

One makes the classical approximation $(a(x, t) \to \varphi(x, t))$ by forgetting the boson field $b(x, t)$ and gets a nonlinear Schrödinger equations for the "wave function" $\varphi(x, t)$ of the condensate

$$i\frac{\partial}{\partial t} \varphi(x, t) = \left(-\frac{\Delta}{2m} + V_e(x) + \frac{g}{2} |\varphi(x, t)|^2 \right) \varphi(x, t) .$$

This equation is called the Gross-Pitaevskii (G.P.) equation [32-34] describing a semi-classical boson liquid in the case one can prove that this equation has a solution different from zero [35-37].

It is a general belief that this equation has a good validity if the scattering length a is small with respect to the average distance between the particles. This is expressed by the inequality $\rho a^3 \ll 1$.

One can consider an alternative derivation of the G.P.-equation namely as the Euler equation minimizing the following energy functional

$$E(\varphi) = \int dx \left(\frac{1}{2m} |\nabla \varphi|^2 + V_e(x)|\varphi|^2 + \frac{g}{2} |\varphi|^4 \right)$$

namely the equation:

$$i\frac{\partial}{\partial t} \varphi(x) = \frac{\delta E(\varphi)}{\delta \varphi(x)} .$$

One should compare the functional $E(\varphi)$ with the Ginzburg-Landau functional, an important tool for the study of superfluidity and superconductivity. Therefore this approach might give insight in the boson condensate as a boson liquid.

References

1. A. Einstein; Quantentheorie des einatsmigen idealen Gases; Sitzungsber. Press. Akad. Wiss. I, 3-14 (1925).

2. G.E. Uhlenbeck, Thesis, Leiden (NL) 1927.

3. B. Kahn, G.E. Uhlenbeck; On the theory of condensation, Physica 5, 399-415 (1938).

128

4. P. Kapitza, Nature 141, 74 (1938).

5. J.F. Allen, A.D. Misener, Nature 141, 75 (1938)

6. F. London; Superfluids, Vol. 2; Macroscopic Theory of Superfluid Helium; Wiley, New york.

7. V.A. Zagrebnov, J.-B. Bru; The Bogoliubov Model of Weakly Imperfect Bose Gas; Phys. Reports 350, 291-434 (2001).

8. A. Pais; Subtle is the lord; Oxford University Press, 1982.

9. M. Girardeau; Relationships between systems of impenetrable bosons and fermions in one dimension; J. Math. Phys. 4, 666-671 (1960).

10. H.B.G. Casimir; One Bose-Einstein condensation; in Fund. Probl. in Stat. Mech. III, North-Holland Publ. Comp. Amsterdam 1968, p. 188-196.

11. M. van den Berg, J.T. Lewis; On generalized condensation in the free boson gas; Physica A 110, 550-564 (1982).

12. M. van den Berg; On boson condensation into a infinite number of low-lying levels; J. Math. Phys. 23, 1159-1161 (1982).

13. M. van den Berg; On condensation in the free boson gas and the spectrum of the Laplacian; J. Stat. Phys. 31, 623-637 (1983).

14. M. van den Berg, J.T. Lewis; On the free boson gas in a weak external potential; Commun. Math. Phys. 81, 475-494 (1981).

15. J.V. Pulé; The free boson gas in a weak external potential; J. Math. Phys. 24, 138-142 (1983).

16. T. Michoel, A. Verbeure; Non-extensive Bose-Einstein condensation model; J. Math. Phys. 40, 1268-1279 (1999).

17. K. Huang; Imperfect Bose gas; in Studies in Statistical Mechanics, Vol. II, North-Holland Publ. Co., Amsterdam p. 3-110, 1964, eds. J. De Boer, G.E. Uhlenbeck.

18. M. Fannes, A. Verbeure; The condensed phase of the imperfect Bose gas; J. Math. Phys. 21, 1809-1818 (1980).

19. M. Schröder; J. Stat. Phys. 58, 1151- (1990).

20. E. Buffet, J.V. Pulé; Fluctuation properties of the imperfect Boson gas; J. Math. Phys. 24, 1608-1616 (1983).

21. M. Broidioi, A. Verbeure; Scaling Behaviour in the Bose gas; Commun. Math. Phys. 174, 635-660 (1996).

22. M. Fannes, J.V. Pulé, A. Verbeure; On Bose Condensation; Helv. Phys. Acta 55, 391-399 (1982).

23. J. Goldstone; Il Nuovo Cimento 19, 154 (1961).

24. T. Michoel, A. Verbeure; Goldstone normal coordinates in interacting boson systems; J. Stat. Phys. 96, 1125-1162 (1999).

25. T. Michoel, A. Verbeure; Goldstone Boson normal coordinates; Commun. Math. Phys. 216, 461-490 (2001).

26. D. Goderis, A. Verbeure, P. Vets; Dynamics of Fluctuations for Quantum Lattice Systems; Commun. Math. Phys. 128, 533-549 (1990).

27. A.S. Perkins, D.F. Walls; The physics of trapped dilute gas Bose-Einstein condensates; Phys. Rep. 303, 1-80 (1998).

28. A. Griffin, D.W. Snoke, S. Stringari; Bose-Einstein Condensation; Univ. Press, Cambridge 1936.

29. F. Dalfovo, S. Giorgini, L.P. Pitaevskii, S. Stringari; Theory of Bose-Einstein condensation in trapped gases; Rev. Mod. Phys. 71, 463-512 (1999).

30. J.V. Pulé; The free boson gas in a weak external potential; J. Math. Phys. 24, 138-142 (1983).

31. J. Messer, A. Verbeure; Free bosons in a scaled external potential; J. Phys. A 15, L111-L114 (1982).

32. E.P. Gross; Structure of a quantized vortex in boson systems; Il Nuovo Cimento 20, 454-466 (1961).

33. L.P. Pitaevskii; vortex lines in an imperfect Bose gas; Sov. Phys.-JETP 13, 451-454 (1961).

34. E.P. Gross; Hydrodynamics of a superfluid condensate; J. Math. Phys. 4, 195-207 (1963).

35. E.H. Lieb, R. Seiringer, J. Yngvason; Bosons in a trap: a rigorous derivation of the Gross-Pitaevskii energy functional; Phys. Rev. A 61, 043602-1–043602-13 (2000).

36. E.H. Lieb, R. Seiringer, J. Yngvason; A rigorous derivation of the Gross-Pitaevskii energy functional for a two-dimensional Bose gas; Commun. Math. Phys. 224, 17-31 (2001).

37. E.H. Lieb, R. Seiringer; Proof of Bose-Einstein Condensation for Dilute Trapped Gases; Texas arXiv math-ph 102-115.

Acknowledgements

To all my co-workers on the topics touched in these notes during all these years, in particular V.A. Zagrebnov for carefully reading the manuscript.

PART II

Invited Conferences
and
Seminars

ON THE TOPOGRAPHIC RECTIFICATION OF OCEAN FLUCTUATIONS

by

Alberto Álvarez, Emilio Hernández-García, Joaquín Tintoré

Instituto Mediterráneo de Estudios Avanzados IMEDEA[1]
CSIC-Universitat de les Illes Balears, E-07071 Palma de Mallorca (Spain).

Abstract

Stochastic fluctuations acting on a model of quasigeostrophic fluid motion on a rotating frame are shown to be rectified giving rise to large-scale noise-sustained average currents. As in other noise rectification phenomena, the effect requires nonlinearity and absence of detailed balance to occur. We apply an analytical coarse-graining procedure to obtain insight into the phenomenon. Relevance of the effect in the context of ocean modeling is briefly discussed.

1 Introduction

Nonlinear interactions can rectify random inputs of energy organizing them into coherent motion. This noise-rectification phenomenon has been discussed in several contexts ranging from biology to physics or engineering [1]. Three ingredients are needed to obtain this kind of noise-sustained directed motion: nonlinearity, random noise lacking the property of detailed balance, and some symmetry-breaking feature establishing a preferred direction of motion.

It has been shown numerically [2] that directed motion sustained by noise also appears in a model of large-scale fluid dynamics on a rotating frame, namely the vorticity equation describing quasigeostrophic forced turbulence [3]. A large amount of rotating fluid problems concerning planetary atmospheres and oceans can be described to some degree of approximation by this model. It uses the fact that, in a planet or frame in fast rotation, vertical velocities are small and slaved to the horizontal motion, so that flow patterns can be described in terms of two horizontal coordinates. The vertical depth of the fluid becomes a dependent variable. The fluid displays many of the unique properties of two-dimensional turbulence, but some of the aspects of three-dimensional dynamics are still important. In particular, topographic features of the bottom wall above which the fluid is flowing appear explicitly in the model. In terms of the streamfunction $\psi(\mathbf{x}, t)$, with $\mathbf{x} \equiv (x, y)$, the model reads:

$$\frac{\partial \nabla^2 \psi}{\partial t} + \lambda \left[\psi, \nabla^2 \psi + h \right] = D + F . \tag{1}$$

D is the dissipation term. Here we will use the standard viscous damping, $D = \nu \nabla^4 \psi$, where ν is the viscosity. $F(\mathbf{x}, t)$ is any kind of relative-vorticity external forcing, and $h = f \Delta H / H_0$,

---

[1]URL: http://www.imedea.uib.es/

O. Descalzi et al. (eds.), Instabilities and Nonequilibrium Structures VII & VIII, 133–139.
© 2004 *Kluwer Academic Publishers. Printed in the Netherlands.*

with f the Coriolis parameter, H_0 the mean depth, and $\Delta H(\mathbf{x})$ the local deviation from the mean depth. λ is a bookkeeping parameter introduced to allow perturbative expansions in the interaction term. The physical case corresponds to $\lambda = 1$. The Poisson bracket or Jacobian is defined as

$$[A, B] = \frac{\partial A}{\partial x}\frac{\partial B}{\partial y} - \frac{\partial B}{\partial x}\frac{\partial A}{\partial y} \ . \tag{2}$$

The streamfunction provides the horizontal components of the fluid velocity $(u(\mathbf{x}), v(\mathbf{x}))$ from

$$u = -\frac{\partial \psi}{\partial y} \ , \quad v = \frac{\partial \psi}{\partial x} \tag{3}$$

Equation (1) represents the time evolution of the relative vorticity subjected to forcing and dissipation. In addition to the context of rotating fluids, this kind of quasi-twodimensional dynamics appears also in the study of drift-wave turbulence in plasmas under strong magnetic fields (perpendicular to the plane of $\mathbf{x}$) [4, 5, 6, 7]. In this case ψ is related to the electrostatic potential, and $h = \ln(\omega_c/n_0)$, where ω_c and n_0 are the cyclotron frequency and plasma density respectively. Eq. (1) is also the limiting case of the more general Charney-Hasegawa-Mima equation when the scales are small compared to the ion Larmor radius or the barotropic Rossby radius[5, 6, 7].

2   The noise-induced currents

The results in [2] imply (Rayleigh friction is used in that paper, but the results apply also to the viscous damping used here) that, when the forcing in Eq.(1) is a white Gaussian noise, the average flow at large scales approaches a state with currents following the large-scale features of the underlying topography. Energy is extracted from the noise forcing and concentrated into these large-scale currents. If noise is switched-off, viscosity dissipates all the energy in the system and currents stop. More interestingly, if nonlinear terms are eliminated from the equation, the average flow is again zero, corresponding to the intuitive idea that noise with zero average would not induce mean flow. But the interaction between noise and nonlinearity induces nonvanishing currents through the phenomenon of noise rectification. The direction of the currents was cyclonic (i.e. counterclockwise in the Northern hemisphere) around depressions and anticyclonic around elevations. An example of average flow induced by noise over the topography of Fig. 1 is shown in Fig. 2. It is obvious that contour levels of the average streamfunction, giving the average velocity field, closely follow topographic contours. Detailed inspection of the simulation data reveals that the correlation is higher for the large-scale features, the small scales being less correlated with topography.

Appearance of currents following topography is not a new issue. Since the work of Salmon et al. [8] it is known that the statistical mechanics of the Euler equations predicts such kind of currents. In addition, there seem to be observations of them in several places of the world ocean [9]. What is quite surprising and new is that they appear here in a forced and dissipative model, and without the property of detailed balance. Thus, the equilibrium statistical mechanics ideas

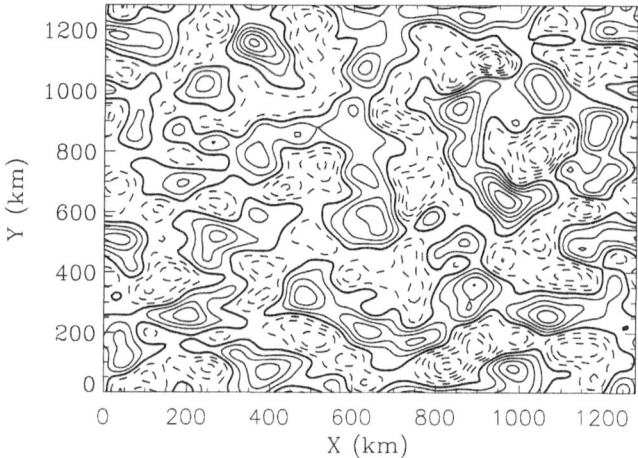

Figure 1: Depth contours of a randomly generated bottom topography. Maximum depth is 381.8m and minimum depth $-381.8m$ over an average depth of 5000m. Levels are plotted every 63.6m. Continuous contours are for positive deviations with respect to the mean, whereas dashed contours are for negative ones.

presented in [8] cannot explain the phenomenon in (1). A review of the different contexts where topographic currents appear can be found in [10].

In the next Section we perform analytic calculations trying to understand the origin of the currents in the present forced and dissipative case [11]. The observation of distinct behavior at different scales suggests that analysis of the relationship between small and large scales would give clues about the process.

3 Coarse-graining approach

Here we analyze how the dynamics of the large scales in (1) is affected by the small scales, when F is a random forcing. A useful choice of F, flexible enough to model a variety of processes, is to assume F to be Gaussian stochastic process with zero mean and correlations given by

$$\left\langle \hat{F}_{\mathbf{k}}(\omega)\hat{F}_{\mathbf{k}'}(\omega') \right\rangle = Dk^{-y}\delta(\mathbf{k} + \mathbf{k}')\delta(\omega + \omega') . \tag{4}$$

$\hat{F}_{\mathbf{k}}(\omega)$ denotes the Fourier transform of $F(\mathbf{x}, t)$, $\mathbf{k} = (k_x, k_y)$, and $k = |\mathbf{k}|$. The process is then white in time but has power-law correlations in space. $y = 0$ corresponds to white-noise also in space, which is the case studied in [2] and displayed in Fig. 2. This value of y has been observed for wind forcing on the Pacific ocean [12]. Thermal noise corresponds to $y = -4$ [13]. In this

136

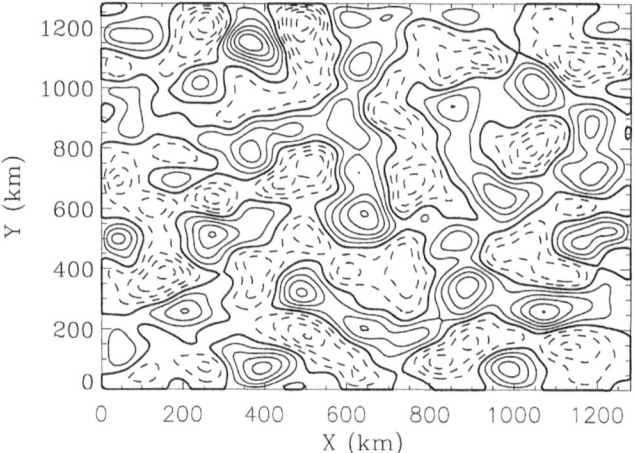

Figure 2: Mean streamfunction computed by time averaging when a statistically stationary state has been achieved. Continuous contours denote positive values of the streamfunction, whereas dashed contours denote negative ones; $\lambda = 1$, $f = 10^{-4}s^{-1}$, $\nu = 200m^2s^{-1}$, and forcing is white noise in space and time with intensity $D = 10^{-9}m^2s^{-3}$. Maximum and minimum values are 991.864 and -991.864 m^2/s, and levels are plotted every 165.31 m^2/s.

case there is a fluctuation-dissipation relation between noise and the viscosity term, so that the fluctuations satisfy detailed balance.

To obtain the desired large-scale closed equation we have applied a coarse-graining procedure to the investigation of the dynamics. For our problem it is convenient to use the Fourier components of the streamfunction $\hat{\psi}_{\mathbf{k}\omega}$ or equivalently the relative vorticity $\zeta_{\mathbf{k}\omega} = -k^2\hat{\psi}_{\mathbf{k}\omega}$. This variable satisfies:

$$
\begin{aligned}
\zeta_{\mathbf{k}\omega} &= G^0_{\mathbf{k}\omega}F_{\mathbf{k}\omega} + \\
&\lambda G^0_{\mathbf{k}\omega}\sum_{\mathbf{p},\mathbf{q},\Omega,\Omega'} A_{\mathbf{k}\mathbf{p}\mathbf{q}}\left(\zeta_{\mathbf{p}\Omega}\zeta_{\mathbf{q}\Omega'} + \zeta_{\mathbf{p}\Omega}h_{\mathbf{q}}\right) \quad,
\end{aligned}
\tag{5}
$$

where the interaction coefficient is:

$$
A_{\mathbf{k}\mathbf{p}\mathbf{q}} = (p_x q_y - p_y q_x)p^{-2}\delta_{\mathbf{k},\mathbf{p}+\mathbf{q}} \quad,
\tag{6}
$$

the bare propagator is:

$$
G^0_{\mathbf{k}\omega} = (-i\omega + \nu k^2)^{-1} \quad,
\tag{7}
$$

and the sum is restricted by $\mathbf{k} = \mathbf{p} + \mathbf{q}$ and $\omega = \Omega + \Omega'$. $\mathbf{p} = (p_x, p_y)$, $p = |\mathbf{p}|$, and similar expressions hold for $\mathbf{q}$. $0 < k < k_0$, with k_0 an upper cut-off. Following the method in Ref. [14], one can eliminate the modes $\zeta^>_k$ with k in the shell $k_0 e^{-\delta} < k < k_0$ and substitute their expressions into the equations for the remaining low-wavenumber modes $\zeta^<$ with $0 < k < k_0 e^{-\delta}$. To second order in λ, the resulting equation of motion for the modes $\zeta^<$ is:

$$\frac{\partial \nabla^2 \psi^<}{\partial t} + \lambda \left[\psi^<, \nabla^2 \psi^< + h^< \right] = \nu' \nabla^4 (\psi^< - gh^<) + F' , \tag{8}$$

where

$$\nu' = \nu \left(1 - \frac{\lambda^2 S_2 D(2+y)\delta}{32(2\pi)^2 \nu^3} \right) , \tag{9}$$

$$g(\lambda, D, \delta, \nu, y) = \frac{\lambda^2 D S_2 (y+4)\delta}{16(2\pi)^2 \nu^3} . \tag{10}$$

$F'(\mathbf{x}, t)$ is an effective noise which turns out to be also a Gaussian process with mean value and correlations given by:

$$< F'(\mathbf{x}, t) > = -\frac{\lambda^2 D S_2 (4+y)\delta}{16(2\pi)^2 \nu^2} \nabla^4 h^<, \tag{11}$$

$$\left\langle \left(\hat{F}'_k(\omega) - \left\langle \hat{F}'_k(\omega) \right\rangle \right) \left(\hat{F}'_{k'}(\omega') - \left\langle \hat{F}'_{k'}(\omega') \right\rangle \right) \right\rangle =$$
$$Dk^{-y}\delta(k + k')\delta(\omega + \omega') \tag{12}$$

S_2 is the length of the unit circle: 2π. Equations (8)-(12) give the dynamics of long wavelength modes $\psi^<$. They are valid for small λ or, when $\lambda \approx 1$, for small width δ of the elimination band. The effects of the eliminated short wavelengths on these large scales are described in the new structure of the viscosity operator and the corrections to the noise term F'. The action of the dressed viscosity term $\nabla^4(\psi^< - gh^<)$ is no longer to drive large scale motion towards rest, but towards a motion state ($\approx gh^<$) characterized by the existence of flow following the isolevels of bottom perturbations $h^<$. This ground state would characterize the structure of the mean pattern. The energy in this ground state is determined by the function $g(\lambda, D, \delta, \nu, y)$ which measures the influence of the different terms of the dynamics (nonlinearity, noise, viscosity). Relation (10) shows that while nonlinearities and noise increase the energy level of the ground state, high values of the viscosity parameter would imply a reduction of the strength of the ground state motion due to the damping that viscosity exerts over small scales. The other mechanism that reinforces the existence of average directed motion comes from the fact that the dressed noise has got a mean component as a result of the small scale elimination.

A most interesting fact in (10) and (11) is the presence of the factor $y+4$. It implies that the tendency to form directed currents reverse sign as y crosses the value -4, and that it vanishes at $y = -4$ which is the value for thermal noise satisfying detailed balance. In consequence noise rectification does not occur when detailed balance holds, a result of general validity [1]. As an illustration we show in Fig. 3 the average streamfunction in a case with $y = -4$. An irregular flow is seen which does not follow the topographic contours. Increasing the number of configurations included in the average, the amplitude of the mean streamfunction structures is seen to decrease, converging towards zero flow.

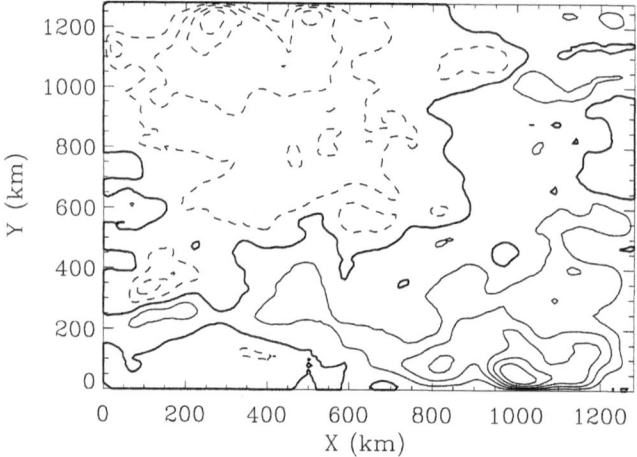

Figure 3: Mean streamfunction (average is over a period of 250 years) for the same parameters as in Fig. 2, except that noise correlations are of the form (4) with $y = -4$. Features are uncorrelated with topography. Averaging for longer times makes the features to disappear.

4   Conclusions

We have shown that quasigeostrophic flows develop mean patterns in the presence of noisy perturbations, and that the origin of these patterns is related with nonlinearity and lack of detailed balance. Nonlinear terms couple the dynamics of small scales with the large ones and provide a mechanism to transfer energy from the fluctuating component of the spectrum to the mean one. This mean spectral component, that is inexistent in purely two-dimensional turbulence [15], is controlled by the shape of the bottom boundary and characterizes the structure of the pattern. The existence of these noise-sustained structures has a wide range of implications in the fields of fluid and plasma physics. First because it highlights the important and organizing role that noise can play in these systems. Second, it establishes the need to modify not only the value of the parameters (as usually done in eddy-viscosity approaches) when performing large eddy simulations with insufficient small-scale resolution, but also the structure of the equations in a way determined by topography. This last statement has been previously suggested from a heuristic point of view in the context of large-scale ocean models [9, 10, 16]. Our results represent a step forward towards the justification of such approaches. The implications of these changes in presently existing general circulation ocean models open new ways for investigating the crucial role of oceanic circulation on climate variability.

Acknowledgements

Financial support from FEDER and MCyT (Spain) (IMAGEN REN2001-0802-C02-01/MAR, CONOCE BFM2000-1108) is greatly acknowledged.

References

[1] M.O. Magnasco , Phys. Rev. Lett. **71**, 1477 (1993); J. Maddox, Nature **369**, 181 (1994); C. R. Doering, W. Horsthemke, J. Riodan, J. Phys. Rev. Lett. **72**, 2984 (1994); J. Maddox, Nature **368**, 287 (1994); J.K. Douglass, Lon Wilkens, Eleni Pantazelou, Frank Moss, Nature **365**, 337 (1993); S. M. Bezrukov, Igor Vodyanoy, Nature **378**, 362 (1995); J. Rousselet, L. Salome, A. Ajdari, J. Prost, Nature **370**, 446 (1994).

[2] A. Álvarez, E. Hernández-García, J. Tintoré , Physica **A 247**, 312 (1997).

[3] J. Pedlosky , *Geophysical fluid dynamics*, Springer-Verlag (New York, 1987).

[4] A. Hasegawa, K. Mima , Phys. Fluids **21**, 87 (1978).

[5] A. Hasegawa, C. G. Maclennan , Phys. Fluids **22**, 11 (1979).

[6] A. Hasegawa , Advances in Physics **34**, 1 (1985).

[7] W. Horton , Phys. Rep. **192**, 1 (1990).

[8] R. Salmon, G. Holloway, M.C. Hendershot J. Fluid. Mech. **75**, 691 (1976).

[9] G. Holloway , J. Phys. Oceanogr. **22**, 1033 (1992).

[10] A. Álvarez and J. Tintoré, in *Ocean Modeling and Parametrization*, edited by E.P. Chassignet and J. Verron (Kluwer, Dordrecht, 1998).

[11] A. Álvarez, E. Hernández-García, J. Tintoré , Phys. Rev. **E 58**, 7279 (1998); *ibid*, Phys. Lett. A **261**, 179 (1999); *ibid.*, Geophys. Res. Lett., **27**, 739 (2000).

[12] M.H. Freilich and D.B. Chelton , J. Phys. Oceanogr. **16**, 741 (1985).

[13] L.D. Landau and M. Lifshitz , *Fluid Dynamics, 2nd Edition. Course of Theoretical Physics, vol. 6* (Pergamon, New York, 1987).

[14] D. Foster, D. Nelson, M. Stephen, Phys. Rev. A **16**, 732 (1977).

[15] P.D. Thompson , J. Fluid Mech. **55**, 711 (1972).

[16] A. Álvarez, J. Tintoré, G. Holloway, M. Eby, J.M. Beckers , J. Geophys. Res., **99**, 16053 (1994).

Dynamics of conical singularities: S type d-cones

Victor Apablaza and Francisco Melo

Departamento de Física de la Universidad de Santiago de Chile and Center for Advanced Interdisciplinary Research in Materials

Av. Ecuador 3493, Casilla 307 Correo 2 Santiago-Chile

Conical singularities or tips of developable cones (d-cones) arising when thin sheets are subjected to large homogeneous deformations are very common but poorly understood structures. This paper reports on the nucleation and motion of these singularities in a cylindrical sheet deformed following a direction perpendicular to the principal cylinder axis. We show that in this geometry two opposite d- cones are generated. As deformation is increased d-cone tips move apart producing a change in the d-cone topology; from a regular d- cone to a type S d-cone. For metallic sheet, a moving tip gives rise to a wrinkle which is in fact one of the wings of a S type d-cone. By studying the motion of single d-cones tips in simple configurations, we conclude that any tip motion can be decomposed into two elementary processes: climbing and gliding. Both processes occur when forces along the d-cone generators, straight lines ending at d-cone tip, are not equilibrated. Thus, S d-cones can be seen as a natural consequence of a motion of d-cone tip involving both climbing and gliding.

Point singularities are common to crumpled structures of two-dimension objects as metal plates, sheet of papers or even textiles. When a plane elastic plate is crumpled because of the action of external forces, the resulting structure, even if it is complex, has the tendency to be a surface with zero Gaussian curvature or on other words a developable surface. This means that the length of a line drawn along the plane surface doesn't change due to the deformation. One can explain this fact as a competition between the two forms of elastic energies which are bending and stretching. The first one is energetically the more convenient when the thickness of the plate goes to zero and, differently to stretching, it does not change

O. Descalzi et al. (eds.), Instabilities and Nonequilibrium Structures VII & VIII, 141–148.

lengths. In technical words, a final surface due only to bending can be described by using "generators". These former are straight lines along directions which have zero Gaussian curvature [1]. Since there is no change in the lengths along the surface and the action of crumpling is strongly related to a reduction of the volume of a plate, the last scenario seems contradictory. However, to satisfy both requirements singularities are necessary. These are regions of large curvature which permit a fast change in the direction of the generators in order to decrease the volume occupied by the surface. Because large curvature exists in the singularities, stretching is present to avoid an expensive energetically bending deformation. Consequently, Gaussian curvature is non zero and double curvature can be observed at those regions. The simplest situation to obtain a crumpled surface having the characteristics given above is to push a plate into a cylinder of radius slightly smaller than the plate radius [2,3]. In this case, a conical developable surface is selected because of the geometric constrains. Such a structure can be obtained as a solution of von Karman equations and has been named a "d-cone" by M. Ben Amar and Y. Pomeau [4].

More recent studies have been addressed to the topology of an experimental d- cone [2]. Theoretically, it has became necessary to include not only bending energy, that account well for the structure of the plate far away from the singularity, but also stretching energy which in turn accounts for singularity regions. Equilibrating bending and stretching energies a realistic scaling for the singularity size has been obtained [5]. This scaling resulted similar to the one introduced earlier by A. Lobkovsky *et. al.* [1995 Science **270** 1482] [6].

In this paper, we address the question of motion of conical singularities in a well defined cylindrical geometry. In our experiment, the d-cones are obtained on a thin cylinder (coke can) made of Aluminum, $0.1mm$ wall thickness, mm radius and mm length, by pushing a round tip ($0.5mm$ diameter) centered perpendicular to the cylinder principal axis. In order to allow the d-cones to form, when pushing the tip, we keep the sheet border free to move in a circular rigid frame whose radius is slightly larger than the cylinder radius. The displacement d of the tip is measured by a $10^{-2}mm$ precision micrometer. A miniature load cell is mounted under the pushing tip to perform force measurement. The method used to measure the topology of the d-cone consists on a profilometric tip, mounted on the active part of a position sensor transducer that enables us to measure the sheet surface topology with a precision of $10^{-2}mm$. A precision positioning system scans the whole d-cone surface. A schema of our experimental apparatus is found in Fig 1.

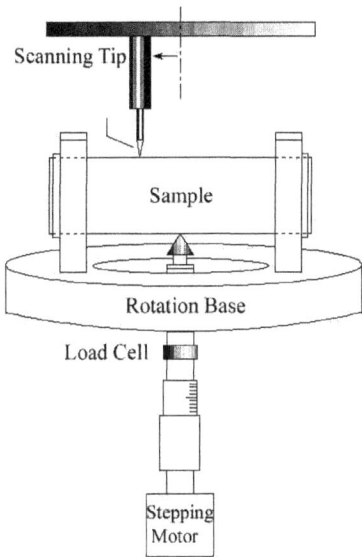

Scanning Tip

Sample

Rotation Base

Load Cell

Stepping Motor

FIG. 1. Experimental setup.

In Fig. 2, a view from above of the deformed cylindrical sheet. For small deformation, upper panel, two symmetric d-cones have nucleated on the sheet. Each d-cone has a convex part and a concave part limited by the curved shape in Fig 2a. More precisely, the generatrics close to the d-cone tip have curved forms, with a well defined radius of curvature. As it was shown in [5], this length is an appropriate measure of the singularity size. As deformation is increased, d-cones tips move apart, following the straight line that join them. This line is in fact a common generator of the two d-cones produced by the panel deformation.

144

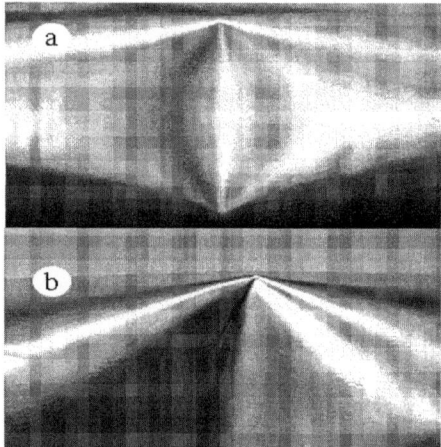

FIG. 2. Nucleating d-cone tips. a) Low deformation regime, two d-cones in opposition whose tips move apart following a line perpendicular to the principal cylinder axis, as deformation is increased. b) One of the above tips moving inclined with respect to the previous motion, as deformation is increased further.

An interesting tip dynamics arises when deformation is increased further, the tips trajectories incline themselves as shown, for one tip, on Fig. 2b. We will show below that such tip motion is a result of non balanced stretching forces along the d- cones generators. In the following, we present d-cones topology obtained by profilometric measurements. In Fig. 3 the profiles z obtained by scanning the surface, for several distances from the d-cone tip, as a function of azimuthal angle θ, for the cases presented in Fig 2. To obtain such profiles, first the cylinder is rotated around its principal axis in such a way that one of the d-cones tips is on the z axis of the scanning tip. This axis is the vertical symmetry axis of the undeformed cylinder. The zero reference for the z axis is taken when the tip touches the undeformed surface cylinder. Thus, in our experiment, the distance from the tip is defined as the radial distance r of the cylindrical coordinates and it is measured by the positioning system sketched in Fig. 1 of the scanning tip. For the case of small deformation, as shown in Fig. 3a, the d-cone profiles are nearly symmetric, however, for large deformation (see

Fig. 3b), they become asymmetric forming S type shapes. Since this type of profiles are multivaluated they can not be measured accurately with our experimental method. Inset in Fig. 3b shows a typical profile of a S type d-cone obtained in sections by tilting the cylinder axis.

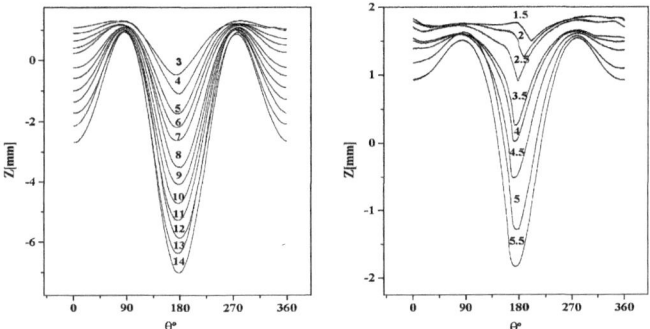

FIG. 3. D-cones profiles as a function of the distance for the tip singularity for two values of tip displacement. a) A symmetric profile, $d = 7.3mm$. b) An asymmetric profile, obtained from a S-type d-cone, $d = 8.4mm$. On the figures r is indicated in mm.

Force applied on the deforming tip can be measured accurately during deformation. Unfortunately, the transition to S type d-cone can not be detected from the data of the force applied on the panel. From Fig. 4, the force acting on the panel increases with the vertical displacement. When deformation reaches an approximate value of $\epsilon \sim 0.08$ the force start to increased faster. However, this change is rather related to plastic deformation taking place at d-cone tip. This is conclusion is well supported by the hysteresis observed in the force for $\epsilon > 0.08$. Interesting information about the energy stored on the cylindrical panel can be obtained by integrating the hysteresis loop, this results will be presented elsewhere. When the d-cone inclines itself, which occurs for $\epsilon \sim 0.22$, the force continue to increase, no changes are observed regarding the forces.

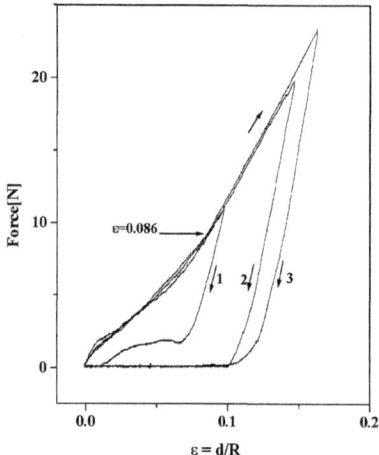

FIG. 4. Force on the d-cone tip as a function of dimensionless displacement, $\epsilon = d/R$.

To understand the S-type d-cone formation let us first to study the ideal motion of a d-cone tip. The ideal motion can be understood by performing two simple experiments. In the first one, a symmetric d- cone is submitted to forces along the generators, as illustrated in Fig. 5a. Of course, complementary forces are required to avoid d-cone translation and rotation. However, these forces can be supposed to act perpendicular to the generators. The reader is invited to perform this ideal experiment in a piece of paper by hand. As indicated in the Fig. 5a, the result is that the d-cone tip moves downward. The upward motion of the d-cone tip can be also obtained by reversing the forces. The second experiments is sketched in Fig. 5b, where forces are applied now along the generators on the two wings of the d-cone. The resulting motion, in this case, is the lateral displacement of the d-cone tip. This motion can be considered as the analogous of the dislocation gliding due to an elastic shear field. Once we have accepted the importance of the two elementary d-cones tip motions, we conclude that the transition from symmetric to non symmetric shapes, reported in Fig.

2b, is not related to only one type of motion alone. It is rather linked to a combination of the two elementary process described previously. In fact, the d-cone gliding motion, necessary to break the symmetry of the moving tip in our sample, Fig. 2a, is a result of non equilibrated forces acting along the wings d-cone generators. When the tip of the d-cone start to move laterally these forces are not equal. In our case, these stretching forces appear as a consequence of the geometric constrain on the cylindrical sample. It is important to note that, our sample has two flat lateral walls that keep the radius nearly constant at the cylinder ends. These constrains produce forces, along the d-cones wings, that compete at the d-cone tip. As reported in a recent work [10], where two flat walls are not present, the singularity motion is rather different than the ones reported here. In this case, similar to our experiment two d-cones nucleates for small deformations. However, as the deformation is increased, two additional d-cones can nucleate. All d-cones are mobile in such a way that they are ejected at the free end of the cylindrical plate.

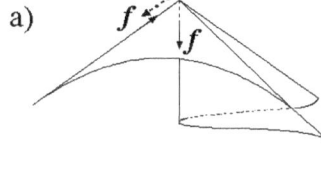

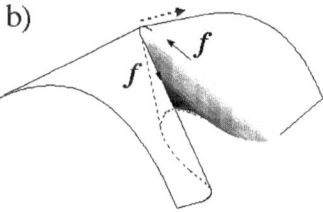

FIG. 5. Two modes of tip motion a) Climbing mode. b) Gliding mode.

In conclusion, we have reported an experimental configuration that simply illustrate the d-cone tip motion. A complex motion of d-cone tip can be qualitatively understood as a

148

decomposition of two elementary process: climbing and gliding. In a simplified view each process is governed by forces along generators, i.e. only stretching. Thus, in our experiment, the S type cone is formed when the two elementary process are induced by external forces on the d-cone tip. However, since in our experiment plastic deformation occurs at the d-cone singularity, more experiments need to be carried out in order to conclude if the mechanism described here is still generic for material where plastic limit is not reached at d- cone tip.

This work was supported by the Chilean Cátedra Presidencial en Ciencias.

[1] Y. Pomeau, Comptes rendus de l'Académie des sciences, **320**, série I, 975 (1995).

[2] S. Chaieb and F. Melo, Phys. Rev. Lett. **80**, 2354 (1998).

[3] E. Cerda and L. Mahadevan, Phys. Rev. Lett. **80**, 2358 (1998).

[4] M. Ben Amar and Y. Pomeau, Proc. R. Soc. London A **453**, 729 (1997).

[5] E. Cerda, S. Chaieb, F. Melo and L. Mahadevan, Nature, **401**, September (1999).

[6] A. Lobkovsky, S. Gentges, H. Li, D. Morse and T. Witten, Science **270**, 1482 (1995).

[7] In this case, our photoelastic method is not sensitive to the bending strain which is zero when average on the sample thickness. Details of this method will be discussed elsewhere.

[8] T. Witten and H. Li, EuroPhys. Lett **23**, 51 (1993).

[9] L. Landau and E.M. Lifshitz, Theory of Elasticity (Pergamon, New York, 1986), 3rd ed.

[10] Arezki Boudaoud, Pedro Patricio, Yves Couder and Martine Ben Amar. Nature, **407**, October (2000).

NONLINEAR DYNAMIC ANALYSIS OF SCALP EEG EPILEPTIC SIGNALS

by

Susana A. Blanco[1], Judith Creso[1], Alejandra Figliola[1],
Rodrigo Quian Quiroga[2], Osvaldo A. Rosso[1,2]

[1]: Instituo de Cálculo, Facultad de Ciencias Exactas y Naturales,
Universisdad de Buenos Aires, Pab.II, Ciudad Universitaria,
1428 Buenos Aires, Argentina.
[2]: Institute of Physiology, Medical University Lübeck,
Ratzeburger Alle 160. D-23538 Lübeck 1, Germany.

Abstract

Noisy signals obtained during a tonic-clonic epileptic seizure, are usually neglected for visual inspection by the physicians due to the presence of muscle artifacts. Although noise obscures completely the recording, information about the underlying brain activity can be obtained by filtering, through the Orthogonal Wavelet Transforms, those frequencies bands associated with muscle activity. After generating a "noise free" signal by removing the muscle artifacts with wavelets, a dynamical analysis of the brain behavior will be performed by using nonlinear dynamics methods. The values for nonlinear metric invariants, like the correlation dimension and the maximum Lyapunov exponent, confirm that the brain dynamical behavior is more ordered during the epileptic seizure than pre-seizure stage.

1   Introduction

The electroencephalogram (EEG) can be raftly defined as the mean electrical activity of the brain in different sites of the head. EEG patterns are correlated with functions, dysfunctions and diseases of central nervous system based on empirical basis. Although visual inspection of the EEG is quite useful, it is subjective and hardly allows any systematization. In order to overcome this fact, quantitative EEG analysis introduces objective measures reflecting the characteristics of the brain activity. One of the new branches of quantitative EEG analysis is the application of the concepts related with nonlinear dynamic theory. The nonlinear metric invariants (correlation dimension, D_2, and Lyapunov exponent, λ_1) proved to be very useful for characterize the brain dynamics in

O. Descalzi et al. (eds.), Instabilities and Nonequilibrium Structures VII & VIII, 149–157.
© 2004 *Kluwer Academic Publishers. Printed in the Netherlands.*

different states, like during sleep, doing mental tasks, etc. Furthermore, nonlinear parameters were compared among normal subjects and patients with different pathologies like epilepsy, schizophrenia, Creutzfeld-Jakob coma among others.

A scalp EEG signal is essentially a nonstationary time series that presents artifacts mainly due to eye movements and muscle activity. Artifacts related with muscle contractions are specially troublesome in the case of tonic-clonic or "Grand Mal" epileptic seizures, where they reach very high amplitudes contaminating the whole seizure recording. In fact, they limit the traditional visual analysis to the pre- and post-seizure periods. Furthermore, they also restrict the application of mathematical methods like ones of nonlinear dynamics. The elimination of the high frequency muscle activity with the use of traditional filters have some disadvantages: *a)* filtering frequencies related with muscle artifacts also affects the morphology of the remaining ones and, *b)* the filtering process alter the nonlinear metric invariants.

In order to overcome the above limitations, a new filtering method based on a multiresolution analysis using orthonormal wavelets is proposed. In this way a nonlinear study of a "Grand Mal" epileptic seizures, avoiding noise related limitations by using a wavelet filtering, can be made and compared with those from more clean records like corresponding to depth electrodes.

2 Subject and Data recordings

Samples of EEG time series corresponding to generalized tonic-clonic epileptic seizures were analyzed. Scalp and sphenoidal electrodes were applied following the 10-20 international system. Each signal was digitized at 409.6 Hz through a 12 bit A/D converter and filtered with an antialiasing eight pole lowpass Bessel filter with a cutoff frequency of 50 Hz. Then, signals were digitally filtered with a $1 - 50$ Hz bandwidth Butterworth filter and stored, after decimation, at 102.4 Hz in a PC hard drive. Recordings were done under video control in order to have an accurate determination of the different stages of the seizure. The different stages of EEG signals were determined by the physicians team.

As an example, in Fig. 1.a we present an scalp EEG signal, corresponding to a tonic-clonic epileptic seizure recorded in a central right location ($C4$ channel). We chose this electrode after visual inspection of the EEG records as the one with the leased amount of artifacts. Seizure starts at second 50, with a "discharge" of slow waves superposed by fast ones with lower amplitude. This discharge lasts approximately 8 *sec* and has a mean amplitude of 100 μV. Afterwards, seizure spreads making the analysis of the EEG more complicated due to muscle artifacts; however, it is possible to establish the beginning of the clonic phase at around second 93 and the end of the seizure at second 125 where there is an abrupt decay of the signals amplitude.

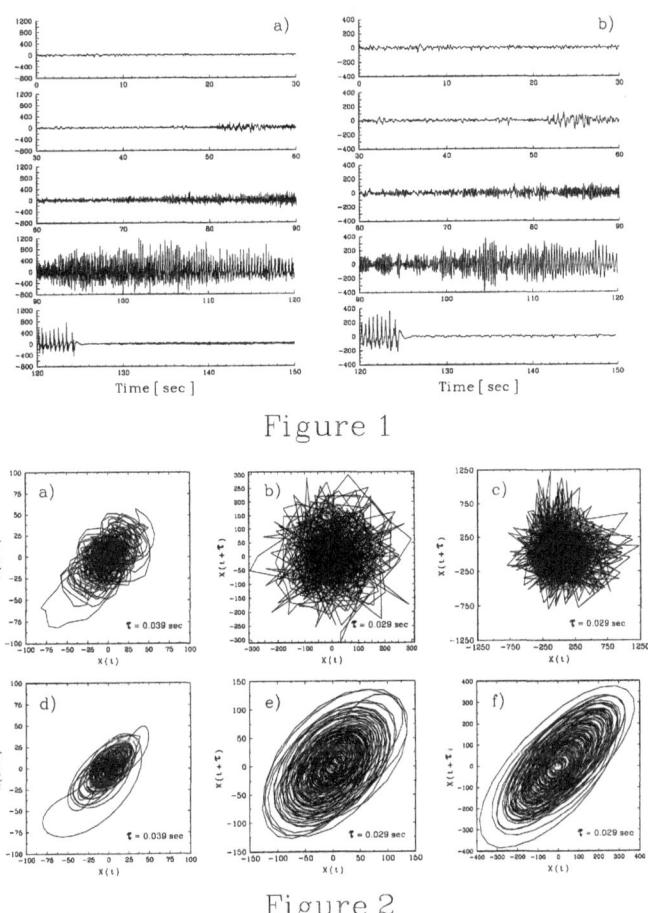

Figure 1

Figure 2

Figure 1: *a*) Scalp EEG signal corresponding to a tonic-clonic epileptic seizure recorded in a central right location (*C*4 channel). In this record, seizure starts at second 10, with a "discharge" of high frequencies. The seizure clearly ends at second 85. *b*) Cleaned signal (without high frequency components).

Figure 2: Two dimensional projection of the phase portrait corresponding to signal segments $20 - 50$ *sec*, $60 - 90$ *sec*, $90 - 120$ *sec* for the original EEG signal (top) and for the cleaned signal (botton) respectively. τ^* is the optimal time lag employed.

3   Wavelet Transform: filtering of the signal

Wavelet transform give us a powerful tool to analyze from time-frequency point of view very diverse problems in applied sciences or pure mathematics The *wavelet* is a smooth and quickly vanishing oscillating function with good localization both in frequency and time. It can be interpreted as single short times signal with oscillating structure. A *wavelet family* $\psi_{a,b}$ is the set of elemental functions generated by dilations and translations of a unique admissible *mother wavelet* $\psi(t)$:

$$\psi_{a,b}(t) \; = \; |a|^{-1/2}\psi\left(\frac{t-b}{a}\right) \; , \tag{1}$$

where $a, b \in \mathcal{R}$, $a \neq 0$ are the scale and translation parameters respectively, and t is the time. As a increases, the wavelet became more narrow. Thus we have a unique analytic pattern and its replicas at different scales and with variable localization in time.

Given a finite energy signal $S(t)$, the different correlations $< S, \psi_{a,b} >$ indicates how precisely the wavelet function locally fits the signal at every scale a. This correlation operation define the transformation that synthesizes the numerical information obtained in this way. From a different viewpoint, the wavelets of a family play the roles of elemental functions, representing the functions as a superposition of wavelets correlated with the function for different scales (different a's). This makes it possible to organize the information in some particular structure to distinguish, for example, trends, or the shape associated with long scales of the local details from corresponding short scales.

The *continuous wavelet transform* (CWT) of a signal $S(t) \in L^2(\mathcal{R})$ is defined as the correlation between the function $S(t)$ with the family wavelet $\psi_{a,b}$ for each a [1]:

$$(W_\psi S)(a,b) \; = \; |a|^{-1/2} \int_{-\infty}^{\infty} S(t)\, \psi^*(\frac{t-b}{a})\, dt \; = \; < S, \; \psi_{a,b} > \; . \tag{2}$$

For special selections of the function, ψ and a discrete set of parameters $a_j = 2^{-j}$ and $b_{j,k} = 2^{-j}k$, with $j, k \in \mathcal{Z}$ and the scale 2^{-j}, give us the shift parameter. The subfamily

$$\psi_{j,k}(t) \; = \; 2^{j/2}\, \psi(\, 2^j\, t \, - \, k\,) \qquad j, k \in \mathcal{Z} \, , \tag{3}$$

constitutes an orthonormal basis of the Hilbert space $L^2(\mathcal{R})$. In this way, we can obtain discrete transformations, and it is possible to expand the signal in series of wavelets. Then, we can join the advantages of the wavelet transform with atomic decomposition of $S(t)$.

The *decimated discrete wavelet transform* (DWT) associated with ψ is simply seen as a restriction of the continuos wavelet transform at the parameters set $\{a_j, b_{j,k}\}$. In this case, as it is well known, the information given by the discrete wavelet transform can be organized according a hierarchical scheme of nested subspaces called multiresolution analysis in $L^2(\mathcal{R})$.

In the present analysis we used a multiresolution scheme based on cubic orthogonal spline functions as mother wavelet, with a discretized version of the integral wavelet transform given by eq. (2). We selected this wavelet due to the fact that it forms a base in $L^2(\mathcal{R})$ with a very convenient characteristic of symmetry and simplicity. Moreover, the smoothness of its derivates are very suitable for representing natural phenomena.

In the following we will assume that the EEG signal is given by sampled values $\{s_0(n)\}$, $n = 1, \cdots, M$, which correspond to an uniform time grid with sampling time Δt and sample frequency $\omega_s = 2\pi/\Delta t$. If the decomposition is carried out over all resolutions levels, $N = Ln_2(M)$, then the wavelet expansion will be:

$$S(t) \;=\; \sum_{j=-N}^{-1} \sum_{k} C_j(k)\, \psi_{j,k}(t) \;=\; \sum_{j=-N}^{-1} r_j(t)\,, \tag{4}$$

where wavelet coefficients $C_j(k)$ can be interpreted as the local residual errors between successive signal approximations at scales j and $j + 1$, and $r_j(t)$ is the *detail signal* at scale j. It contains the information of the signal $S(t)$ corresponding with the frequencies $2^{j-1}\omega_s \leq |\omega| \leq 2^j\omega_s$.

Natural time series are usually the desired "clean" signal with effects of contamination by the environment through which it passes on its way from the source to measurement device or contamination by properties of the measurement process itself. Many applications, like nonlinear dynamics, requires the separation of signal and noise. Otherwise, when nonlinear invariants are evaluated the contaminating noise can give spurious results, and the obtained values will underestimate or overestimate the real complexity of the system under study.

When noise is present only in specific frequency bands, a filtering process based on orthonormal wavelet transform can be implemented. Supposing, as in the case of tonic-clonic seizures, that we are interested in eliminating high frequency noise, using eq. (4) at scale j_0 we can obtain a smoothed version of the original signal, $\tilde{S}(t)$ by

$$\tilde{S}(t) \;=\; = \sum_{j=-N}^{-j_0} r_j(t)\,. \tag{5}$$

This smoothed signal have less amount of high frequencies in comparison to level $j_0 + 1$ but it will have half number of data than the previous level. Then the questions are how to obtain the remaining data in order to recover the original signal sample ratio and which is the best form to do that ?. Which are answering in optimal way if we use an *undecimated discrete wavelet transform* (UWT).

Following the work of Serrano and Fabio [2], we implemented the undecimated discrete wavelet transform from orthogonal spline wavelets and used the efficient algorithm, proposed by these authors, to compute it from the sampling values of the given signal. Moreover, they also proposed an extended multiresolution analysis structure, to organize the available numerical information of the signal. In particular, eqs. (4) and (5) will be still valid but now the coefficients $C_j(k)$ are the wavelets coefficients of the UWT obtained with the above mentioned algorithm.

Then, the undesimated discrete wavelet transform associated with orthonormal spline wave- lets can be accurately used for separation of noise and the cleaned signal can be obtained as a sum of residual components corresponding to the interesting frequency bands. In summary, this method allows the elimination of non desired frequency bands that hidden other more interesting or unknown effects. Due to the orthogonality of the wavelet functions employed we can assure that only the unwanted and previously selected frequency bands were extracted, without needing of the assumption of linearity necessary for making the traditional Fourier-based digital filtering and without affect the dynamics associated with the remaining frequency bands.

4   Nonlinear Dynamics Metric Tools

In recent years the theory of nonlinear dynamics has developed methods for quantitative analysis of experimental time series. The methods usually employed in nonlinear dynamic analysis (metric approach) are based on distances, and assume the stationary of the data sets. Distances between points in appropriate embedding of the data are used to compute a set of metric parameters, such as fractal dimensions and Lyapunov exponents. These quantities are difficult to compute, require large data sets and degrade rapidly with additive noise.

The methods employed in this work are independent of any modeling of brain activity. They rely solely on the analysis of data obtained from a single variable time series. The insight gained by the concept of deterministic chaos for the EEG is that a seemingly disordered process may be governed by a relatively few simple laws, which could be determined.

In order to characterize the global brain dynamics using the nonlinear dynamic metric tools, we implemented the following protocol: *i)* Select the stationary subsets of the signal in order to satisfy the mathematical hypothesis of the metrical methods [3]. *ii)* Use the attractor reconstruction expansion as a geometry-based framework for choosing proper times delays [4]. *iii)* Determine the minimum embedding dimension for phase space reconstruction using a geometrical algorithm [5]. In addition this method provide a measurement of the noise-signal ratio. *iv)* Evaluate the Correlation Dimension and largest

Lyapunov exponent simultaneously [5, 6]. The convergence of these metrics invariants is checked for increasing embedding dimensions for which the noise-signal ratios is less than 0.5%. *v)* Evaluate the maximum expected correlation dimension according with the length of the data sets. Estimation of the minimum number of data points necessary to extract accurate dynamical information represented by the maximum lyapunov exponent [7]. In case of use shorter data sets, their convergence with increasing number of data must be carefully analyzed.

5 Result and Discussion

EEG spectral analysis is traditionally performed by studying different frequency bands, whose boundaries are well defined but could have some small variations according to the particular experiment being considered. Absolute and relative intensities of these bands are analyzed and correlated with different pathologies. In this work, we chose seven frequency bands associated with the resolution levels appropriate for the wavelet analysis in the scheme of multiresolution proposed. We denoted these band-resolution levels by B_j ($|j| = 1, \cdots, 7$).

Owing that one of the goals of this work is to analyze the middle and low frequencies brain activity during an Tonic-Clonic epileptic seizure, we eliminated B_1 and B_2 bands, with wavelet resolution level $j = -1$ and -2 respectively ($\omega > 12\ Hz$), both containing high frequency artifacts that obscures the EEG. Although high frequency brain activity will be also eliminated, it is well accepted that the components are not important in the study of an epileptic seizure [8]. Then applying undecimated discrete wavelet transform from orthogonal spline wavelet we can filter the signal by subtracting these bands.

Using the power of the wavelet transform, now we can analyzed the dynamic of the noise-free signal shown in Fig. 1.b (signal without high frequency artifacts). For this propose the signal was divided in representative epochs of lengths of 30 *sec* ($N = 3072$ data). These intervals were: *a)* $20 - 50\ sec$, *b)* $60 - 90\ sec$, *c)* $90 - 120\ sec$ corresponding to the pre-seizure, seizure begging and tonic stage, and clonic stage respectively.

In Fig. 2 we present two dimensional projection of the reconstructed attractor of the original signal (Fig. 1.a) and the noise-free signal (Fig. 1.b) for pre-seizure and for the different seizure stages. Note that the attractors corresponding to the pre-seizure (Fig. 2.a and 2.d) are quite similar indicating that in this stage the contribution of the fast frequencies are not so important. On the other hand, the remaining phase portraits (Fig. 2.e and 2.f) show an underlying low dimensional dynamic for the different stages of the seizure, which are completely blurred by noise in the original signal (Fig. 2.b and 2.c). It is interesting to note that attractors obtained with the noise-free signal are similar to those corresponding to EEG signals obtained using surgically implanted electrodes where there is no influence of muscle activity [9, 10].

In order to evaluate the correlation dimension D_2 and the biggest Lyapunov exponent λ_1 we determined the minimum suitable embedding dimension for each seizure stage following the method of false nearest neighbors [5]. We consider that an $\%FNN \leq 0.5$ corresponds to an embedding dimension in which the attractor will be unfolded enough, and in consequence, it will be define the minimum embedding dimension. The minimum embedding dimensions for the three signal portions were $D_e^{(min)} = 8, 6$ and 7 respectively.

The different nonlinear metrical invariants (averaged over 3 consecutive increasing embedding dimensions, starting from $D_e^{(min)}$) obtained for the different interval were $< D_2 >= 4.23$ and $< \lambda_1 >= 2.15$ for the pre-ictal stage; $< D_2 >= 4.02$ and $< \lambda_1 >= 1.53$ for the seizure beginning and tonic stage; $< D_2 >= 4.09$ and $< \lambda_1 >= 2.04$ for the clonic stage. We can observe that the correlation dimension, D_2, and the biggest Lyapunov exponent, λ_1, present lower values in the starting of the seizure and increasing values in the other seizure stages.

Even when this values of D_2 and λ_1 must be taken only for comparative proposes. However, the obtained values for λ_1 are indicatives that a chaotic behavior can be associated with all the EEG signal portions considered and these values are consistent with the behavior observed with the Information Cost Function analysis performed in previous work [8].

As in previous works (using EEG time series from depth electrodes), from the present nonlinear analysis of wavelets filtered tonic-clonic epileptic time series, we confirm the validity of the basic assumption that: at the seizure onset, a transition takes place in the dynamical behavior of the neural network from a complex behavior to a simpler one. That means, during an epileptic seizure the dimension of the attractor and the biggest Lyapunov exponent present lower values compared with the ones evaluated at pre seizure stage. Also the reconstructed attractor in phase space looks more simple and ordered.

Acknowledgments

This work was partially supported by the Consejo Nacional de Investigaciones Científicas y Técnicas (CONICET, Argentina) and the International Office of BMBF, Germany.

The authors wish to thank H. Garcia and A. Rabinowicz of the Instituto de Investigaciones Neurológicas Raúl Carra (FLENI), Argentina, for useful comments and the providing of the EEG recordings used in this work.

The authors (J.C., A.F. and O.A.R.) are very grateful to the organizer of the workshop for their very kind hospitality during their stay in Chile.

References

[1] A. Aldroubi and M. Unser (edt). *Wavelets in Medicine and Biology,* CRC Press, Boca Raton, 1996.

[2] E. P. Serrano and M. Fabio. Undecimated wavelet transform from orthogonal spline wavelets, in *Wavelet Theory and Harmonic Analysis in Applied Sciences,* E. M. Fernandez-Berdaguer and C. E. D'Attellis (edt), Birkhouser Publishers, 1997, p. 33-72.

[3] S. Blanco, H. García, R. Quian Quiroga, L. Romanelli and O. A. Rosso. *Stationarity of the EEG series,* IEEE Eng. Med. and Biol. 14 (1995) 395-399.

[4] M. T. Rossestein, J. J. Collins and C. J. De Luca. *Reconstruction expansion as a geometry-based framework for choosing proper delay times,* Physica D 73 (1994) 82-98.

[5] H. D. I. Abarbanel, R. Brown, J. J. Sidorowich and L. S. Tsimring. *The analysis of observed chaotic data in physical systems,* Rev. Mod. Phys. 65 (1993) 1331-1392.

[6] M. T. Rossestein, J. J. Collins and C. J. De Luca. *A practical method for calculating largest Lyapunov exponents from small data sets,* Physica D 65 (1993) 117-134.

[7] J. P. Eckmann and D. Ruelle. *Fundamental limitations for dimensions and Lyapunov exponents in dynamical systems,* Physica D 56 (1992) 185-187.

[8] S. Blanco, A. Figliola, R. Quian Quiroga, O. A. Rosso and E. Serrano. *Time-frequency analysis of electroencephalogram series (III): Wavelet Packets and Information Cost Function,* Phys. Rev. E 57 (1998) 932-940.

[9] S. Blanco, R. Quian Quiroga, O. A. Rosso and S. Kochen. *Time-frequency analysis of electroencephalogram series,* Phys. Rev. E 51 (1995) 2624-2631.

[10] S. Blanco, S. Kochen, R. Quian Quiroga, L. Riquelme, O. A. Rosso and P. Salgado (1997). Characterization of epileptic EEG time series (I): Gabor Transform and nonlinear dynamics methods, in *Wavelet Theory and Harmonic Analysis in Applied Sciences,* E. M. Fernandez-Berdaguer and C. E. D'Attellis (edt), Birkhouser Publishers, 1997, p. 179-226.

A theorem on wave packets.

Mario Castagnino, Jorge. Guerón
Instituto de Astronomía y Física del Espacio
Casilla de correos 67 sucursal 28
1428 Buenos Aires, Argentina.

Adolfo Ordoñez
Instituto de Física Rosario.
Av. Pellegrini 250,
2000 Rosario, Argentina.

November 26, 1999

Abstract

When $t \to \infty$ a wave packet becomes a Hardy class function from below.

1. Introduction.

Many authors consider that time asymmetry has a global origin [1]. The Reichenbach branched diagram has being used to explain this global phenomenon [2]. This diagram is the combination of all the scattering processes within the universe, beginning at the initial global instability of the universe, which is considered as the source of all energy, and then following this energy flux trough all the diagram. In each scattering process the incoming energy is used to produce unstable states that decay originating the outgoing one. So the outgoing lines of the diagram can be considered as evolutions from an unstable state towards equilibrium. Essentially the diagram can be considered as having just outgoing lines. In fact, the incoming lines in each scattering are outgoing lines of a previous process. Moreover the incoming lines as such cannot be consider as spontaneous evolutions, since they just show the pumping of energy from a previous system,

159

O. Descalzi et al. (eds.), Instabilities and Nonequilibrium Structures VII & VIII, 159–167.

that really in this phase it is coupled with the scattering system making the incoming lines *non spontaneous* or forced evolutions. On the contrary the outgoing lines clearly corresponds to *spontaneous* ones. As a whole the diagram symbolize the evolution of the universe from its unstable beginning up to its final equilibrium state, resolved as a sum of scattering process and showing a clear time asymmetry. Then time asymmetry (which cannot be explained as a consequence of the time-symmetric physical laws) can be explained as a consequence of the time asymmetry of the object-universe.

On the other hand time asymmetry can be consider the consequence of the existence of a time-asymmetric space of physical admissible states [3]. For causality reasons this space was associated with the space of Hardy class functions from below [4]. If all the lines in the Reichenbach diagram would be electromagnetic waves this choice would be natural, since the outgoing waves of electromagnetic scattering can be represented as states belonging to the just mentioned Hardy space [5]. This conclusion can be extended to all hyperbolic scattering but not to the parabolic equation of non relativistic quantum mechanics. Nevertheless we will show that, far from the scatterer, the outgoing lines belong to a Hardy space, if the physical admissible states are wave packets. Then all the spontaneously evolving states of the universe would belong to such a type of space, that we will call Ψ_-. The unphysical non spontaneous time inverted states would belong to a space $\Psi_+ = K\Psi_- \neq \Psi_-$, where K is Wigner time inversion operator [6] [1]. The arrow of time would be the consequence of this asymmetry.

2. Two examples.

Two example show that what we are looking is not so unusual.

2.1. Baker's transformation.

Let us consider a square 1×1 and the Baker's transformation:

$$B(x, y) = \begin{cases} (2x, \frac{1}{2}y) & \text{if } 0 \leq x \leq \frac{1}{2} \\ (2x - 1, \frac{1}{2}y + \frac{1}{2}) & \text{if } \frac{1}{2} \leq x \leq 1 \end{cases} \tag{2.1}$$

---

[1]In the usual popularization language Ψ_- would be the space of the spontaneous evolutions: the sugar lump solving in the coffee, or the elephant breaking the crystal shop, while Ψ_+ will be the space of impossible (or better non-spontaneous) evolutions: the sugar lump concentrating in the coffee, or the elephant reconstructing the crystal shop.

Let us consider the space of functions $f\ (x,y)$ on 1×1 of support:

$$S = \bigcup_i h_i \cup \bigcup_j v_j \tag{2.2}$$

where h_i are "horizontal" rectangles $x_i \times y_i$, i. e.: such that $x_i \geq y_i$, and v_j are "vertical" rectangles $x_j \times y_j$, i. e.: such that $x_j \leq y$, being all these rectangles disjoint. Let us decompose $f\ (x,y)$ as:

$$f(x,y) = f_h(x,y) + f_v(x,y) \tag{2.3}$$

such that:

$$\sup f_h(x,y) = \bigcup_i h_i, \ \ \sup f_v(x,y) = \bigcup_j v_j \tag{2.4}$$

We can define two spaces of functions H and V such that:

$$f_h(x,y) \in H, \ f_v(x,y) \in V \tag{2.5}$$

Then it is obvious that:

$$BH \subset H, \ BV \subset H \oplus V \neq V \tag{2.6}$$

and that:

$$\lim_{n\to\infty} B^n(H \oplus V) = H \tag{2.7}$$

So we will say that the supports of the functions in H are "stable fibers", while the supports of the functions in V are "unstable fibers", in such a way that any function of $H \oplus V$, if multiply by B^n, in the "far future" will end in space H, precisely when $n \to \infty$.

2.2. Beurling theorem.

The quantum version of the example above will be the following: Let us consider a quantum system with quantum states ρ that belong to a space $\mathcal{L}$. Let be the Liouville operator $\mathbb{L}$, endowed with a uniform Lebesgue spectrum $\mathbb{R}$, let be $l \in \mathbb{R}$ an eigenvalue and $|l)$ the corresponding eigenvector. Then $\{|l)\}$ is an orthonormal eigenbasis. Let us define two spaces $\Phi_\pm$:

$$\rho \in \Phi_\pm \Leftrightarrow (l|\rho) \in H^2_\pm \tag{2.8}$$

where $H^2_\pm$ are the Hardy classes from $\{^{above}_{below}\}$. Then

$$\mathcal{L} = \Phi_+ \oplus \Phi_- \tag{2.9}$$

Now, from the inverse of the Beurling theorem [7] we know that:

$$e^{-i\mathbb{L}t}\Phi_- \subset \Phi_-, \text{ for } t > 0 \tag{2.10}$$

so we can say that Φ_- is an stable space towards the future, while

$$e^{-i\mathbb{L}t}\Phi_+ \subset \mathcal{L} = \Phi_- \oplus \Phi_+ \neq \Phi_+, \text{ for } t > 0 \tag{2.11}$$

so we can also say that Φ_+ is an unstable space towards the future (of course it would be an stable space towards the past). The last two equation are the quantum version of the Bakers's equations (2.6). We would like to find the condition that allows to say that also the quantum version of equation (2.7) is satisfy, namely that:

$$\lim_{t\to\infty} e^{-i\mathbb{L}t}\mathcal{L} = \Phi_- \tag{2.12}$$

We will see that this equation is satisfy by the wave packets. We will try to explain these theorem in the simplest way both for mathematician and for physicist.

3. Pure states and the hamiltonian.

Let us begin considering just pure states $|\psi\rangle$, belonging to a Hilbert space $\mathcal{H}$, of a quantum system with hamiltonian H, such that:

$$H|\omega, n\rangle = \omega|\omega, n\rangle \tag{3.1}$$

where $0 \leq \omega < \infty$ or $\omega \in \mathbb{R}^+$ and n belong to a set of indices N, which is the same for any ω (for didactical reasons we may consider that this set is numerable and therefore the index n is discrete). Thus H can be consider a typical scattering hamiltonian just endowed with an absolutely continuous energy spectrum. So the pure states read:

$$|\psi\rangle = \sum_n \int_0^\infty d\omega |\omega, n\rangle\langle\omega, n|\psi\rangle \tag{3.2}$$

Moreover, let us consider that the real physical states are wave packets. We adopt, for each value of n, as the mathematical model of these wave packets the Schwarz functions of $\omega \in \mathbb{R}^+$, so:

$$f(\omega) = \langle\omega, n|\psi\rangle \in \mathcal{S}^+ = \theta(\mathcal{S}) \tag{3.3}$$

where $\mathcal{S}$ is the Scharz space of infinite derivable functions that converge to zero towards $\pm\infty$, better that the inverse of any polynomial and θ the Heaviside symbol. Taking into account all the values of n we can say that:

$$f(\omega) = \langle \omega, n | \psi \rangle \in \bigoplus_n \mathcal{S}_n^+ \tag{3.4}$$

The mathematical model above is adopted for the following reasons:

1.- It is clear that we do not find infinities energies in nature so $\langle \omega, n | \psi \rangle$ must somehow go to zero when $\omega \to \infty$.

2.- In order to use derivatives in our calculations it is not enough that the states belong to a Hilbert space. They must be represented by derivable functions. We postulate that they are infinitely derivable

3.- But since these functions must de square integrable they must go fast to zero when $\omega \to \infty$. We postulate that they go faster that the inverse of any polynomial.

Of course we are free to chose other spaces than $\mathcal{S}^+$, but it is evident that $\mathcal{S}^+$ is the simplest space endowed with the properties of the usual wave packets (moreover the same choice is made in [8]).

4. Mixed states and the liouvillian.

We will use the notation of ref. [9]. Then the Liouville operator reads.

$$\mathbb{L} = [H, .] = H \times I - I \times H \tag{4.1}$$

Let us consider the space of density matrices $\mathcal{L} = \mathcal{H} \otimes \mathcal{H}$ and the basis:

$$|\omega, n\rangle\langle\omega', n'| \tag{4.2}$$

Let us define the Riezs indices [9]:

$$\nu = \omega - \omega', \quad -\infty < \nu < \infty$$

$$\sigma = \frac{1}{2}(\omega + \omega'), \quad \frac{|\nu|}{2} \leq \sigma < \infty \tag{4.3}$$

Then we can label basis (4.2) as:

$$|\omega, n\rangle\langle\omega', n'| = |\nu, \sigma, n, n') \tag{4.4}$$

Then:

$$\mathbb{L}|\nu,\sigma,n,n') = \nu|\nu,\sigma,n,n')$$ (4.5)

So $\{|\nu,\sigma,n,n')\}$ is an eigen basis of the liouvillian, being ν the corresponding eigen value and σ, n, n', degeneration indices. From (4.3) we see that $\nu \in \mathbb{R}$ while for the properties of H introduced in the previous section n and $n' \in N$, and $\sigma \in \mathbb{R}^+$, and these spaces are the same for any ν. So $\mathbb{L}$ has the homogeneous Lebesgue spectrum $\mathbb{R}$.

In the basis $\{|\nu,\sigma,n,n')\}$ the ν "wave function reads":

$$\rho(\nu) = (\rho|\nu,\sigma,n,n')$$ (4.6)

Following the ideas of the previous section these functions can be considered as physical if they are sums of products of functions $f(\omega) \in \mathcal{S}^+$, namely:

$$f(\omega)g(\omega') = f\left(\sigma + \frac{\nu}{2}\right) g\left(\sigma - \frac{\nu}{2}\right)$$ (4.7)

This function has infinite derivatives with respect to ν since f and g are infinite derivable. $\nu \in \mathbb{R}$ since $\omega, \omega' \in \mathbb{R}^+$ and $\rho(\nu)$ goes to zero when $\nu \to \pm\infty$ faster than the inverse of any polynomial since this is a property of functions f and g. Thus, for any σ, n, n' :

$$\rho(\nu) = (\rho|\nu,\sigma,n,n') \in \mathcal{S}$$ (4.8)

and taking into account all the values of σ, n, n' we can say that:

$$(\rho|\nu,\sigma,n,n') \in \bigoplus_{\sigma,n,n'} \mathcal{S}_{\sigma,n,n'}$$ (4.9)

5. Age operator.

For the physical states we can define an "age operator" [10] as:

$$\mathbb{A}\rho(\nu,\sigma,n,n') \doteq i\frac{\partial}{\partial\nu}\rho(\nu,\sigma,n,n')|_{\sigma,n,n'=const.}$$ (5.1)

Since $\mathbb{L}$ has a homogeneous Lebesgue spectrum $\mathbb{R}$ we can prove that:

$$[\mathbb{A},\mathbb{L}] = i$$ (5.2)

since A and $\mathbb{L}$ are in the same relation as the usual operators q and p. Then $\widehat{\rho}(a,\sigma,n,n') = (\rho|a,\sigma,n,n')$, the Fourier transform in variables (ν,a) of $\rho(\nu,\sigma,n,n')$, is an eigen vector of A precisely:

$$A\widehat{\rho}(a,\sigma,n,n') = a\widehat{\rho}(a,\sigma,n,n')$$

and A has also a homogeneous Lebesgue spectrum $\mathbb{R}$. Moreover $\widehat{\rho}(a,\sigma,n,n') \in \mathcal{S}$ in the variable a since it is the Fourier transform of $\rho(\nu,\sigma,n,n')$. Then the time evolution of $\widehat{\rho}(a,\sigma,n,n')$ reads:

$$e^{-i\mathbb{L}t}\widehat{\rho}(a,\sigma,n,n') = e^{t\frac{\partial}{\partial a}}\widehat{\rho}(a,\sigma,n,n') =$$

$$\widehat{\rho}(a,\sigma,n,n') + t\frac{\partial}{\partial a}\widehat{\rho}(a,\sigma,n,n') + \frac{t^2}{2!}\frac{\partial^2}{\partial a^2}\widehat{\rho}(a,\sigma,n,n') + ... =$$

$$\widehat{\rho}(a+t,\sigma,n,n') \tag{5.3}$$

thus, as could be specked, $\mathbb{L}$ is the generator of the time translations, and $\widehat{\rho}(a,\sigma,n,n')$ increase its age as $a \to a+t$ becoming $\widehat{\rho}(a+t,\sigma,n,n')$. This fact justify the name given to A, the "age operator". But in its time evolution $\widehat{\rho}(a,\sigma,n,n')$ do not change its shape. As it is a wave packet, since $\widehat{\rho}(a,\sigma,n,n') \in \mathcal{S}$, it continue to be a wave packet of exactly the same shape. So in basis $\{|a,\sigma,n,n')\}$ all physical states are wave packets at any time. Moreover

$$\lim_{t\to\infty}\widehat{\rho}(a+t,\sigma,n,n') = 0 \tag{5.4}$$

and this function goes to zero faster than the inverse of any polynomial.

6. The theorem.

We can decompose the Hilbert space $\mathcal{L}$ as in (2.9). Analogously the space of physical states as can be decompose as $\Psi_+ \oplus \Psi_-$ where $\Psi_\pm = \mathcal{S} \cap \Phi_\pm$ are wave packet spaces that also are Hardy class functions, for the variable ν, from $\{{}^{above}_{below}\}$. Then our theorem states that.

The limit of any physical states, when $t \to \infty$, belongs to Ψ_-.

Demonstration:

We can decompose any $\widehat{\rho}(a)$ as:

$$\widehat{\rho}(a) = \widehat{\rho_+}(a) + \widehat{\rho_-}(a) \tag{6.1}$$

such that:

$$\widehat{\rho_+}(a) = \widehat{\rho}(a) \text{ for } a > 0, \ \widehat{\rho_+}(a) = 0 \text{ for } a < 0$$

$$\widehat{\rho_-}(a) = \widehat{\rho}(a) \text{ for } a < 0, \ \widehat{\rho_-}(a) = 0 \text{ for } a > 0 \qquad (6.2)$$

Then from eq. (5.4) we see that when $t \to \infty$ $\widehat{\rho_-}(a + t) \to 0$ and therefore $\widehat{\rho}(a+t) \to \widehat{\rho_+}(a+t)$. So when $t \to \infty$ these two functions are zero for $a < 0$, and therefore, from Tichmar theorem [11] the Fourier transform of $\widehat{\rho}(a) : \rho(\nu)$ belongs to H^2_-. q. e. d.

So we have proved eq. (2.12) for the wave packets as announced.

7. Conclusion.

As the characteristic distance among the scatterers is much bigger that the characteristic dimension of the scatterers, most of the states can be considered as far from these scatterers. Therefore most of the physical states do belong to space Ψ_-, explaining time-asymmetry (see the introduction). Moreover using space of physical admissible states Ψ_- most of the irreversible phenomenon of nature can be foresee, obtaining the same results as those of well established formalisms (as coarse-graining, Lindblad, etc. [12]).

References

[1] R. P. Feynman, R. B. Leighton, M. Sand, *The Feynman lectures on physics*, vol. 1, Addison-Wesley, New York, 1964.

D. Layser, *The arrow of time*, Scientific America, Dic. 1975.

M. Castagnino. *The global nature of the arrow of time and the Bohm-Reichenbach diagram,* in "Irreversibility and Causality" (Proceeding G. 21, Goslar 1996) A. Bohm, H. Doebner, P. Kielarnowski, eds., p. 282, Springer-Verlag, Berlin, 1998.

[2] H. Reichenbach, *The direction of time,* University of California press, 1956.

P. C. Davies, *Stirring out trouble,* in "Physical origin of time asymmetry", J. Halliwell et al. eds. Cambridge University press, Cambridge, 1994.

M. Castagnino, Phys Red. D, **57**, 750, 1998.

M. Castagnino, M. Gadella, F. Gaioli, R. Laura, *Gamow vectors and time asymmetry,* Int. Jour. Theo. Phys. in press, 1999.

M. Castagnino, A. Korol, L. Lara, *The global nature of the arrow of time,* submitted to Europ. Journ . Phys., 1999.

[3] M. Castagnino, R. Laura, Phys. Rev A, **56,** 108, 1997.

R. Laura, M. Castagnino, Phys. Rev. A, **57,** 4140, 1998.

[4] I. Antoniou, A. Bohm, P. Kelanowski, J. Math Phys.

[5] P. D.Lax, R. S. Phyllips, *Scattering theory,* Academic press, New York, 1979.

[6] A Messiah, *Quantum mechanics,* vol. II, Noeth-Holland, Pub. Co., Amsterdam, 1962.

[7] Koosis

[8] N. N. Bogolubov, A. A. Logunov, I. J. Thodorov, *Introduction to axiomatic quantum field theory,* Benjamin, London, 1975.

[9] M. Castagnino, F. Gaioli, E. Gunzig. Found Cos. Phys. **16,** 221, 1996.

[10] M. Courbage.

[11] Tichmar

[12] R. Laura, M. Castagnino, Phys. Rev E, **57,** 3948, 1998.

D. Arbó, M. Castagnino, F. Gaioli, S. Iguri, *Minimal irreversible quantum mechanics,* Physica A, in press, 1999.

THE KOLMOGOROV-LAX-PHILLIPS SYSTEMS
AS BRANCH SYSTEMS OF THE REICHEMBACH MODEL

by

Mario Castagnino[1], Sergio Iguri[1], Edgard Gunzig[2], Adolfo Ordóñez[3].

[1]: Instituto de Astronomía y Física del Espacio, Casilla de Correos 67, Sucursal 28, 1482 Buenos Aires, Argentina

[2]: Instituts Internationaux de Physique et de Chimie, 50 Av. F.D.Roosevelt, 1050 Bruxelles, Belgium.

[3]: Instituto de Física de Rosario, Av. Pellegrini 250, 2000 Rosario, Argentina.

Abstract

Based in some mathematical theorems, related to Lax-Phillips scattering theory, a unified (classical-quantum) formalism is introduced to explain the appearance of irreversibility in some class of systems (KLPS-systems). These systems are endowed with a unitary evolution that maps densities into densities, and they have a fixed subsystem in which this evolution is strongly contracting, yielding a Lyapounov variable and a unique equilibrium state. A global model of these systems is also studied.

1 Introduction

We will try to explain the main ideas about what is usually called "the arrow of time", namely irreversibility and time asymmetry.

First of all: *the arrow of time is global.* In fact, let us suppose that the arrow of time would be local. Thus, it would be possible to consider two laboratories and to define, in each one, an arrow of time independently. Precisely, let us suppose that the two laboratories are perfectly isolated. Then, we can ask ourselves: Are the two arrows of time pointing to the same direction or not? Of course, this question has no answer, since as the two laboratories are isolated, it is impossible to compare one arrow with the other. If we would like to compare the two arrows of time, an interaction must be introduced between the laboratories, and we would be forced to consider the global system of the two laboratories and the interaction. Then, perhaps we could coordinate the two arrows of time to have juts one arrow in our enlarged laboratory. But we know, based on the astronomical observation, that this coordination can be done (in fact, there is only one arrow of time in the Universe). We can repeat the same story adding a third isolated laboratory, and so forth. Eventually, *the* arrow of time will only be well defined and studied if we consider *all* possible laboratories and the interactions among each other, namely the whole universe. (See the coincident opinion of Feynman in [1], see also [2]). So, *to study the problem of time asymmetry, we need to go to a global cosmological model,* and we know that cosmological models are far from being completely understood (e.g. quantum cosmological

O. Descalzi et al. (eds.), Instabilities and Nonequilibrium Structures VII & VIII, 169–181.
© 2004 *Kluwer Academic Publishers. Printed in the Netherlands.*

models). Moreover, they are not formalized in a rigorous way. Nevertheless, what we need is not a *complete* cosmological model, but a model that represent the *relevant* physical facts for our problem, precisely: *the global Reichenbach system.*

The global arrow of time is best represented by the phenomenological global Reichenbach system [3],[4]: the system of all irreversible processes within the universe, such that any process within any subsystem (branch system) begins in an unstable state that was produced using energy coming from another process of the global system. E. g.: the famous Gibbs ink drop in the glass of water (initial unstable state), evolves towards a final equilibrium state, the homogeneous mix of ink and water (final stable state), showing that we are dealing with an irreversible process. But the ink drop was not produced by an extremely improbable fluctuation that concentrates the ink in the glass. This unstable states was obtained from an ink factory where, to get the necessary energy, coal (initial unstable state) was burnt in an oven until it became ashes (final equilibrium state). Furthermore the system "ink diffusing in the glass of water" *only exists as such after the instant when we put the ink drop into the water.* Before this instant, a much more complex system exists, that eventually contains the ink factory, the oven, the coal burning, etc. Again, the system "burning coal" exists only after we light the oven. In turn, coal was not produced by a fluctuation, quite on the contrary, it was produced using the energy coming from the sun in geological ages. The necessary energy was provided by the light of the sun, where H (initial unstable state) was burnt until it became He, and finally Fe (final equilibrium state). Finally, H was produced using the energy coming from the initial state of the whole global system: a cosmological initial instability. This initial unstable state can be explained, after decoupling time, by the effect of the gravitational field that takes the gas and radiation, in equilibrium before that time, into a state of hot condensed clouds of matter (where nuclear reactions take place) surrounded by cold radiation, in an expanding geometry [4], [5]. If we want to go beyond decoupling time we must consider Big-Bang quantum cosmological models, which also have an unstable initial state [6], [7]. Thus, *through this hierarchical chain, that begins in the cosmological instability and contains all the irreversible processes, where each spontaneous process begins where the corresponding creation device has finished its task, the irreversible nature of the universe and the origin of any irreversible process in it can be explained.* Therefore, Gibbs ink drop only exists because there was a primordial cosmological instability and *Irreversible Statistical Mechanics can not be fully explained without Irreversible Cosmology.* The global system can be symbolized as in fig. 1, which has a clear time asymmetry: the branch arrow of time (BAT), which points in opposite direction to the initial cosmological instability and follows the evolution of the hierarchical chain towards equilibrium. In each box of this diagram (branch system), the energy coming from the left is produced by other boxes, while part of the energy going to the right is used to produce new unstable states, the rest is degraded. This is also an asymmetry of the diagram of fig. 1. All the arrows of the diagram symbolizes spontaneous processes, and thus they are outgoing, with origin in the left box of each arrow. Global Reichenbach system is clearly a realistic model of the set of irreversible processes within the universe. Of course, if we go to a General Relativity model, all the arrows of the branch system must be contained in the corresponding future light semicones.

The global system allow to introduce the notion of *causality* in the universe (i.e., *global causality)*, since, using fig. 1, we can say that events A and B are not causally related, while C is the (partial) cause of D, and D is the (partial) effect of C. Also we can state causality as the law that say that *no cause can occur before its effect*. This statement is only meaningfully because now we have the BAT that defines the word "before". Now, we also know the nature of the physical "messenger" that mediates the causality, is the energy coming from an unstable state and creating a new unstable state, namely unstable structures that are created pumping energy from sources in the past.

Trying to find the mathematical structure corresponding to the phenomenological description given above, we can make the following reasoning. Since all the arrows of fig. 1 (if considered as emerging of the corresponding boxes), are outgoing states, let us use the best mathematical model for these states, namely, Lax-Phillips scattering theory [12] [13] [36] [37].

The boxes in fig. 1 represent systems where some irreversible processes take place. These systems have a set of common properties that we shall use:

i.-They evolve according to a unitary evolution U in a Hilbert space $\mathcal{H}$, that maps densities into densities, in such a way that all the process is time-symmetric. Therefore, all mathematical structures related with the future (as the future equilibrium states, the outgoing states space, etc.) are isomorphic (i.e., essentially identical) to the corresponding mathematical structures in the past (as the past equilibrium state, the incoming states space, etc.). There is only a *conventional* difference between the future and the past structures. If this symmetry would be absent, past would be *substantially* different than future, and the origin of time-asymmetry would be not a problem.

ii.-Nevertheless, if the evolution in any branch system (in Reichembach's terminology), is restricted to a certain subspace, a non-unitary, time-asymmetric evolution T appears. If it would be not so, it would be impossible to transform the reversible system into an irreversible one, and the system would be reversible and trivial, like a harmonic oscillator or a free particle. We will consider only the case where T is a contractive operator in $\mathcal{H}$, that can be strongly dilatated to U (even if, more generally linear non-markovian processes, or even non-linear ones, could be the origin of irreversibility). More complex models will be considered elsewhere.

iii.- The evolution will have only one final equilibrium state. This is only a technical and provisional simplification.

iv.- The system will have a continuous infinity of particles or modes (e.g. it will contain a field)

v.- The system will have finite characteristic times. Namely, its irreversibility can be experimentally verified within a finite period of time.

For these systems, that we will call Kolmogorov-Lax-Phillips systems (KLPS), we will develop a general theory that coincide with our "minimal irreversible quantum mechanics" in the quantum case [8] [9].

We end this introduction explaining how the paper is organized. In section II we state a set of mathematical definitions and theorems that will be used all through the paper. In section III we analyze the characteristic properties of an irreversible system, and found the corresponding mathematical structure. Section IV is devoted to list the systems where the formalism can be used. This contains the scattering (quantum and classical) system, the K-fluxes, etc. In section

V, we define the notion of entropy. In section VI we prove the existence of dispersion relations in KLPS, showing its time-asymmetrical nature, and also the experimental physical consequences of the theory.

2 Mathematical definitions and theorems.

Let A be a linear operator, $A \in \mathcal{L}(\mathcal{H})$ (the set of linear continuous operators over the Hilbert space $\mathcal{H}$), which non necessarily is neither unitary nor an isometry (see [10] and [11]). Let $E \subset \mathcal{H}$ be a closed subspace of $\mathcal{H}$ which it is not invariant under A, i. e.: $E \notin Lat(A)$ (where "Lat"=lattice, symbolized the set of all closed invariant subsets of $\mathcal{H}$ under A, endowed with the lattice structure given by the set inclusion "$\subset$"). Then, about $A|_E$, the restriction of the operator A to the subspace E, we can only say that $A|_E : E \to \mathcal{H}$, because the elements of E do not always remain in this space under the mapping A. If we want to remain in E we must project the vector obtained by the application of $A|_E$ on the space E. Precisely, let P_E be the orthogonal projector over E, then if $B = P_E A|_E$ this operator goes from E to E :

$$P_E A|_E : E \to E \tag{1}$$

Then, we will say that $B = P_E A|_E$ is the *compression* of A in E. Its characteristic property is that:

$$(B\xi|\eta)_\mathcal{H} = (A\xi|\eta)_\mathcal{H} \ , \ \forall \xi, \eta \in E \tag{2}$$

Viceversa, we will say that A is the *dilatation* of B in $\mathcal{H}$.

We will say that A is a *strong dilatation* of B if and only if:

$$P_E A^n|_E = B^n, \ \forall n \in \mathbb{N} \tag{3}$$

B is also called a *strong compression* of A. Symbol n represents time, precisely, the discontinuous time jumps (or *cascades*) that will be used in this section, going latter to the continuous case. Then, we can state the following [10]:

Lemma 1 (Sarason): Let $A \in \mathcal{L}(\mathcal{H})$ and let E be a subspace $E \subset \mathcal{H}$. In order that A would be a strong dilatation of $B = P_E A|_E$, it is necessary and sufficient that $\mathcal{H}$ would have the following decomposition:

$$\mathcal{H} = E_- \oplus E \oplus E_+ \tag{4}$$

where E_+ is invariant under A, and E_- is invariant under $A^\dagger$, namely:

$$A(E_+) \subset E_+ \ , \ A^\dagger(E_-) \subset E_- \tag{5}$$

If A is unitary, then B is a *contraction* (i. e.$\|B\| \le 1$).

Provisionally, we could call E_- the **incoming space** and E_+ the **outgoing space**. The physical motivation for these names is obvious: if A is the evolution operator towards the future, the outgoing space, e.g. the space of states after the scattering must be invariant under repeated application of A. But, if A is unitary, $A^\dagger$ is the evolution operator towards the past, and the incoming space, e.g. the space of states before the scattering, must be invariant under the repeated application of $A^\dagger$. Nevertheless more properties must be added to these spaces in order to consider them as real physical incoming and outgoing spaces. Let us observe that *an extra space E remains to be interpreted. In fact, it is the space where dissipative phenomena take place*, since within this space, the evolution operator is not A, but B, and if A is unitary, B is a contraction, and $||B^n\rho||$, where $\rho \in E$, turns out to be a Lyapunov variable.

The following result is also very important [10]:

Theorem 1 (fundamental): If $T : \mathcal{K} \to \mathcal{K}$ is a contraction and $U : \mathcal{H} \to \mathcal{H}$ its minimal unitary dilatation, then:

1.- $\mathcal{K}$ is a closed space in $\mathcal{H}$, $T^n = P_\mathcal{K} U^n|_\mathcal{K}$ for $\forall n \in \mathbb{N}$, and:

$$\mathcal{H} = \mathcal{H}_- \oplus \mathcal{K} \oplus \mathcal{H}_+ \tag{6}$$

where: $\mathcal{H}_- \in Lat(U^{-1})$, $\mathcal{H}_+ \in Lat(U)$, $U^{-1}|_{\mathcal{H}_-}$ and $U|_{\mathcal{H}_+}$ are *pure isometries*, i.e. unilateral shifts, and, if we call: $N_- = \mathcal{H}_- \ominus U^{-1}(\mathcal{H}_-)$ and $N_+ = \mathcal{H}_+ \ominus U(\mathcal{H}_+)$ we have:

$$\mathcal{H}_- = \bigoplus_{n \in \mathbb{N}_0} U^{-n}(N_-), \quad \bigcap_{n \in \mathbb{N}_0} U^{-n}(\mathcal{H}_-) = \{0\} \tag{7}$$

$$\mathcal{H}_+ = \bigoplus_{n \in \mathbb{N}_0} U^n(N_+), \quad \bigcap_{n \in \mathbb{N}_0} U^n(\mathcal{H}_+) = \{0\} \tag{8}$$

2.- If: $\bigvee = \overline{\bigcup}$ symbolizes the closed subspace generated by the union, then:

$$K_- = \bigvee_{n \in \mathbb{Z}} U^n(\mathcal{H}_-), \quad K_+ = \bigvee_{n \in \mathbb{Z}} U^n(\mathcal{H}_+) \tag{9}$$

where

$$\mathcal{H} = K_+ \iff \forall \xi \in \mathcal{K} \ : \ T^n \xi \to 0 \text{ when } n \to \infty \tag{10}$$

$$\mathcal{H} = K_- \iff \forall \xi \in \mathcal{K} \ : \ (T^\dagger)^n \xi \to 0 \text{ when } n \to \infty \tag{11}$$

A contraction that has all these properties, is called a $(0,0)$-**contraction**. This is the typical contraction in the *non-dissipative* Lax-Phillips scattering theory [12].

In the case of a $(0,0)$-contraction, the properties of spaces $\mathcal{H}_+$ and $\mathcal{H}_-$ are those physically necessary to consider them as the real outgoing and incoming spaces. So, we will call to $\mathcal{H}_+$ and $\mathcal{H}_-$, the outgoing and incoming spaces, from now on.

Let us now state two theorems about the representation of the spaces just introduced:

Theorem 2 (Sinai) [12] [13]. If $\mathcal{H}_+$ is the outgoing space of a $(0,0)$-contraction, then:

(a) $\mathcal{H}$ can be represented isometrically as $L^2(-\infty, \infty, N)$ (see appendix) where $N \sim N_+ \sim N_-$ (since, due to time-symmetry, postulated in section I (i), $N_+ \sim N_-$), and

(b) U acts as a translation or shift to the right by a unit over a *shift variable* $a \in (-\infty, \infty)$, while

(c) $\mathcal{H}_+$ is isometrically represented by $L^2(0, \infty, N)$, and, via a Fourier transform, by $\mathcal{H}^2_+(N)$, the Hardy space from above, of functions from $[0, \infty)$ to the Hilbert space N. Analogously, $\mathcal{H}_- \sim L^2(-\infty, 0, N) \sim \mathcal{H}^2_-(N)$.

If N_1 and N_2 are Hilbert spaces, and θ is a measurable function in $(-\infty, +\infty)$, such that for almost all $\nu \in (-\infty, +\infty)$, that is to say, with the possible exception of a set of measure cero, $\theta(\nu) : N_1 \to N_2$ is a unitary operator, then $\theta(\nu)$ is called an *interior function*.

Now we can formulate the following [10]:

Theorem 3 (standard model for (0,0)-contractions).

Let $T : \mathcal{K} \to \mathcal{K}$ be a (0,0)-contraction, then, $T = P_{\mathcal{K}} U|_{\mathcal{K}}$ is unitarily equivalent to the *model operator* $P_L S|_L$, where S is the shift $S : \mathcal{H}_+^2(N) \to \mathcal{H}_+^2(N)$, relative to the auxiliary space:

$$N = \text{Closure of } (1 - T^\dagger T)^{\frac{1}{2}} \mathcal{K} \tag{12}$$

and there exists an interior function $\theta(\nu) : N \to N$, such that $L \subset \mathcal{H}_+^2(N)$, is the subspace:

$$L = \mathcal{H}_+^2(N) \ominus \theta \mathcal{H}_+^2(N) \tag{13}$$

being

$$\theta \mathcal{H}_+^2(N) := \left\{ g \ / \ g(\nu) = \theta(\nu) \left[f(\nu) \right] \text{ for certain } f \in \mathcal{H}_+^2(N) \right\}$$

If we are dealing with a scattering model, S will be the scattering matrix, but the mathematical structure just introduced is more general.

3 Characteristic properties of an irreversible system.

Physical models are reversible (like the harmonic oscillator that oscillates eternally), or irreversible (like a damped harmonic oscillator, that inevitably will reach a final equilibrium state). A real physical system is usually composed of several parts, some of them reversible and some of them irreversible. Then, the essential feature of an irreversible system is that, at least one of its components, will reach final equilibrium state ρ_* under an irreversible non unitary evolution T (here, we will only address the case where there is *only one* final equilibrium state, and *only one* irreversible T, leaving the problem of many equilibria or many T, or general non-markovian T, for future papers). Since there is only one equilibrium, in this sense, our system will be *ergodic* [1]. Moreover, we know that, in almost all models in the literature about the subject, there exists a unitary dilatation U of T, acting on a Hilbert space $\mathcal{H}$, and preserving densities (classical or quantum ones), then according to Sarason's lemma, T is a contraction.

Let us translate these ideas, and those of the introduction, into a mathematical language. Let $\mathcal{K}$ be the space corresponding to the irreversible subsystem, and let $T : \mathcal{K} \to \mathcal{K}$. If $\rho \in \mathcal{K}$, then there is a unique final equilibrium state $\rho_* \in \mathcal{K}$ such that:

$$T\rho_* = \rho_*, \quad \lim_{n \to \infty} T^n \rho = \rho_* \tag{14}$$

---

[1] In paper [14], an example is shown of a non ergodic system that reaches a final equilibrium state. But this system has a *finite* number of particles or modes. Our models always contain a field, namely infinite modes. To see how a system with a finite number of modes behaves in the limit of infinite modes see [15].

Then, let us define an equilibrium state towards the past:

$$\lim_{n \to \infty} (T^{-1})^n \rho = \rho'_*$$ (15)

for any $\rho \in K$. We have that $\rho_* = \rho'_*$, i.e. the past and future equilibria are equal, as a consequence of the time-symmetry of our system.

On the other hand, in usual physical systems, the unitary dilatation U of the operator T, is $U = e^{-iLt}$, where L is the Liouville operator. Let ρ_0 be such that $L\rho_0 = 0$, and let ρ_* be the orthogonal projection of ρ_0 onto K. Obviously, $L\rho_* = 0$, being therefore constant in time. Then, we can define the space:

$$\mathcal{F}_0 = \mathcal{K} \ominus \{\rho_*\}$$ (16)

where the equilibrium state is the null vector 0. We will call $\xi = \rho - \rho_*$. In general, the states ρ and ρ_* are functionals, the spaces $\mathcal{H}$ and $\mathcal{K}$ are functional spaces [9], and the limit of eq. (14) is a weak limit. But it can be proved [23] that $\mathcal{F}$ and $\mathcal{F}_0$ are Hilbert spaces, and the limit becomes a strong one.

Using the fundamental Theorem 1, the irreversible contractive evolution $T : \mathcal{F}_0 \to \mathcal{F}_0$ has a minimal strong dilatation $U : \mathcal{F} \to \mathcal{F}$. Therefore,

1.- $\mathcal{F}_0$ is a closed subspace of some Hilbert space $\mathcal{F}$, and $T^n = P_{\mathcal{F}_0} U^n|_{\mathcal{F}_0}$ for any $n \in \mathbb{N}$, and:

$$\mathcal{F} = \mathcal{F}_- \oplus \mathcal{F}_0 \oplus \mathcal{F}_+$$ (17)

where: $\mathcal{F}_+ \in Lat(U)$, $\mathcal{F}_- \in Lat(U^{-1})$, $U^{-1}|_{\mathcal{F}_-}$ and $U|_{\mathcal{F}_+}$ are pure isometries, i.e. unilateral shifts in spaces $\mathcal{F}_-$ and $\mathcal{F}_+$. This, in turn, implies that these two spaces are isometrically isomorphic to Hardy class functions spaces $\mathcal{H}^2_-, \mathcal{H}^2_+$, respectively.

If we call: $N_- = \mathcal{F}_- \ominus U^{-1}(\mathcal{F}_-)$ and $N_+ = \mathcal{F}_+ \ominus U(\mathcal{F}_+)$ we have:

$$\mathcal{F}_- = \bigoplus_{n \in \mathbb{N}_0} U^{-n}(N_-), \quad \bigcap_{n \in \mathbb{N}_0} U^{-n}(\mathcal{F}_-) = \{0\}$$ (18)

$$\mathcal{F}_+ = \bigoplus_{n \in \mathbb{N}_0} U^n(N_+), \quad \bigcap_{n \in \mathbb{N}_0} U^n(\mathcal{F}_+) = \{0\}$$ (19)

2.-If:

$$K_- = \bigvee_{n \in \mathbb{Z}} U^{-n}(\mathcal{F}_-) \,, \quad K_+ = \bigvee_{n \in \mathbb{Z}} U^n(\mathcal{F}_+)$$ (20)

then:

$$\mathcal{F} = K_+ \text{ since } \forall \xi \in \mathcal{F}_0 \,, \ T^n\xi \to 0 \text{ when } n \to \infty$$ (21)

$$\mathcal{F} = K_- \text{ since } \forall \xi \in \mathcal{F}_0 \,, \ (T^\dagger)^n\xi \to 0 \text{ when } n \to \infty$$ (22)

because the equilibrium state in space $\mathcal{F}_0$ is the null vector.

Therefore, we are dealing with a $(0,0)$-contraction, and we have obtained a Lax-Phillip structure [12] or Kolmogorov structure in our system. Then, we will call them *Kolmogorov-Lax-Phillips systems* (KLPS). In fact, its incoming and outgoing states have the same abstract properties of the Lax-Phillips scattering theory, and also of the Kolmogorov flows [16].

We conclude that any KLPS, which essentially is an irreversible system, with a unique equilibrium state and a unique T, can be studied in a space:

$$\mathcal{H} = \mathcal{F}_- \oplus \mathcal{F}_0 \oplus \mathcal{F}_+ \bigvee \{\rho_*\} \tag{23}$$

As the system described in space $\mathcal{H}$ with the unitary evolution U is time symmetric, the difference between past and future is just conventional there. Thus, all the minus subspaces are unitarily isomorphic to the plus subspaces, and as we already said, $N_- \sim N_+ \sim N$, etc.

We have the following isometric representations:

$$\mathcal{F}_+ \sim \mathcal{H}_+^2(N), \ \mathcal{F}_- \sim \mathcal{H}_-^2(N) \tag{24}$$

and, using eqs. (16) and the fact that

$$\mathcal{F}_- \oplus \mathcal{F}_0 \oplus \mathcal{F}_+ \sim \mathcal{H}_+^2(N) \oplus \mathcal{H}_-^2(N)$$

we have also the isometric representations:

$$\mathcal{F}_+ \oplus \mathcal{F}_0 \sim \mathcal{H}_+^2(N), \ \mathcal{F}_- \oplus \mathcal{F}_0 \sim \mathcal{H}_-^2(N) \tag{25}$$

Using Theorem 3

$$\mathcal{F}_0 \sim \mathcal{H}_+^2(N) \ominus \theta\mathcal{H}_+^2(N) \sim \mathcal{H}_-^2(N) \ominus \theta\mathcal{H}_-^2(N) \tag{26}$$

where in a scattering theory, θ will be determined by the scattering matrix. In order to have a non trivial system, it is necessary that $\theta \neq 1$. Moreover, the imaginary part of the eventual poles of the scattering matrix, will provide us with the characteristic decaying times of the system.

4   Examples in the literature

The KLPS just described are extremely frequent in the literature. In this subsection we are going to list and explain briefly the main examples, referring the reader to the corresponding literature.

Going now closer to the physical examples, let us suppose that $\mathcal{L} = \mathcal{H}$ is the classical (quantum) Liouville-Hilbert space, $\rho \in \mathcal{L}$ is a classical density (density matrix) and $U_t = e^{-i\mathbb{L}t}$ is the time evolution operator, where $\mathbb{L}$ is the classical (quantum) Liouville operator (in this section we consider a continuous time, so $n \to t$). As it is well known [12], for a system with a Lax-Phillips or Kolmogorov structure, the spectrum of $\mathbb{L}$ is a uniform Lebesgue spectrum over $\mathbb{R}$, namely the spectrum of $\mathbb{L}$ is $\mathbb{R}$ with Lebesgue measure, and the same degeneracy for any value of $\mathbb{R}$. Then, our space $\mathcal{F}$ is necessarily endowed with a continuous spectrum. This justify the symbol "$\mathcal{F}$" for "field".

It is also known [16] [17], that a classical time operator (quantum time superoperator) $\mathbb{T}$, can be defined, such that:

$$[\mathbb{T}, \mathbb{L}] = i\mathbb{I} \text{ and that } U_t^{-1}\mathbb{T}U_t = \mathbb{T} + t\mathbb{I} \tag{27}$$

Thus, the existence of $\mathbb{T}$, and the just described type of spectrum, are characteristic properties of the systems under study.

1.- **Scattering processes**. Many usual scattering processes, as those of reference [24], [35], and also models like the Lee-Friedrichs model [8],[21] [30] (namely models with variable particle number, that allows non-elastic scattering), if studied at the density matrices level, are KLPS. If there is no bound state, the equilibrium state will be just the null vector, since after the scattering nothing remains in the scattering zone (in a moment, we will discuss what happens with the poles of S).

If there are real poles (namely the hamiltonian discrete eigenvalues), these poles must be considered as bound states that constitute the equilibrium state. If there is just one pole, with an energy ω_0, and eigenvector $|0>$, the equilibrium $\rho_* = |0><0|$ must be substracted from the beginning (remember we are considering just *one* equilibrium state in this paper), and space $\mathcal{F}$ must be obtained projecting $\mathcal{H}$ into the orthogonal complement of $\{\rho_*\}$. Now, in $\mathcal{F} = \mathcal{H} \ominus \{\rho_*\}$ the equilibrium is 0, and the situation is the same as above.

2.- **System plus bath**. Let us analyze the usual model to study irreversible quantum process, i.e. an oscillator (q, p) coupled to a field (or infinite set of oscillators (q_ω, p_ω)) like the one in paper [20], also studied in great detail (but for the pure state case) in paper [21].

The classical free hamiltonian of these models reads:

$$H_0 = \frac{1}{2}(p^2 + \Omega^2 q^2) + \frac{1}{2}\int_0^\infty (p_\omega^2 + \omega^2 q_\omega^2)d\omega \tag{28}$$

In the quantum version, the oscillator (q, p) has an infinite discrete set of energy eigenvalues, corresponding to stable states. If the interaction is such that all these states, but the fundamental one, $|0>$, becomes unstable, only this last state produces a final equilibrium state for the oscillator. The fundamental state of the oscillator will be the only remaining discrete state, while all other ones (turned into unstable states by the interaction), will appear as poles of the S-matrix (all these facts are explained in paper [21] in great detail). The total hamiltonian H will have the eigenvalue ω_0 corresponding to the fundamental state plus a continuous spectrum $\mathbb{R}^+$ that corresponds to the field (if the field has not a self-interaction, which, in fact, is the usual case for these models). Going to the Liouville space of density matrices, we will find an equilibrium state ρ_* corresponding to the oscillator ground state, and a field $\mathcal{F}$ with a basis $|\omega><\omega'|$. If $\mathcal{F}$ is studied as in the preceding subsection, it can be proved that it contains a Kolmogorov-Lax-Phillips structure, and $\mathcal{F}$ can be decompose as in eq. (17).

3.- **Classical Lax-Phillips scattering**. Are KLPS, as hard balls scattering, electromagnetic wave scattering, acoustic wave scattering, etc. [12]. This statement is evident since we have developed all our formalism based in this theory.

4.- **Kolmogorov flows**. These and also all higher flows in the chaotic hierarchy [18], are KLPS. The definition of Kolmogorov flows can be found in ref. [18]. In fact, in the case of K-flows, the spectrum of $\mathbb{L}$ is an uniform Lebesgue spectrum over $\mathbb{R}$ with infinite multiplicity. Since these flows are ergodic, it is evident that they are KLPS I. Antoniou and B. Pavlov showed explicitly the Lax-Phillips structure in Baker's transformation, but this paper is not yet published.

So, we have a large number of KLPS in the literature. We hope, this fact could convince the reader that the mathematical effort to understand the theory is worthwhile (after all it explains from the **A**nosov flows, which are K-flows, to the **Z**urek quantum models).

5 Thermodynamics.

Let us now see the thermodynamical consequences of our formalism We will consider only the classical case. As we have postulated an evolution U (or U^t) unitary and densities-preserving, then $T = P_\mathcal{K} U \mid_\mathcal{K}$ (or T^t) will be a contraction by Sarason's lemma, and positiveness preserving. Considering a phase space Γ with a measure μ such that $\mu(\Gamma) = \int_\Gamma d\mu = 1$ (so that 1 is integrable), and writing $L^2 = L^2(\Gamma, \mu)$ and $L^1 = L^1(\Gamma, \mu)$, by the Cauchy-Schwarz inequality in $\mathcal{K} \subset L^2$:

$$
\begin{aligned}
\left\| T^t \rho - \rho_* \right\|_{L^1} &= \int_\Gamma \left| T^t \rho - \rho_* \right| d\mu = \int_\Gamma \left| T^t \rho - \rho_* \right| .1 d\mu \\
&\leq \left\| T^t \rho - \rho_* \right\|_{L^2} \to 0 \text{ , as } t \to \infty
\end{aligned}
\tag{29}
$$

Thus, even if T could be not a Markov operator according to [18], it is "*exact*" in the sense of eq. (29). Then, by almost the same argument of ref. [18], it can be proved that the *conditional entropy*:

$$
H_C(T^t \rho | \rho_*) = - \int_\Gamma T^t \rho(x) \log \left(\frac{T^t \rho(x)}{\rho_*(x)} \right) d\mu
\tag{30}
$$

is always non decaying, and:

$$
\lim_{t \to \infty} H_C(T^t \rho | \rho_*) = 0
\tag{31}
$$

Then, we have defined a completely satisfactory unique entropy out of equilibrium in our formalisms (even if uniqueness comes from the fact that we are considering a unique T).

6 Causality and dispersion relations.

We will see how the Hardy class functions space motivates the dispersion relations. See the theorems of Tichmarsh and of Toll in [34] [33]. From these theorems, we know that *a dispersion relation appears when an input is causally related with an output, if the corresponding functions are endowed of adequate properties, e.g. belong to the Hardy class functions. Since in our theory we have global causality and Hardy functions, it is very easy to find dispersion relations.*

Let us consider any one of the squares of fig. 1. The arrows coming from the left are the input that we will call $\rho_{in}(t')$. The arrows going to the right are the output that we will call $\rho_{out}(t)$. The input is related with the output through a Green function $F(t)$. Now:

i.- Since the input and the output belongs to the global Reichenbach system of fig. 1, they are both outgoing fields and belong to some $\mathcal{H}_+^2$. In fact, in any scattering process the S-matrix maps $\mathcal{H}_+^2$ in $\mathcal{H}_+^2$.

ii.- From global causality (section I-B), we now that the relation between the input and the output is causal, then $F(t) = \widehat{f(t)} = 0$ for $t < 0$.

Therefore, from:

$$\rho_{out}(t) = \int_{-\infty}^{\infty} F(t - t')\rho_{in}(t')dt' \tag{32}$$

and using the *converse theorem*, it follows that F is a causal factor, and this fact originates the dispersion relations [34].

Usually, in the literature some additional assumptions are made to obtain this result (see [34] p. 269, where it is postulated, with no proof, that the input is a causal transform). Now the result is a natural consequence of the formalism and can be used for all models (e.g. that of paper [20]). The systematic appearance of dispersion relations in our formalism is a suggestion toward its validity. We can also see the final mathematical meaning of the global system of fig. 1: it is *a hierarchical chain of outgoing fields (the arrows) connected with causal factors (the boxes)*. We can also remark that, even if we start with one markovian process T, linear non-markovian process, as those described by eq. (32) appear into the play.

7   Acknowledgments

We are very grateful to Boris Pavlov that, in several lectures and talks, introduce us in the subject, and also to Maurice Courbage for many interesting discussions. This paper is partially supported by Grants CI1*-CT94-0004 and PSS*-0992 of the European Community and the OLAM Foundation.

References

[1] Feynman R., Leighton R., Sands N., *The Feynman lectures on physics*, Addison-Wesley Pub Co. Inc., Readings, 1964.

[2] Haag R., Commun. Math. Phys., **132**, 245, 1990.

[3] Reichenbach H., *The direction of time*, Univ. of Calif. Press, Berkeley, 1956.

[4] Davies, P. C., *Stirring up trouble*, in *Physical Origin of Time Asymmetry*, Halliwell et al. eds. Cambridge Univ. Press, Cambridge, 1994.

[5] Aquilano R., Castagnino M., Astrophys. and space Sci., **238**, 159, 1996 / Modern Phys. Lett. A, **11**, 755, 1996.

[6] Hartle J. and Hawking S., Phys. Rev. D, **28**, 2960, 1983.

[7] Vilenkin, A., Phys. Rev. D, **33**, 3560, 1986 / **37**, 388, 1988.

[8] Castagnino M., Laura R. Phys. Rev. A, **56**, 108, 1997

180

[9] Laura R., Castagnino M., *Minimal irreversible quantum mechanics: the mixed states and the diagonal singularity*, Phys. Rev. A, in press 1997.

[10] Cotlar M., *Teoremas Espectrales, Modelos Funcionales y Dilataciones de Operadores en Espacios de Hilbert*, I.A.M., Cursos de Matemática No 5, Buenos Aires, 1991.

[11] Rosenbaum M. Rovnyck J, *Hardy classes and operators theory*, Oxford. Univ. Press, 1985.

[12] P.D.Lax, R.S.Phillips, *Scattering Theory*, Academic Press (1967).

[13] M.Reed and B.Simon *Methods of Modern Math.Phys.III*, Academic Press (1979)

[14] Bricmont J., Physicalia Mag., **17**, 159, 1995.

[15] Gaioli F., García Alvarez E., Guevara J., Int. Jour. Theor. Phys. **36**, 2167, 1997.

[16] M.,B.Misra, I.Prigogine, M.Courbage. Physica 98 A (1979) 1-26 / Courbage M.

[17] M.Courbage, *Dynamical Systems and Microphysics*, CISM Courses and Lectures No 261, Int. Centre for Mech. Sciences, Springer-Verlag, 1980, 225-232.

[18] IM.Mackey, *Times Arrow: The Origins of Thermodynamic Behaviour*, Springer Verlag (1992)

[19] Unruh W., Wald R., Phys. Rev. D **40**, 2598, 1989.

[20] Hu B., Paz J. P., Zhang Y., Phys. Rev. D **45**, 2843, 1992.

[21] Arbó D., Castagnino M., Gaioli F., Iguri S., *Minimal irreversible quantum mechanics: the decay of instable states*, submitted to Phys. Rev. A, 1997.

[22] Castagnino M., Gaioli F., Gunzig E., Found. Cosmic Phys. **16**, 221, 1997.

[23] Castagnino M., Gadella M., Gaioli F., Laura R., *Gamow vectors and time asymmetry*, submitted to Progress of Physics, 1997.

[24] Bohm A., *Quantum Mechanics: Foundations and Applications*, Springer-Verlag (1986) / Bohm A., Phys. Rev. A, **51**, 1996.

[25] Castagnino M., Gunzig E. *Minimal irreversible quantum mechanics: the axiomatic formalism*, submitted to Int. Jour. Theor. Phys. 1998.

[26] Castagnino M., *The global nature of the arrow of time and the Bohm-Reichenbach diagram*, Proceeding Group 21, Goslar 1996, Bohm A. ed. World Scientific, 1998.

[27] Castagnino M, Gunzig E., Lombardo F., Gen. Rel. and Grav., **27**, 257, 1995.

[28] Castagnino M., Lombardo F., Gen. Rel. and Grav., **28**, 263, 1996.

[29] Castagnino M., *The Mathematical structure of quantum superspace as a cosequence of time asymmetry*, Phys. Rev. D., in press 1997.

[30] Antoniou I., Prigogine I., Physica A, **192**, 443, 1993.

[31] Penrose R., *Singularities and time asymmetry, in General Relativity*, Hawking S., Israel S. eds., Cambridge Univ. Press, Cambridge, 1979.

[32] Sach R. G., *The physics of time reversal*, Univ. of Chicago Press, Chicago, 1987.

[33] Tichmarsh E. C., *Theory of Fourier integrals*, The Clarendom Press, Oxford, 1948.

[34] Roman P., *Advanced quantum theory*, Addison-Wesley Pub. Co., Reading, 1965.

[35] Sudarshan E., Phys. Rev. A, **46**, 37.

[36] Pavlov B., *Nonselfadjoint operators: between complex analysis and the geometry of Hilbert space*, Inaugural lecture of Professor at the Univ. of Auckland, New Zeland, 27/7/95.

[37] Pavlov B., Irreversibility, Lax-Phillips approach to Resonance Scattering and Spectral Analysis of Nonselfadjoint Operators in Hilbert Space, preprint, 1997.

Computing the quantum Boltzmann equation from a Kossakowski-Lindblad generator

A. Karina Chattah[*]

Facultad de Matemática Astronomía y Física, Universidad Nacional de Córdoba, 5000 Córdoba, Argentina

Manuel O. Cáceres[†]

Centro Atómico Bariloche e Instituto Balseiro, CNEA y Universidad Nacional de Cuyo, 8400 Bariloche, Argentina.

Abstract

We obtain the quantum Boltzmann equation for the occupation number of electronic states, for the *polaron* model. The calculation is performed under the picture of a quantum dynamical semigroup. Starting from a microscopic dynamics, we also discuss the formal aspects of these quantum semigroups deduced from second order perturbation theory. Our formalism allows us to obtain the mean value and the fluctuations of any physical observable for the system of interest.

I. INTRODUCTION

Classically the Boltzmann kinetic equation for the particle density in phase space (or for the electronic density), arises from the hypothesis of molecular chaos when dealing with

---

[*]Postdoctoral Fellow of Fundación Antorchas

[†]Senior Independent research associated at CONICET. Email: caceres@cab.cnea.gov.ar

O. Descalzi et al. (eds.), Instabilities and Nonequilibrium Structures VII & VIII, 183–195

the two-body probability. This fact introduces the irreversibility and gives a Markovian approach to the dynamics of the system [1]. For open quantum systems, the Markovian description of the dynamics is based on the concept of quantum dynamical semigroups. Only with these semigroups are the properties of the reduced density matrix of the quantum system of interest, preserved during the whole time evolution (positivity, trace and hermiticity) [2,3]. From a microscopic description (considering the total Hamiltonian of the system of interest and the environment) it is possible to derive a picture involving the quantum dynamical semigroups. In this work we first describe some formal aspects and also different representations of quantum dynamical semigroups. We analyze the second order approximation obtained for a quantum open system weakly coupled to the environment. We describe some general conditions on the system of interest, the environment and the interaction Hamiltonian to obtain a well behaved semigroup. After these steps we apply our formalism to a particular problem: *the polaron model*, i.e.: a set of independent electrons in contact with an equilibrium phonon thermal bath. Extending the formal analysis of quantum dynamical semigroups for a particular system with an infinite Hilbert space, we obtain the evolution equation for the number of electrons in a given state under the Markovian approximation, and reobtain the non-linear quantum Boltzmann equation. The fluctuations of the one-body density can also be studied in the same framework proposed here.

II. QUANTUM DYNAMICAL SEMIGROUPS

Quantum dynamical semigroups are the generalization of Markov semigroups for non-commutative algebras [2,3]. In the Markovian approximation, Kossakowski and Lindblad established the form of the Quantum Master Equation (QME) in order that the evolution of the system of interest corresponds to a quantum dynamical semigroup (these semigroups are also called "completely positive semigroups" CPS). In the *structural theorem*, Lindblad [4] established that the *generator* of a CPS, acting on the reduced density matrix of the system ρ, has the form: $L[\bullet] \equiv -\frac{i}{\hbar}[H_{eff}, \bullet] + \frac{1}{2}\sum_\alpha \left[V_\alpha\bullet, V_\alpha^\dagger\right] + \left[V_\alpha, \bullet V_\alpha^\dagger\right]$, where V_α are bounded operators. If the system has a finite number of degrees of freedom, the expression for $L[\bullet]$ can alternatively be written in a different way. Consider for example the algebra of the $N \times N$ complex matrices, then assume that the set of operators $\{V_\alpha\}_{\alpha=1}^{N^2-1}$ fulfilling $Tr[V_\alpha^\dagger V_{\alpha'}] = \delta_{\alpha,\alpha'}$, is a basis in that space (i.e.: the basis is orthonormal with respect to the scalar product: $< A|B > = Tr[A^\dagger B]$). In term of this basis, the generator of the CPS is [5]

$$L[\bullet] = -\frac{i}{\hbar}[H_{eff}, \bullet] + \frac{1}{2}\sum_{\alpha,\gamma=1}^{N^2-1} a_{\alpha\gamma}([V_\alpha\bullet, V_\gamma^\dagger] + [V_\alpha, \bullet V_\gamma^\dagger]), \tag{1}$$

where the matrix of elements $a_{\alpha\gamma}$ is positive-definite. This generator is written in the Schrödinger representation and acts on the density matrix of the system of interest. In the Heisenberg representation the dual generator $L^*[\bullet]$, defined as $Tr[L^*[A]\rho] = Tr[AL[\rho]]$ acts on any physical observable A (Hermitian operator). Now we define the superoperator

$$F[\bullet] = \sum_{\alpha,\gamma=1}^{N^2-1} a_{\alpha\gamma} V_\alpha \bullet V_\gamma^\dagger, \tag{2}$$

and we consider its dual $F^*[I] = \sum_{\alpha,\gamma=1}^{N^2-1} a_{\alpha\gamma} V_\gamma^\dagger V_\alpha$ evaluated in the identity operator I. Then the generator (1) can be written in the compact form

$$L[\bullet] = -\frac{i}{\hbar}[H_{eff}, \bullet] + F[\bullet] - \frac{1}{2}\{F^*[I], \bullet\}_+. \tag{3}$$

Equations (1) and (2) allow us to define the *structure matrix* $[a_{\alpha\gamma}]$. This matrix contain the information concerning the relaxation times of the dynamical system in contact with a thermal bath. We say that a generator with the structure (3), with an Hermitian matrix $[a_{\alpha\gamma}]$ has the *form* of a Kossakowski-Lindblad (KL) [6]. We note that the corresponding semigroup is completely positive *if and only if* $[a_{\alpha\gamma}]$ is a positive matrix, and it is equivalent to say that the generator $L[\bullet]$ fulfills the structural theorem. Alternatively, we will say that when $[a_{\alpha\gamma}] \geq 0$ the generator is a *well defined* KL.

The formal solution of (3) for the density matrix $\rho(t)$ is

$$\rho(t) = \exp\left\{(-\frac{i}{\hbar}[H_{eff}, \bullet] + F[\bullet] - \frac{1}{2}\{F^*[I], \bullet\}_+)t\right\}\rho(0). \tag{4}$$

This expression can be put in the form [6,7]

$$\rho(t) = \sum_{m=0}^{\infty} \int_0^t dt_m \int_0^{t_m} dt_{m-1} \cdots \int_0^{t_2} dt_1$$
$$\times \{S(t-t_m)F[\bullet]S(t_m - t_{m-1}) \cdots F[\bullet]S(t_1)\}\rho(0).$$

In this way the dynamics of the system can be interpreted as if it were composed by quantum jumps (associated to the superoperator $F[\bullet]$) and in between them there is a smooth non-unitary evolution determined by

$$S(t)\rho = \exp\left\{\left(-\frac{i}{\hbar}[H_{eff}, \bullet] - \frac{1}{2}\{F^*[I], \bullet\}_+\right)t\right\}\rho = N(t)\rho N^\dagger(t),$$

where $N(t) = \exp(-\frac{i}{\hbar}tH_{eff} - \frac{t}{2}F^*[I])$ characterizes an exponential decay. This representation is very suitable for describing the decoherence of the off-diagonal elements of the density matrix [5,8].

III. THE QUANTUM MASTER EQUATION AND THE SECOND ORDER APPROXIMATION

In this section we show that the QME arising from second order perturbation theory has, in general, the KL *form* (3). We assume the total Hamiltonian is of the form: $H_T = H_S + H_B + \theta H_I$, and that the system $\mathcal{S}$ interacts with a equilibrium thermal bath $\mathcal{B}$ through the term θH_I (θ is the coupling intensity). Now consider the Liouville equation for the total density matrix and trace out the bath variables, keeping only up to the second order $\mathcal{O}(\theta^2)$. This gives a QME for the reduced density matrix of the system ρ having a KL *form* where the generator is defined through an effective Hamiltonian H_{eff} and the superoperator $F[\bullet]$ [6],

$$H_{eff} = H_S - i\frac{\theta^2}{2\hbar} \int_0^\infty d\tau \ Tr_B \left([H_I, H_I(-\tau)] \rho_B^e\right) \tag{5}$$

$$F[\rho(t)] = \left(\frac{\theta}{\hbar}\right)^2 \int_0^\infty d\tau \ Tr_B \left(H_I \ \rho(t) \otimes \rho_B^e \ H_I(-\tau) + H_I(-\tau) \ \rho(t) \otimes \rho_B^e \ H_I\right). \tag{6}$$

Here $H_I(-\tau) = e^{-i\tau(H_S + H_B)/\hbar} H_I \, e^{i\tau(H_S + H_B)/\hbar}$ and ρ_B^e is the equilibrium density matrix of the bath. We remark that this structure for the generator is independent of any particular system

$\mathcal{S}$ under consideration; it is also valid for finite or infinite dimensional Hilbert space. Now we particularize for a Hilbert space of dimension N, and consider the interaction Hamiltonian to be characterized by the product of operators:

$$H_I = \sum_{\beta}^{n} V_\beta \otimes B_\beta, \qquad n \le N^2 - 1. \tag{7}$$

Then using explicitly the Hermitian condition of H_I, H_{eff} and the superoperator $F[\bullet]$ can be written as:

$$H_{eff} = H_S - i\frac{\theta^2}{2\hbar} \sum_{\alpha\beta} \int_0^\infty d\tau \left(\chi_{\alpha\beta}(-\tau) V_\alpha^\dagger V_\beta(-\tau) - \chi_{\alpha\beta}^*(-\tau) V_\beta^\dagger(-\tau) V_\alpha \right) \tag{8}$$

$$F[\bullet] = \left(\frac{\theta}{\hbar} \right)^2 \sum_{\alpha\beta} \int_0^\infty d\tau \left(\chi_{\alpha\beta}(-\tau) V_\beta(-\tau) \bullet V_\alpha^\dagger + \chi_{\alpha\beta}^*(-\tau) V_\alpha \bullet V_\beta^\dagger(-\tau) \right). \tag{9}$$

Here we have introduced the correlation functions of the bath: $\chi_{\alpha\beta}(-\tau) \equiv Tr_B \left(\rho_B^e \ B_\alpha^\dagger B_\beta(-\tau) \right)$, where $B_\alpha(-\tau) \equiv \exp(-i\tau H_B/\hbar) B_\alpha \exp(i\tau H_B/\hbar)$. Because the bath is stationary the correlation function fulfills the symmetry condition: $\chi_{\alpha\beta}(-\tau) = \chi_{\beta\alpha}^*(\tau)$. The KL *form* allow us to recognize the matrix $[a_{\alpha\gamma}]$ and analyze its possible positivity. First we assume that the system operators belong to a particular basis $\{V_\beta\}_{\beta=1}^{N^2-1}$, then their dynamical evolution can be written in terms of the coefficients $C_{\beta\gamma}(-\tau)$ determined by

$$V_\beta(-\tau) \equiv \exp(-i\tau H_S/\hbar) V_\beta \exp(i\tau H_S/\hbar) = \sum_{\gamma=1}^{N^2-1} C_{\beta\gamma}(-\tau) V_\gamma \tag{10}$$

In this way the equation for the reduced density matrix $\rho(t)$ can be written in the form (1) where

$$a_{\alpha\gamma} = \left(\frac{\theta}{\hbar} \right)^2 \sum_{\beta} \int_0^\infty d\tau \left(\chi_{\gamma\beta}(-\tau) C_{\beta\alpha}(-\tau) + \chi_{\alpha\beta}^*(-\tau) C_{\beta\gamma}^*(-\tau) \right) \tag{11}$$

This expression indicates that the matrix $[a_{\alpha\gamma}]$ is Hermitian; nevertheless it is not possible to prove in general that it is a positive-definite matrix. Therefore, even when the QME up to the second order approximation can be written in a KL *form*, it is not possible to assure that the semigroup will be completely positive. On the other hand the condition $[a_{\alpha\gamma}] \ge 0$

188

implies the following inequalities for the matrix elements; which are known as Sylvester's criterium,

$$a_{\alpha\alpha} \geq 0$$

$$a_{\alpha\alpha} a_{\gamma\gamma} \geq |a_{\alpha\gamma}|^2, \quad \forall \alpha \neq \gamma. \tag{12}$$

These equations allows us to analyze $[a_{\alpha\gamma}]$ and introduce a *necessary condition* on the Hamiltonian H_I in order to arrive to a *well defined* KL. Assuming, as before, that the interaction Hamiltonian can be written (in a particular basis) in the form: $H_I = \sum_{\beta=1}^{n} V_\beta \otimes B_\beta$ with $n \leq N^2 - 1$. The set $\{V_\beta\}_{\beta=1}^{n}$ *must be closed* in the Heisenberg representation, i.e.:

$$V_\beta(-\tau) = \sum_{\gamma=1}^{m} C_{\beta\gamma}(-\tau) V_\gamma \quad \text{with} \quad m \leq n; \tag{13}$$

otherwise the matrix $[a_{\alpha\gamma}]$ will not be positive-definite. This proposition can be proved by writing the elements of (11), and noting that if $m > n$ Sylvester's criterium is not fulfilled [6].

A. Davies' average

In the context of a second order QME it is well known that applying Davies' average, the generator goes to a *well defined* KL, i.e.: with a positive-definite structure matrix $[a_{\alpha\gamma}]$. Davies' average is particular useful when the spectrum of the system is discrete. If the obtained generator from the second order approximation is K, Davies' average is defined by the operation [2,3,9]

$$K^{\#} = \lim_{T \to \infty} \frac{1}{2T} \int_{-T}^{T} \exp(-t\mathcal{L}_S) K \exp(t\mathcal{L}_S) \, dt, \tag{14}$$

where $\mathcal{L}_S = -i/\hbar [H_S, \bullet]$ is the Liouville operator of the system $\mathcal{S}$. Now let us apply the procedure to each term appearing in the generator $L[\bullet]$ of the previous section. To do this, let us define the operators Q_ω^β by the relation

$$V_\beta(-\tau) = \sum_\omega Q_\omega^\beta \exp(-i\tau\omega), \qquad Q_\omega^\beta = \sum_{\varepsilon_n - \varepsilon_{n'} = \hbar\omega} \langle n| V_\beta |n'\rangle \, |n\rangle \, \langle n'|, \tag{15}$$

here $\{\varepsilon_n\}$ are the eigenenergies of H_S and $\sum_{\varepsilon_n - \varepsilon_{n'} = \hbar\omega}$ means the sum over all n, n' that fulfill the restriction $\hbar\omega = \varepsilon_n - \varepsilon_{n'}$. We are also assuming that the system is non-regular and all the $\{\omega\}$ are different. Then the superoperator $F[\bullet]$ adopts the form

$$F[\bullet] = \left(\frac{\theta}{\hbar}\right)^2 \sum_{\alpha,\beta} \sum_{\omega,\omega'} \int_0^\infty d\tau \left(\chi_{\alpha\beta}(-\tau)\exp\left(-i\tau\omega'\right) + \chi_{\beta\alpha}^*(-\tau)\exp\left(i\tau\omega\right)\right) Q_{\omega'}^\beta \bullet Q_\omega^{\dagger\alpha}. \quad (16)$$

With this notation it is easy to see that when we apply Davies' average (14), all the terms with different $\omega' \neq \omega$ cancel out. The operation (14) is also known as the Rotating Wave Approximation in the context of Quantum Optics [10], or secular approximation in the jargon of magnetic systems [11]. The interpretation is that if the coupling θ is weak, then the evolution time-scale is much longer than any free dynamics of the system. Now invoking the stationary property of the thermal bath, $\chi_{\alpha\beta}(-\tau) = \chi_{\beta\alpha}^*(\tau)$, and eliminating the terms with $\omega' \neq \omega$ we obtain

$$F^\#[\bullet] = \left(\frac{\theta}{\hbar}\right)^2 \sum_{\alpha,\beta,\omega} \int_{-\infty}^\infty d\tau\, \chi_{\alpha\beta}(-\tau)\exp\left(-i\tau\omega\right) Q_\omega^\beta \bullet Q_\omega^{\dagger\alpha}. \quad (17)$$

From this expression it is possible to realize that the elements of the structure matrix are $h_{\alpha\beta}(\omega) \equiv \int_{-\infty}^\infty d\tau \chi_{\alpha\beta}(-\tau)\exp\left(-i\tau\omega\right)$. On the other hand Bochner' theorem assures its positivity because $\chi_{\alpha\beta}(-\tau)$ are correlation functions of a thermal bath. Then it is also possible to see that Detailed Balance condition is also guaranteed on the KL generator [3]. Therefore the *form* of the generator can be written in terms of the superoperator

$$F^\#[\bullet] = \left(\frac{\theta}{\hbar}\right)^2 \sum_{\omega>0} Q_\omega \bullet Q_\omega^\dagger + e^{-\beta\omega}\, Q_\omega^\dagger \bullet Q_\omega, \quad (18)$$

where $Q_\omega(t) = e^{-i\omega t}Q_\omega$ (and $\beta = 1/k_B T$), which assures the validity of Detailed Balance.

If the spectrum of H_S is non degenerate, and after taking Davies' average, the generator (14) leads to a Pauli Master Equation for the diagonal elements of the reduced density matrix (independent of the off-diagonal elements, which in fact decay to zero). The gain terms in the QME are associated to $F[\bullet]$, while the loss terms are characterized by $\{F^*[I], \bullet\}_+$. Then the dynamics of the diagonal elements of the reduced density matrix becomes classical by taking Davies' average.

IV. BOLTZMANN EQUATION FOR THE POLARON MODEL

Now we analyze the QME in the second order approximation for the *polaron* model, i.e.: the system of interest is a set of free electrons interacting with phonons in a harmonic solid. This system is associated to an infinite dimensional Hilbert space, however we can extract information from the different terms in the KL *form* and its associated structure matrix. Taking into account the evolution equation for the reduced density matrix, we get the Boltzmann kinetic equation for the occupation number of electronic states of the system.

The unitary microscopic dynamics is governed by a Frölich Hamiltonian describing a model of independent electrons in contact with a thermal bath of phonons [12]. The system and bath Hamiltonians are:

$$H_S = \sum_k \hbar \varepsilon_k \alpha_k^\dagger \alpha_k, \qquad H_B = \sum_\beta \hbar \omega_\beta b_\beta^\dagger b_\beta,$$

here $\alpha_k, \alpha_k^\dagger$ and $b_\beta, b_\beta^\dagger$ are fermionic and bosonic operators destroying and creating energy states ε_k and ω_β. The interaction Hamiltonian is deduced from the quantization of the electromagnetic interaction between the Coulomb field of the electrons and the polarization waves of the solid; then it takes the form

$$H_I = \sum_{k,\beta} \hbar (v_\beta \, b_\beta \, \alpha_{k+\beta}^\dagger \, \alpha_k + v_\beta^* \, b_\beta^\dagger \, \alpha_k^\dagger \, \alpha_{k+\beta}) \equiv \sum_\beta (B_\beta V_\beta + B_\beta^\dagger V_\beta^\dagger). \tag{19}$$

This interaction Hamiltonian is written in terms of the bath operators $B_\beta = \hbar v_\beta \, b_\beta$, and the system operators $V_\beta = \sum_k \alpha_{k+\beta}^\dagger \, \alpha_k$ producing transitions between different electronic states. Taking into account this total Hamiltonian and using the corresponding QME in the second order approximation, the KL *form* has the terms (5) and (6):

$$H_{eff} = H_S - \frac{i\theta^2}{2\hbar} \sum_{k,k'} \sum_\beta \left(\left(n_\beta(-w_{k,\beta}) \mathcal{V}_{k',\beta}^\dagger \mathcal{V}_{k,\beta} + \tilde{n}_\beta(-w_{k,\beta}) \mathcal{V}_{k',\beta} \mathcal{V}_{k,\beta}^\dagger \right) - h.c \right) \tag{20}$$

$$F[\bullet] = \frac{\theta^2}{\hbar^2} \sum_{k,k'} \sum_\beta \left(\left(n_\beta(w_{k,\beta}) \mathcal{V}_{k',\beta} \bullet \mathcal{V}_{k,\beta}^\dagger + \tilde{n}_\beta(w_{k,\beta}) \mathcal{V}_{k',\beta}^\dagger \bullet \mathcal{V}_{k,\beta} \right) + h.c \right) \tag{21}$$

(*h.c.* means Hermitian conjugate). In this expression we can see the appearance of the operators $\mathcal{V}_{k,\beta} = \alpha_{k+\beta}^\dagger \alpha_k$. Note that the non-unitary evolution of the reduced density matrix (irreversible dynamics) is characterized by $F[\bullet]$ and $\{F^*[I], \bullet\}_+$. Then we can recognize the occurrence of molecular chaos as the result of using a second order approximation, and in addition, the particular structure that we have used for the weak interaction H_I [see also the formal solution for the reduced density matrix equation (4)].

Now, to determine the possible positivity of the KL generator, we define the following time integrals of the bath correlation functions, evaluated at the difference of electronic energies $\hbar w_{k,\beta} = \varepsilon_{k+\beta} - \varepsilon_k$

$$n_\beta(w) = \int_0^\infty d\tau \exp(iw\tau) Tr_B[B_\beta^\dagger B_\beta(-\tau)]$$

$$\tilde{n}_\beta(w) = \int_0^\infty d\tau \exp(-iw\tau) Tr_B[B_\beta(-\tau) B_\beta^\dagger],$$

Therefore using eqs. (20) and (21) we will able to analyze the corresponding structure matrix using the Sylvester criterium (12).

A. Structure Matrix

Here we analyze the positivity condition of the structure matrix arising from expression (21). Considering the operators $\{\mathcal{V}_\mu\}$ (where $\mu = \{k, \beta\}$) and the scalar product: $<A|B> = Tr(A^\dagger B)$, we see that these operators satisfie $Tr(\mathcal{V}_{k,\beta}^\dagger \mathcal{V}_{k',\beta'}) = \delta_{k,k'} \delta_{\beta,\beta'}$, so forming an orthogonal set. Then these operators can be taken as a basis in the space of operators of finite trace and we can write the superoperator $F[\bullet]$ in terms of them:

$$F[\bullet] = \sum_{\mu,\mu'} a_{\mu,\mu'}^A \mathcal{V}_\mu^\dagger \bullet \mathcal{V}_{\mu'} + a_{\mu,\mu'}^B \mathcal{V}_\mu \bullet \mathcal{V}_{\mu'}^\dagger \tag{22}$$

This structure is similar to (2); in that case the Hilbert space had finite dimension. Now we want to extend the analysis of the structure matrix to the present model characterized by a Hilbert space of infinite dimension. Although, in this case, there is no formal proof of the positivity criterium for the structure matrix, we assume the validity of Sylvester's inequalities (12). We support this kind of procedure from experience of previous literature where it has been proved that a Lindblad generator satisfying the *structural theorem* is connected with a matrix $[a_{\mu,\mu'}]$ fulfilling Sylvester's inequalities in infinite dimensional spaces [13–15].

From expression (22) we can see that the KL *form* has two contributions associated to the structure matrices $a^A_{\mu,\mu'}$ and $a^B_{\mu,\mu'}$. These matrices, defined as:

$$a^B_{\mu,\mu'} = \theta^2[n^*_\beta(w_{k,\beta}) + n_\beta(w_{k',\beta'})]\delta_{\beta\beta'}/\hbar^2$$

$$a^A_{\mu,\mu'} = \theta^2[\tilde{n}^*_\beta(w_{k,\beta}) + \tilde{n}_\beta(w_{k',\beta'})]\delta_{\beta\beta'}/\hbar^2,$$

must be positive-definite in order to have a *well defined* KL. The diagonal elements of these matrices are positive

$$a^B_{\mu,\mu} = 2Re[n(w_{k,\beta})]/\hbar^2 \geq 0 \;, \qquad a^A_{\mu,\mu} = 2Re[\tilde{n}(w_{k,\beta})]/\hbar^2 \geq 0.$$

This fact is guaranteed because these elements are the Fourier transforms of the correlation functions for the thermal bath and their positivity is proved by Bochner theorem [3]. Using Sylvester's criterium we can see that the off-diagonal elements do not satisfy the required inequalities. Thus we conclude that the generator of the QME is not a *well defined* KL. On the other hand, assuming periodic boundary conditions, the energy levels are well defined; this fact allows us to apply Davies' formalism.

B. Boltzmann equation

Considering a discrete set of energy levels for the system, we apply the Davies average to the KL generator determined by (20) and (21). Then by this procedure we expect to get a generator of a CPS. Applying the procedure given by (14) to the superoperator (21) we obtain:

$$F^\sharp[\bullet] = \sum_\mu a^A_{\mu,\mu}\mathcal{V}^\dagger_\mu \bullet \mathcal{V}_\mu + a^B_{\mu,\mu}\mathcal{V}_\mu \bullet \mathcal{V}^\dagger_\mu, \tag{23}$$

in a similar way it is possible to obtain $H^\sharp_{eff}$. Note that in the superoperator $F^\sharp[\bullet]$ there remains only the diagonal elements of the structure matrix, which are positive, and thus the resulting generator is a *well defined* KL.

Now, for any operator A of the system, Davies' averaged generator allow us to obtain the dynamical equation in the Heisenberg picture:

$$\dot{A} = L^*[A] = \frac{i}{\hbar}[H_{eff}^\sharp, A] - \frac{1}{2}\left\{F^\sharp[I]^*, A\right\}_+ + F^\sharp[A]^*, \tag{24}$$

where $F^\sharp[\bullet]^*$ denotes the dual operator calculated from (23). In particular we work out the evolution equation for the operator measuring the number of electrons in a certain state k', which is defined as

$$f_{k'} = \alpha_{k'}^\dagger \, \alpha_{k'}.$$

Now taking into account the commutation relations between fermionic operators we can see that $[H_{eff}, f_{k'}] = 0$. Finally the evolution equation for $f_{k'}$ is ($\theta = 1$):

$$\begin{aligned}\dot{f}_{k'} = (2/\hbar^2)\sum_\beta Re[\tilde{n}_\beta(w_{k',\beta})](1-f_{k'})f_{k'+\beta} - Re[n_\beta(w_{k',\beta})]f_{k'}(1-f_{k'+\beta})\\ + Re[n_\beta(w_{k'-\beta,\beta})](1-f_{k'})f_{k'-\beta} - Re[\tilde{n}_\beta(w_{k'-\beta,\beta})]f_{k'}(1-f_{k'-\beta})\end{aligned} \tag{25}$$

This is the non-linear Boltzmann equation for the occupation number of the electronic states. Note that the QME for the density matrix, and also equation (24) are linear; nevertheless the structure of $L^*[A]$ mixes terms of the form $\mathcal{V}_\mu^\dagger A \mathcal{V}_\mu$ then producing a non-linear equation (25) for $f_{k'}$. The terms containing combinations of the form $(1-f_{k'})f_{k'+\beta}$ are typical for fermionic systems.

The non-equilibrium statistical mechanics Boltzmann kinetic equation starts from the "molecular chaos" hypothesis. Then irreversibility is introduced leading to a Markovian dynamics. In our case we have reobtained this approach, from the microscopic quantum weak interaction H_I, through the second order approximation and Davies formalism. The quantum Boltzmann kinetic equation can be associated to a semigroup as the result of weak interaction with the bath. On the other hand, the QME for the reduced density matrix and the Heisenberg representation (24), are starting points to study the behavior of higher moments of the quantum observables. This fact could be useful to study the fluctuations of the mean value of the electronic density. Classically, a related analysis can be found in chapter XIV of Ref. [1].

V. CONCLUSIONS

In this work we have analyzed the second order approximation for the QME (also called Born-Markov equation) for the polaron model starting from the microscopic dynamics. We have characterized the different terms of the KL generator of the quantum semigroup, which is the starting point to analyze the positivity condition of the structure matrix using Sylvester's inequalities. Applying Davies's formalism to the generator, a completely positive semigroup is obtained and the Heisenberg evolution for any operator can be studied. In particular, the evolution equation for the occupation number of electronic states leads to the quantum Boltzmann equation, representing then the Markovian description of an irreversible dynamics. On the other hand this approach can be used to get information about the fluctuations and higher moments of the fermionic occupation number, and work along these lines is in progress.

REFERENCES

[1] N.G. van Kampen, *Stochastic processes in Physics and Chemistry*, 2^a ed. (North Holland, Amsterdam, 1992).

[2] H. Spohn; Rev. Mod. Phys. **52,** 569 (1980); R. Dumcke, H. Spohn, Z. Phys. B, **34,** 419, (1979).

[3] R. Alicki and K. Lendi, *Quantum Dynamical Semigroups and Applications,* Lectures Notes in Physics 286, Springer-Verlag, Berlin, (1987).

[4] G. Lindblad, Commun. Math. Phys. **48,** 119 (1976).

[5] V. Gorini and A. Kossakowski; J. Math. Phys. **17,** 821 (1976).

[6] A.A. Budini, A.K. Chattah and M.O. Cáceres; J. Phys. A Math. and Gen. **32,** 631 (1999).

[7] M.B. Plenio, P.L. Knight; Rev. Mod. Phys, **70,** 101 (1998)

[8] J. Dalibard,Y. Castin and K. Molmer; Phys. Rev. Lett., **68,** 580 (1992)

[9] E.B. Davies; Commun. Math. Phys. **39,** 91 (1974).

[10] C.W. Gardiner, *Quantum Noise* (Springer, Berlin, 1991).

[11] K. Blum, *Density Matrix Theory and Applications* (Plenum Press, New York, 1981).

[12] H. Haken, *Quantum Field Theory of Solids* (North-Holly, Amsterdam, 1976).

[13] A. Sandulescu and H. Scutaru, Annals of Physics **173,** 277 (1987); A. Isar, A. Sandulescu, W. Scheid, J. Math. Phys. **34,** 3887 (1993)

[14] H. Dekker and M. C. Valsakumar; Phys Lett A **104,** 67 (1984).

[15] A.K. Chattah and M.O.Cáceres, Cond. Matt. Phys. **3,** pags. 51-73 (2000).

A COMBINED DENSITY FUNCTIONAL AND SEMICLASSICAL APPROACH TO DESCRIBE ATOMIC/IONIC RADII

M.Cristina Donnamaria [1]

[1]: Instituto de Fisica de Liquidos y Sistemas Biologicos (IFLYSIB)-Comision de Investigaciones Cientificas de la Provincia de Buenos Aires (CIC)
IFLYSIB-CIC-CONICET-UNLP,C.C. 565, 1900 La Plata ,Argentina, donna@lpsat.com

1 Abstract

A combination of a Thomas Fermi Amaldi (TFA) density functional and a semiclassical Z-optimization criterion is apply in this work, to neutral atoms and their first positive ions to calculate their effective atomic radii.

2 Introduccion

In Density Functional Theory (DFT), the focus is on the electronic density, rather than on the associated wavefunction. This is the aim of DFT, which simplest formulation is embodied in the Thomas-Fermi (TF) theory [1,2]. The TF model works well in situations in which the electronic density is slowly varying. On the other hand, properties which depend on the details of the electronic structure, such as shell structure, are not properly represented by this formalism. The Amaldi correction [3] enhanced the TF results, removing the spurious self interaction among the electrons, as it was proved previously by Donnamaria et al.[4-7].Besides using the Thomas-Fermi-Amaldi procedure ions can be easily included [5-7], while TF formalism is useful only for neutral atoms. In this work, within the context of the TFA density functional, the effective nuclear charges for 44 neutral atoms and their first positive ions are calculated, using a trial electronic density function and a particular nuclear charge (Z) optimization criterion. These values are used to predict their effective atomic/ionic radii. This treatment also brings out some interesting features about atomic shell structure.

3 Method

In the Amaldi correction [3] to the TF theory the spurious self interaction among electrons is removed by a simple factor in the energy density functional (N-1/N)[5-7]

$$E\left[\rho\right]_{TFA} = 2.8712 \int \rho\left(r\right)^{5/3} dv + \int \rho\left(r\right) V_N.dv + \frac{1}{2}\left(\frac{N-1}{N}\right) \int \int \frac{\rho\left(r\right).\rho\left(r\right)'.dvdv'}{|r-r|} \quad (1)$$

ρ being the electronic variational density, N is the number of electron and Z is the nuclear charge . The three terms respectively are the electronic kinetic energy, the interaction between

O. Descalzi et al. (eds.), Instabilities and Nonequilibrium Structures VII & VIII, 197–203.

the electrons and the nucleus and the interaction among the electrons. This ρ, obtained by minimizing the energy is the same that can be achieved from the numerical solution of the TF equation, [1]. Instead of using the "exact" TF solution, ϕ_{TF}, which decreases too slowly with distance to the origin, this method allows to use a density function with an adequate distance dependence, particularly that of Wu exhaustively analyzed in previous papers [4-7]

$$\rho\left(x\right)_{Wu} = \frac{\left(1 + mx^{1/2} + nx\right)^3 (x)^{-3/2} e^{-3mx^{1/2}}}{4\pi a^3} \tag{2}$$

The optimum (m,n) parameters are obtained by minimizing , Eq. (1).

3.1 Charge optimization criterion

On the Hasee Kirkwood Vinti [10] approach, the atomic electric dipole polarizability,α is

$$\alpha = 16 \left(\frac{m_o.c^2}{N_A.e_o^2}\right)^2 . \left(\frac{S^2}{a_B.Z}\right) = \frac{0.1058.S^2}{Z} \tag{3}$$

N_A is de Avogadro's number, the other quantities have their conventional meaning, and S is the diamagnetic susceptibility [10]. Replacing S by the expression obtained with the Wu function [6-9] the α expression in the TFA formalism is

$$a_{Wu} = \frac{0.163125.P_{nm}N^2}{Z^{7/3}} \tag{4}$$

$$P_{nm} = \frac{10880}{729m^7} + \frac{76160n}{729m^9} + \frac{67200n^2}{2187m^{11}} + \frac{1971200n^3}{656m^{13}} \tag{5}$$

In another work [11] we found that the minimizing of the energy, Eq.(1), by variation of N and keeping constant the atomic number Z, gives poor results for polarizabilities. To tackle this problem, we suggest the following optimization criterion to define a charge optimum value:

$$\alpha_{Wu(Z_{op},N,n,m)=\alpha_F(HF-SCF)} \tag{6}$$

where a_{Wu} is given by Eq.(4) assuming that Z = $Z_{op}$and $\alpha_F(HF - SCF)$ is the Self-Consistent-Field atomic dipole polarizability evaluated by Fraga et. al.[12], then

$$\frac{0.163125.P_{nm}N^2}{Z^{7/3}} = \alpha_F(HF - SCF) \tag{7}$$

and

$$Z_{op}(n,m) = \left(\frac{0.163125Pnm^2N^2}{\alpha_F}\right) \tag{8}$$

Z_{op} being the effective nuclear charge which reproduces the Fraga's polarizabilities. The TFA formalism, allows to apply the treatment not only to neutral atoms, but also to ions since each species is described in terms of (Zop,N,n,m),

3.1.1 The effective ionic radius

Redefining an "effective radius" as:

$$r_{eff}\left(Z_{op}\right) = \left\langle r(z)^2 \right\rangle^{1/2} \tag{9}$$

In the TF formalism :

$$\left\langle r^2 \right\rangle = \int \rho\left(r\right) r^2 dv \tag{10}$$

In the previous expression ρ is replaced by $\rho\left(x\right)_{Wu}$, Eq. (2), then, [7-9]

$$\left\langle r^2 \right\rangle = \frac{1.56766 a_B . Pnm.N}{Z^{7/3}} \tag{11}$$

and assuming $Z = Z_{op}$ the following expression is obtained for the "effective radius"

$$r_{eff}\left(Z_{op}\right) = \left(\frac{1.56766 a_B . Pnm.N}{Z_{op}^{7/3}}\right)^{1/2} \tag{12}$$

3.2 Results

In previous papers [6,7] we have minimized the TFA-Wu energy functional, Eq. (1), using the Wu trial function [2]. For 44 neutral atoms and their first positive ions we obtained a particular set of values (mTFA, nTFA) and then each Pnm. These parameters are displayed in Table I.. The corresponding optimum charges, Z_{op}, Eq.(8) and the effective radii, r_{eff}, Eq. (12), calculated by the TFA formalism are also given. The α_F and r_{eff} values correspond to the atomic polarizability and to the radius for the outermost orbital as given by Fraga et. al [12] from HF- SCF calculations respectively. Table I and the Figures 1(a. and b.) show a particular behavior of $r_{eff}\left(Z_{op}\right)$ with Z_{op} for neutral atoms and for their first positive ions. A family of curves is obtained and each curve can be correlated with each period of the periodic table, F1 to F4. In each one, when working with neutral atoms (Fig.1.a) the first element corresponds to an alkaline atom while the last one is the respective noble gas atom of the period. When the description is for ions (Fig. 1.b) the first element corresponds to an ion with an alkaline earth-atom electronic structure while the last one is an ion with the respective noble gas atom electronic structure. The Ne atom family has been omitted since the present formalism does not behave properly here. As regards the discontinuities associated with the origin of each curve, Na, K, Rb and Cs (neutral atoms) and Mg^+, Ca^+, Sr^+, and Ba^+ (ions), from inspection of the Z_{op} values (Table I) could be correlated with the start of a new shell. One way of interpreting the varying Z effect is in terms of the nuclear charge shielding by the electrons [13]. Thus, the electron entering the new shell "sees" a more "compact" electron distribution . The same breakdowns appear in the HF-SCF Fraga's ionization potential [12], IP_F versus N, (Figures 2a.and 2.b), and in the trend of Fraga's polarizabilities α_F vs N, showing that this behavior is also related to the start of a new shell. Moreover, there is a strict correlation between them, the higher the IP_F values, the lower the α_F for each Z_{op} . The largest ionization potentials are related to the lowest polarizabilities of the rare gas atom electronic configuration, while the lowest IP_F correspond to the alkaline-earth atom electronic structure, which has the highest polarizabilities, as expected, due to their physical chemical properties. Some other internal strong shifts within the curves of r_{eff} vs Z_{op} are observed, Zn/Ga, Cd/In, Yb/Lu and Hg/Tl (neutral atoms), and Cu^+/Zn^+, Ag^+/Cd^+, Tm^+/Yb^+ and Au^+/Hg^+ . From the analysis of IP_F and from the Z values (Table I) it can be inferred that these irregularities are related to the filling of a subshell. It is also well known [14] that d^{10} and f^{14} configurations have a shell structure in some atomic properties, and they behave as new shell. The atomic radius contraction of the first or second element in some periods is also a typical shell effect: Rb/Sr Cs/Ba and Sr^+/Rb^+ Ba^+/La^+. As far as neutral atoms and their first positive ions are concerned, the present r_{eff} radii and the r_F sets have similar trends with Z_{op} and reproduce fairly well the desired systematic general behavior of atomi/ionic radii. Thus is: i) the atomic radii tend to decrease while the atomic number increases within a given period of the periodic table by the filling of shells/subshells, ii) the atomic radii tend to increase while the atomic number increases within a given group of the periodic system iii) the radii increase for the first (or the second) element of a new shell/subshell, consistent with the contraction effect. In fact, they are the general conditions that a "good theoretical radius " must fulfill [15]. Taking into account that r_F (orbital representation) and r_{eff} are close related, the latter could be associated with the "size" of the outer atomic/ionic "shell". This assignation is equivalent to associating each shell with a particular region of the coordinate space, in which the atom concentrates.

4   Conclusions

These findings reveal that the present Z optimum criterion in addition to the modified TFA-density functional provide us with a "good" ionic radius set, which reproduces the desired systematic behavior of atomic radii. It also contributes to the atomic shell structure study, giving a reasonable "size" criterion for neutral atoms and their first positive ions.

Acknowledgments:This work has been funded by CIC, CONICET and UNLP of Argentina. MCD is member of of Research Career of Comision de Investigaciones Cientificas de la Provincia de Buenos Aires (CIC). She is also indebted to the Organizers of the 7th International Workshop on Inestabilities and Non-equilibrium Structures for their financial help which allowed her to attend the Symposium.

References

[1] N.H. March, *Theory of the Inhomogeneous Electron Gas* Plenum, New York, 1983.

[2] R.M. Dreizler and E.K. Gross, *Density Functional Theory* Springer Verlag, Berlin 1990.

[3] E. Fermi and E. Amaldi. Mem. Acad. Ital., 1934, 6, 117.

[4] M.C. Donnamaria, E.A. Castro and F.M. Fernandez, Int. J. Quantum Chem, 1982,22,1005, J. Chem. Phys. 1983,78(2),5013.

[5] M.C. Donnamaria, E.A. Castro and F.M. Fernandez, J. Chem. Phys. 1984, 80(3),1179.

[6] D.M. Glossman, M.C. Donnamaria, E.A. Castro and F. M. Fernandez, J. Physique, 46,173,(1985) and Acta Physica Slovaca, 1987, 37(5), 298.

[7] M.C. Donnamaria and A. N. Proto, MATCH 1991, 26, 95 and in *Atomic and Ionic Information Entropies.* Condensed Matter Theories, Vol 7,p.243. Plenum Press, NY, 1992

[8] M.C.Donnamaria, R.E. Cachau and E.A. Castro, Teochem, 1990, 210, 121.

[9] M.C.Donnamaria and E. Castro, Tubitak (Turkish J. of Physics) 1997, 1.

[10] J.O. Hirschfelder, C.P. Curtis and R.P. Bird, *Molecular Theory of Gases and Liquids*, Wiley, New York, 1965.

[11] M.C. Donnamaria , E.A. Castro and F.M. Fernandez, An. Quim. 1983, 79(3), 613.

[12] S. Fraga, J. Karwoski and K.M.S. Saxena, Handbook of Atomic Data, Elsevier, Amsterdam, 1980.

[13] P.L. Pilar, J. Chem. Educ. 1978,55, 2.

[14] L. Pauling, *The Nature of Chemical Bond*, Cornell University Press, Ithaca, N.Y. 1960.

[15] R.J. Boyd, J. Phys. B. 1977, 10, 2283.

Table I: TFA-Atomi radii , I=Neutral Atoms, II= First Positive ions, PolF = α_F

At(I)	Z	Pnm	Zop	$PolF$	reff	r-F	At(II)	Z	Pnm	Zop	$PolF$	reff	r-F
Ar Series													
Na	11	5.28	4.26	18.67	3.11	4.21	Na*	10	4.16	25.33	0.15	1.46	0.31
Mg	12	5.29	5.18	14.13	3.05	3.25	Mg*	11	4.26	7.43	3.22	2.32	1.29
Al	13	5.29	6.19	10.97	2.99	3.43	Al*	12	4.34	7.64	3.85	2.42	1.05
Si	14	5.29	8.10	6.81	2.84	2.75	Si*	13	4.41	9.12	3.09	2.40	1.10
P	15	5.30	10.35	4.42	2.71	3.32	P*	14	4.47	11.11	2.32	2.34	0.91
S	16	5.30	12.18	3.44	2.65	2.06	S*	15	4.53	13.49	1.74	2.29	0.80
Cl	17	5.31	14.45	2.61	2.58	1.84	Cl*	16	4.57	15.60	1.44	2.27	0.75
Ar	18	5.31	17.92	1.98	2.51	1.66	R*	17	4.62	18.10	1.17	2.23	0.67
Kr Series													
K	19	5.31	5.07	37.58	3.87	5.24	K$^+$	18	4.66	20.94	0.95	2.20	0.63
Ca	20	5.33	5.57	33.8	3.86	4.21	Ca$^+$	19	4.69	8.70	8.34	3.04	1.72
Sc	21	5.33	6.41	26.8	3.86	4.21	Sc$^+$	20	4.72	17.32	1.88	2.49	0.61
Tl	22	5.33	7.20	22.4	3.71	3.78	Tl$^+$	21	4.75	19.24	1.64	2.47	0.55
V	23	5.33	8.01	19.1	3.66	3.63	V$^+$	22	4.78	21.61	1.39	2.44	0.49
Cr	24	5.33	8.83	16.6	3.62	3.49	Cr$^+$	23	4.81	24.37	1.16	2.40	0.49
Mn	25	5.33	9.66	14.6	3.59	3.38	Mn$^+$	24	4.80	25.85	1.11	2.41	0.40
Fe	26	5.33	10.60	12.7	3.55	3.26	Fe$^+$	25	4.84	27.94	1.01	2.40	0.38
Co	27	5.33	11.52	11.3	3.52	3.15	Co$^+$	26	4.87	30.22	0.92	2.39	0.37
Ni	28	5.33	12.41	10.2	3.49	3.06	Ni$^+$	27	4.89	32.56	0.84	2.38	0.36
Cu	29	5.33	13.47	9.04	3.46	2.97	Cu$^+$	28	4.90	35.37	0.75	2.36	0.32
Zn	30	5.33	14.52	8.12	3.43	2.90	Zn$^+$	29	4.89	22.84	2.22	2.78	1.16
Ga	31	5.33	14.01	9.43	3.53	3.42	Ga$^+$	30	4.91	20.69	3.01	2.92	1.06
Ge	32	5.33	16.35	7.01	3.41	2.87	Ge$^+$	31	4.92	21.10	3.09	2.96	1.16
As	33	5.33	19.05	5.22	3.29	2.51	As$^+$	32	4.93	22.92	2.73	2.93	0.99
Se	34	5.33	20.71	4.56	3.25	2.30	Se$^+$	33	4.94	25.27	2.32	2.88	0.96
Br	35	5.33	23.49	3.60	3.16	1.95	Br$^+$	34	4.95	26.92	2.13	2.86	0.88
Kr	36	5.33	25.28	3.21	3.12	1.95	Kr$^+$	35	4.98	29.12	1.89	2.83	0.80
Xe Series													
Rb	37	5.33	8.27	45.96	4.60	5.63	Rb$^+$	36	4.90	31.25	1.65	2.79	0.78
Sr	38	5.33	8.54	44.92	4.61	4.63	Sr$^+$	37	4.98	14.17	11.43	3.71	1.94
Y	39	5.33	9.68	35.32	4.48	4.30	Y$^+$	38	4.99	22.08	4.30	3.25	0.95
Zr	40	5.33	10.72	29.30	4.39	4.09	Zr$^+$	39	5.00	24.22	3.66	3.19	0.87
Nb	41	5.33	11.74	24.92	4.31	3.93	Nb$^+$	40	5.01	26.60	3.11	3.14	0.80
Mo	42	5.33	12.73	21.63	4.24	3.79	Mo$^+$	41	5.02	29.37	2.60	3.08	0.73
Tc	43	5.33	13.70	19.13	4.19	3.69	Tc$^+$	42	5.03	31.07	2.40	3.06	0.67
Ru	44	5.33	14.74	16.88	4.14	3.57	Ru$^+$	43	5.04	32.70	2.24	3.05	0.66
Rh	45	5.33	15.76	15.10	4.09	3.47	Rh$^+$	44	5.04	35.84	1.90	2.99	0.66
Pd	46	5.33	16.79	13.64	4.05	3.38	Pd$^+$	45	5.05	38.45	1.69	2.96	0.59

Xe Series(continued)

Ag	47	5.33	17.79	12.42	4.02	3.30	Ag$^+$	46	5.05	41.37	1.49	2.92	0.54
Cd	48	5.33	18.78	11.41	3.98	3.24	Cd$^+$	47	5.06	29.25	3.50	3.31	1.27
In	49	5.33	18.22	12.77	4.07	3.78	In$^+$	48	5.06	26.08	4.78	3.48	1.17
Sn	50	5.33	20.41	10.20	3.96	3.25	Sn$^+$	49	5.07	26.19	4.95	3.51	1.30
Sb	51	5.33	22.96	8.06	3.84	2.90	Sb$^+$	50	5.07	27.70	4.53	3.48	1.20
Te	52	5.33	24.36	7.30	3.80	2.69	Te$^+$	51	5.08	29.74	4.00	3.44	1.10
I	53	5.33	26.36	6.31	3.74	2.50	I$^+$	52	5.08	31.08	3.76	3.42	1.02
Xe	54	5.33	28.68	5.38	3.67	2.34	Xe$^+$	53	5.09	32.89	3.43	3.39	1.00

Rn Series

Cs	55	5.33	10.01	65.23	5.26	6.31	Cs$^+$	54	5.09	35.01	3.08	3.35	0.98
Ba	56	5.33	9.99	67.68	5.31	5.2	Ba$^+$	55	5.09	16.80	17.74	4.33	2.17
La	57	5.33	10.39	64.04	5.29	5.18	La$^+$	56	5.10	34.66	3.40	3.43	0.89
Ce	58	5.33	10.76	61.07	5.27	5.11	Ce$^+$	57	5.11	37.39	2.96	3.38	0.87
Pr	59	5.33	11.13	58.45	5.26	5.05	Pr$^+$	58	5.11	39.26	2.74	3.35	0.86
Nd	60	5.33	11.49	50.06	5.25	5.00	Nd$^+$	59	5.11	40.98	2.57	3.33	0.84
Pm	61	5.33	11.86	53.84	5.24	4.94	Pm$^+$	60	5.12	42.68	2.42	3.32	0.83
Sm	62	5.33	12.22	51.84	5.23	4.69	Sm$^+$	61	5.12	44.53	2.27	3.31	0.82
Eu	63	5.33	12.59	50.01	5.22	4.84	Eu$^+$	62	5.13	45.88	2.19	3.30	0.80
Gd	64	5.33	12.97	48.13	5.21	4.79	Gd$^+$	63	5.13	47.39	2.10	3.29	0.79
Tb	65	5.33	13.35	46.41	5.20	4.75	Tb$^+$	64	5.13	48.97	2.01	3.28	0.78
Dy	66	5.33	13.73	44.84	5.19	4.70	Dy$^+$	65	5.13	51.79	1.82	3.24	0.77
Ho	67	5.33	14.11	43.35	5.18	4.60	Ho$^+$	66	5.14	52.14	1.85	3.26	0.76
Er	68	5.33	14.49	41.95	5.17	4.62	Er$^+$	67	5.14	53.86	1.77	3.25	0.75
Tm	69	5.33	14.87	40.64	5.17	4.58	Tm$^+$	68	5.15	55.53	1.70	3.24	0.73
Yb	70	5.33	15.25	39.43	5.16	4.54	Yb$^+$	69	5.14	26.35	9.96	4.19	1.85
Lu	71	5.33	16.75	32.6	5.04	4.26	Lu$^+$	70	5.15	38.68	4.10	3.72	0.90
Hf	72	5.33	18.18	27.7	4.93	4.08	Hf$^+$	71	5.15	40.90	3.79	3.67	0.90
Ta	73	5.33	19.53	24.1	4.85	3.94	Ta$^+$	72	5.15	43.44	3.39	3.63	0.84
W	74	5.33	20.79	21.4	4.78	3.82	W$^+$	73	5.15	46.59	2.96	3.57	0.78
Re	75	5.33	22.03	19.2	4.72	3.73	Re$^+$	74	5.15	47.93	2.84	3.56	0.77
Os	76	5.33	23.36	17.2	4.66	3.62	Os$^+$	75	5.16	50.30	2.62	3.53	0.71
Ir	77	5.33	24.63	15.6	4.61	3.53	Ir$^+$	76	5.16	52.82	2.40	3.49	0.70
Pt	78	5.33	25.93	14.2	4.56	3.46	Pt$^+$	77	5.16	55.59	2.19	3.46	0.65
Au	79	5.33	27.14	13.1	4.52	3.39	Au$^+$	78	5.17	58.73	1.98	3.42	0.64
Hg	80	5.33	28.38	12.1	4.48	3.33	Hg$^+$	79	5.17	43.55	4.08	3.80	1.34
Tl	81	5.33	26.95	14.00	4.59	3.93	Tl$^+$	80	5.17	38.66	5.53	3.98	1.24
Pb	82	5.33	29.41	11.7	4.49	3.42	Pb$^+$	81	5.17	38.09	5.87	4.03	1.39
Bi	83	5.33	32.39	9.57	4.37	3.08	Bi$^+$	82	5.17	39.54	5.52	4.01	1.29
Po	84	5.33	33.79	8.88	4.33	2.88	Po$^+$	83	5.17	41.68	5.00	3.96	1.20
At	85	5.33	35.93	7.88	4.27	2.70	At$^+$	84	5.18	42.84	4.81	3.95	1.19
Rn	86	5.33	38.46	6.88	4.20	2.54	Rn$^+$	85	5.18	44.66	4.47	3.91	1.10

BOUNDARY-FORCED SPATIAL CHAOS

by

Víctor M. Eguíluz, Emilio Hernández-García, Oreste Piro

Instituto Mediterráneo de Estudios Avanzados IMEDEA[1]
CSIC-Universitat de les Illes Balears, E-07071 Palma de Mallorca (Spain)

Abstract

We show that the presence of undulated boundaries can induce the formation of spatially chaotic, stationary, and stable structures in models as simple as the Fisher-Kolmogorov equation, which does not display any kind of chaos under common boundaries.

1 Introduction

In the past few decades, considerable understanding of the phenomenon of temporal chaos in dynamical systems of few degrees of freedom has been achieved[1]. On the other hand, spatiotemporal chaos in extended dynamical systems with infinitely many degrees of freedom is currently under very active investigation[2]. It is remarkable however, that an area of problems laying somehow between the two extremes has not received so much attention, namely, purely spatial chaos as a stationary attractor of extended dynamical systems [3, 4, 5, 6, 7, 8, 9].

In the context of fluid dynamics, the existence of spatially chaotic, but temporally steady solutions would also fill a conceptual gap between two well-studied complex phenomena: Lagrangian chaos, and Eulerian chaos or turbulence. The former refers to the chaotic motion of a fluid parcel which might occur in flows that are not necessarily chaotic in their Eulerian description. In fact, a laminar flow may induce chaotic motion for the fluid particles [11, 12, 13], and in three dimensions this is even possible if the flow is steady. On the other extreme, the road to turbulence is usually associated to a hierarchy of increasingly spatiotemporally chaotic Eulerian velocity fields $\mathbf{v}(\mathbf{r})$. Frozen spatial chaos would then refer in this context to a third possibility: a stationary velocity field $\mathbf{v}(\mathbf{r})$ spatially chaotic in the Euler description.

Stationary structures in extended nonlinear dynamical systems in one spatial dimension have been considered in some detail in recent investigations, and spatial chaos found. Rigidly travelling waves with spatial chaotic structure can also be considered as a case of spatial chaos, since it is purely spatial in a moving frame of reference [5, 7]. Such one-dimensional systems are specially suitable to analysis because their steady state configurations, which depend only on the unique spatial coordinate, can be described using results of the theory of low-dimensional dynamical systems, by simply interpreting the spatial coordinate as a fictitious time. In the cases previously studied, spatial chaos appears because the nonlinear dynamical system 'evolving' in

---

[1]URL: http://www.imedea.uib.es/Nonlinear

O. Descalzi et al. (eds.), Instabilities and Nonequilibrium Structures VII & VIII, 205–212.

the space coordinate has dimensionality large enough. The high dimensionality of the equations can be due to a) the presence of high-order spatial derivatives in a single evolution equation as in the cases of Kuramoto-Sivashinsky[8] and Swift-Hohenberg equation [4, 9], b) the coupling of several fields each one satisfying lower order differential equation as in the complex Ginzburg-Landau equation[5] which supports chaotic travelling waves, or c) explicit space dependent forcing terms as in the driven KdV-Burgers [10] equation.

Stationary two-dimensional spatial chaos is basically unexplored at present. In this case application of dynamical system concepts to the analysis of steady state solutions cannot be direct, since we have in principle two equally important spatial coordinates. In addition there is a large variety of possible boundary conditions which surely leads to a variety of steady configurations much richer than in the one-dimensional case. We will be able to show, for a extremely simple model, that rather simple undulated strip-like domain shapes can induce the formation of patterns that are both *spatially chaotic* and *temporally attracting*. We believe that the kind of *modulated boundaries* we propose can be easily implemented in standard experimental pattern formation set-up's such as Faraday waves, convection cells, or open flows. In fact, our work was originally motivated by the observation, in the fluid dynamics experimental setup of a periodic array of pipe bents, that the transversal profile of the steady flow does not necessarily repeat itself with the same periodicity of the array [12].

2   The model and its boundary conditions

The connection between the theory of dynamical systems and the study of stationary spatial configurations of one-dimensional extended systems is direct: A stationary pattern satisfies a system of ordinary differential equations with the spatial coordinate as its independent variable which we can think of as a 'time'.

On the contrary, the cases with two or more spatial dimensions can not be tackled along these lines since conventional dynamical systems theory deals with a single time-variable. There are two-dimensional situations, however, where the two independent spatial directions are naturally differentiated by the geometry of the system and we can interpret one of them as playing the role of the time. Thus, the spatial variation in one direction would be interpreted as time evolution of a one-dimensional field that only depends on the remaining spatial coordinate. Symmetries such as parity in a spatial coordinate will appear as time-reversal symmetry after reinterpretation of this coordinate as a time. Particularly suited to our approach will be the case of two-dimensional extended systems evolving in strip-shaped regions much longer (ideally infinite) in the time-like direction than in the space-like one. If the strip is narrow enough, only patterns composed of one or few transverse spatial modes will be allowed and spatial chaos can be readily defined and identified in terms of the usual concepts of dynamical systems theory.

A first remark to be done about this interpretation is that any explicit dependence on the coordinate along the strip amounts to time dependence of the corresponding dynamical system. In particular, if the lateral boundaries of the strip are undulated, the corresponding boundary conditions will translate into time-periodic forcing.

To focus our attention on spatial chaos induced by boundary effects, we consider a very simple model equation containing only up to second order derivatives and a single field variable known as the Fisher-Kolmogorov (FK) equation:

$$\partial_t \psi = \nabla^2 \psi + a\psi - \psi^3 \tag{1}$$

where $\psi(x, y, t)$ is a real field and ∇^2 is the two dimensional Laplacian operator. The real coefficient a in the linear term could be scaled out, but we find convenient to keep it explicit in the equation. The FK equation appears in several contexts ranging from phase transitions (under the name of real Ginzburg-Landau equation, or time-dependent Ginzburg-Landau model [14]) to population dynamics and ecology. Equation (1) has been extensively studied in one and two dimensions.

In one dimension, for systems large enough, most initial field distributions evolve into configurations made of domains where the field takes values near either ψ_+ or ψ_-, where $\psi_\pm = \pm\sqrt{a}$. These domains are separated by kink or anti-kink-type walls that can move into each other leading to mutual annihilation. By this mechanism, small domains disappear feeding the growth of larger domains whose sizes then increase logarithmically in time until only one of the stationary homogeneous solutions, ψ_+ or ψ_-, prevailing by chance, takes over the whole system.

In two dimensions, the evolution of a large system leads to coarsening of domains containing either the ψ_+ or the ψ_- phases, with typical lengths growing as the square root of time[14]. In addition to the weak wall interaction, present as before, the most important driving force for evolution is the tendency to minimize domain wall length. Independently of the dimensionality, the dynamics of (1) is a pure (non inertial) relaxation seeking a minimum of a functional potential[15]. This implies that the asymptotic states can only be fixed points in the functional space, and no limit cycle oscillations nor more complicated attractors can exist.

It was shown by Collet [16] in a more general context that the time evolution and asymptotic states of Eq. (1) in either one or two dimensions are similar to those of an infinite system except within a boundary layer of a size that depends on the a parameter. Therefore, in order to observe the influence of boundaries on pattern evolution we need to consider a domain small enough at least in one of its dimensions. We will consider a strip-shaped domain, elongated in the x direction, which will be called the *longitudinal* direction, so that the small dimension is the y *transversal* direction. To be concrete, our domain will be limited by the curves $y_0(x)$ and $y_1(x)$, where we impose null Dirichlet conditions (that is $\psi(x, y_0(x), t) = \psi(x, y_1(x), t) = 0$). The choice of Dirichlet boundary conditions is just for convenience but it is not essential for our conclusions. In the one-dimensional case, viewing the stationary solutions of Eq. (1) as orbits of a dynamical system, we immediately rule out chaotic configurations since the dynamical system is just an ordinary differential equation of second order. Chaos can appear, however, if some x-dependent periodic forcing is added to the equation. This argument does not apply directly to the two-dimensional case, but it suggests that lateral undulated boundaries (similar to a time-periodic forcing) could induce spatially chaotic structures.

As a particular case we will study the following periodically modulated boundaries:

$$y_1(x) = \frac{d_1}{2}(1 - \cos(\alpha x)), \quad y_0(x) = -1 - \frac{d_0}{2}(1 - \cos(\alpha x + \phi)) \tag{2}$$

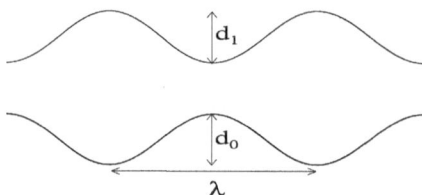

Figure 1: Symmetrically oscillating channel: $\alpha = \frac{2\pi}{\lambda}$; $d_0 = d_1$, $\phi = 0$.

Here d_1, d_0 are the amplitudes of each boundary, α is the spatial frequency, which is the same for both boundaries, and ϕ is the phase lag between both boundaries. The case $d_1 = d_0$ and $\phi = 0$ gives a channel with symmetrically oscillating width (Fig. 1) whilst if $d_0 = 0$ one boundary is flat. Another interesting case is $d_1 = d_0$ and $\phi = \pi$ that corresponds to having the boundaries in phase, so that the domain has a constant transversal width and undulates along the longitudinal coordinate. Here we will concentrate on the symmetrically oscillating channel, shown in Fig. 1. We stress that we try to remain in the simplest situation that will display spatial chaos. Consideration of more complex models such as the Kuramoto-Sivashinsky or the Swift-Hohenberg equations, which show spatial chaos already with simple boundaries, or more complicated boundaries having for example two incommensurate undulation periods for the upper and lower boundaries (corresponding to a quasi-periodic forcing) could only increase the complexity of stationary solutions.

Boundary conditions in the longitudinal x direction remain to be specified. The analogy with a temporal variable would be the strongest if the region were infinite in that direction, with the only requirement of boundness of the solution. However, an infinite domain is not adequate for the numerical studies that will follow. As a consequence we need to impose periodic boundary conditions (of period L) along the x direction. In this way we are restricting the class of solutions to periodic orbits of period L or less in the time-like x-coordinate. However we can still identify as related to spatial chaos configurations that have the maximal period L, provided this period increases as system size L increases, and the periodic orbits approach a chaotic trajectory.

A convenient way to handle the boundaries consists in mapping the region limited by $y_0(x)$ and $y_1(x)$ (and by $x = 0, L$) to a rectangular one: $\tilde{y}_1 = 1$, $\tilde{y}_0 = 0$ and $x = 0, L$. The map $(x, y) \longmapsto (x, \tilde{y})$ for arbitrary functions $y_0(x)$ and $y_1(x)$ transforms Eq. (1) into an equation for $\tilde{\psi}(x, \tilde{y}, t) = \psi(x, y, t)$ satisfying:

$$\partial_t \tilde{\psi} = \partial_{xx}^2 \tilde{\psi} + \mathcal{F}(x)\partial_{\tilde{y}\tilde{y}}^2 \tilde{\psi} + \mathcal{G}(x)\partial_{x\tilde{y}}^2 \tilde{\psi} + \mathcal{H}(x)\partial_{\tilde{y}}\tilde{\psi} + a\tilde{\psi} - \tilde{\psi}^3 \tag{3}$$

where $\mathcal{F}$, $\mathcal{G}$, $\mathcal{H}$ depend on the boundary shape $y_{0,1}$.

The new transversal boundary conditions are

$$\tilde{\psi}(x, \tilde{y} = 0) = \psi_0(x) = 0, \quad \tilde{\psi}(x, \tilde{y} = 1) = \psi_1(x) = 0 \tag{4}$$

Inspection of Eq. (3) tells us that the effect of the boundaries is reflected in the new variables as a parametric forcing.

Making the right hand side of (3) equal to zero to seek for stationary solutions, and thinking of x as time we can look at it as a nonlinear equation for the temporal evolution of a one-dimensional pattern, with a 'time'-periodic parametric forcing due to the boundaries. The present knowledge about spatiotemporal chaos can be used to analyze such dynamics. In particular cases some simple approximations [17] can be carried out to reduce the problem to a simpler one without explicit space dependence.

We could have also performed a conformal transformation, one which conserves the orthogonality of the coordinate lines. The advantage of this kind of transformations is that the Laplacian transforms to itself in the new coordinates but with a space dependent coefficient, i.e. if T is a conformal transformation then $T : \nabla^2 \longmapsto a(x,y)\nabla^2$. Even in this representation the main conclusion would be the same: the non-trivial boundaries affect the dynamical evolution through parametric forcing terms.

3 Numerical Results

We have performed direct two-dimensional simulations of Eq. (1) in the domain of Fig. 1. Periodic boundary conditions in the longitudinal direction restrict the values of the possible values of the frequency of the boundary oscillation: $\alpha = \frac{2\pi}{L}n$, being n an integer.

An example of the typical behavior is shown in Fig. 3, where a two-dimensional frozen structure is presented for parameter values $d = 1.0$, $\alpha = \pi$, $a = 20$, $\phi = 0$. The solution consists in the irregular alternation of two phases $\psi_\pm$ along the 'time-like' longitudinal direction. Calling 'spatially chaotic' this irregular alternation may raise some doubts. In fact since we are using periodic boundary conditions in the longitudinal direction the configuration shown in Fig. (3) is periodic with the maximum period L. However, one can associate to each configuration a sequence of 0's and 1's by taking values of the field at the periodicity $\lambda = \frac{2\pi}{\alpha}$ imposed by the boundaries (in the same way as one would get a stroboscopic map in the case of a 'external' periodic forcing): if ψ is positive, then one takes the value 1, if negative, 0. For example the trajectory in Fig. (2) can be read as '111111100101101000001011'. Our claim is that whatever sequence one could write, it is realizable in this system by only changing the initial condition. Longer sequences would be realized in systems with larger L. This establishes the 'spatially chaotic' character of the typical configurations. Other indicators such as the fractal character of Poincaré maps will be discussed in [17]. In the limit of very large systems one could in principle look for Lyapunov exponents and correlation integrals [9].

Our claim can be proved when the domain is a small perturbation of a rectangular one, so that approximations leading to an evolution equation for a single spatial mode can be obtained [17]. In the general case the following physical argument gives an heuristic justification for it: The tendency of the system is to be in a single phase, so that a Ginzburg-Landau energy is minimized [15]. But initial conditions lead to domains of the two phases in competition, with domain walls between them. The tendency to minimize the free energy has two effects: On the

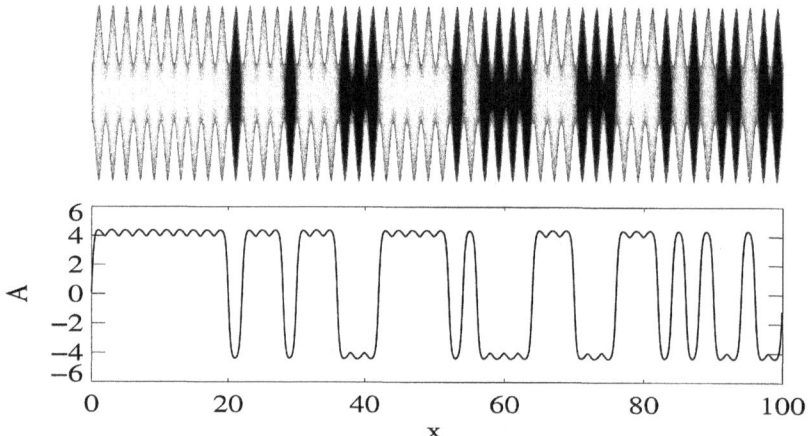

Figure 2: Two-dimensional simulation of the system (1) starting from random initial conditions. The amplitude of the field in the center of the domain is also shown. Parameter values: $d = 1.0, \alpha = \pi, a = 20, \phi = 0$

one hand the walls interact attractively and tend to annihilate by pairs. On the other hand its length tends to be minimal, so that, in our geometry, the walls tend to stay in the places where the distance between the upper and lower boundaries is minimal. If the distance between such places and the amplitude of the boundary oscillations are large enough, this pinning tendency will stop wall annihilation and stabilize configurations such as the one in Fig. 2. This seems to be the origin of the stable disordered structures found. Other geometries, where the distance between both boundaries varies in space, can lead to further interesting behavior.

4 Conclusions

We have shown that an undulated boundary can induce stationary, stable, longitudinally chaotic solutions in a model as simple as the Fisher-Kolmogorov equation, which does not display any kind of chaos under common boundary conditions. We can talk about boundary-forced or boundary-induced structures. Spatial chaos should be present in more complex systems, and changing the shape of the boundaries could be a useful way of controlling it.

We are investigating the possibility that these spatially chaotic structures appear in low Reynolds-number flows. Related work on nonlinear extended dynamics influenced by boundaries can be found in [18].

Acknowledgements

We acknowledge financial support from MCyT (Spain) projects CONOCE BFM2000-1108 and BFM2002-04474-C02-01.

References

[1] H. Bai-Lin, *Chaos* (World Scientific, Singapore 1990).

[2] M.C. Cross, P.C. Hohenberg, Science **263**, 1569 (1994)

[3] P. Coullet, C. Elphick, and D. Repaux, Phys. Rev. Lett. **58**, 431 (1987).

[4] E. Hernández-García, M. San Miguel, R. Toral, and J. Viñals, Physica D **61**, 159 (1992).

[5] R. Montagne, E. Hernández-García, M. San Miguel, Phys. Rev. Lett. **77**, 267 (1996); R. Montagne, E. Hernández-García, A. Amengual, M. San Miguel, Phys. Rev. **E 56**, 151 (1997).

[6] N.J. Balmforth, Annu. Rev. Fluid Mech. **27**, 335 (1996).

[7] H-C. Chang, Annu. Rev. Fluid Mech. **26**, 103 (1994).

[8] Y.A. Demekhin, G.Yu. Tokarev, V.Ja. Shkadov, Physica **D 52**, 338 (1991)

[9] M.I. Rabinovich, A.L. Fabrikant, and L. Sh. Tsimring, Sov. Phys. Usp. **35**, 629 (1992).

[10] Malkov, Physica D **95**, 62 (1996).

[11] S.W. Jones, O.M. Thomas, H. Aref, J. Fluid Mech. **209**, 335 (1989)

[12] Y. LeGuer et al., *Experimental study of chaotic advection in a twisted duct flow*, preprint.

[13] J.H.E. Cartwright, M. Feingold, and O. Piro, in *Proceedings of the NATO ARW "Mixing, Chaos, and Turbulence"*, H. Chaté and E. Villermaux, editors, (1997).

[14] J. D. Gunton, M. San Miguel and P.S. Sahni, *The dynamics of first order phase transitions* in *Phase Transitions and Critical Phenomena, vol. 8*. C. Domb and J. L. Lebowitz, Eds. (Academic Press, New York, 1983).

[15] R. Montagne, E. Hernández-García, M. San Miguel, Physica **D 96**, 47 (1996); M. San Miguel, R. Montagne, A. Amengual, E. Hernández-García, in *Instabilities and Nonequilibrium Structures V*, E. Tirapegui and W. Zeller, Eds. (Kluwer Academic Publishers, Dordrecht, 1996).

[16] P. Collet, Nonlinearity **7**, 1175 (1994).

[17] V.M. Eguíluz, E. Hernández-García, O. Piro, S. Balle, Phys. Rev. E **60**, 6571 (1999).

[18] V.M. Eguíluz, P. Alstrøm, E. Hernández-García, O. Piro, Phys. Rev. E **59** 2822 (1999); V.M. Eguíluz, E. Hernández-García, O. Piro, Int. J. Bif. Chaos **9**, 2209 (1999); *ibid.*, Physica A **283**, 48 (2000); *ibid.*, Phys. Rev. E **64**, 036205 (2001); I. Sendiña-Nadal, V. Pérez-Muñuzuri, V.M. Eguíluz, E. Hernández-García, O. Piro, Phys. Rev. E **64**, 046208 (2001).

CHAOTIC TOY MODEL OF MOBILE BEDS IN NATURAL CHANNELS

by

Pablo M. Jacovkis[1,2], Walter E. Legnani[1] and Osvaldo A. Rosso[1]

[1]: Instituto de Cálculo, Facultad de Ciencias Exactas y Naturales,
Universidad de Buenos Aires, Pab. II, Ciudad Universitaria, 1428 Buenos Aires, Argentina.
[2]: Departamento de Computacíon, Facultad de Ciencias Exactas y Naturales,
Universidad de Buenos Aires, Pab. I, Ciudad Universitaria, 1428 Buenos Aires, Argentina.

Abstract

The aim of this work is to formulate a dynamical model of mobile beds in natural channels. The proposed model is based on a simplified form of the rivers hydrodynamic equations and is similar to the generalized logistic map. As a natural consequence, the two parameters of the model are the control variables and they are related to properties of the real system. The qualitative regime of the phenomena has been totally reproduced by the proposed model.

1   Introduction

To study the unsteady hydrodynamic flow in open shallow-water channels we must often consider that the bed is not fixed. In general, the time evolution of the bed height of a natural channel, at each point of the channel, will be a function of the characteristics of the bed material, the fluid velocity, the water depth and the geometry of the channel, among other variables. Besides, the dynamics of the channel bed will be directly influenced by the transport, the settling and the resuspension of the bed particles. The physical processes involved in this phenomenon are very complicated and are not well understood at the moment, so that many descriptions of it include empirical o semi-empirical laws. For a review in this field we suggest the work of Raudkivi [1].

The main research lines in the field may be identified as: *i)* du Bois type models based on the shear stress; *ii)* formulations based on probability concepts, and *iii)* others which consider as starting point the work done by the fluid on the mobile bed channel. However, all of these formalisms include parameters which must be fitted to each particular case and for each flow regime.

In the present work we propose a simplified nonlinear dynamical model for the behavior of the mobile bed in natural channels. This model, that had not been applied yet, allows us to describe all the observed dynamical behaviors in real systems by choosing the values of their parameters in appropriate form. The simplicity of the model and its similarity to the general logistic map, whose characteristics are well known, render this model a powerful tool to obtain with expediency the main characteristics of the mobile bed in a channel.

O. Descalzi et al. (eds.), Instabilities and Nonequilibrium Structures VII & VIII, 213–218.

2 The Model

We consider a one-dimensional open channel flow, with surface width B and length L ($L \gg B$) as shown in Fig. 1. We denote by h the water depth measured from the bed of the channel; by S the cross section to the flow; by Q the discharge and by e the height of the mobile bed with respect to an arbitrary plane of reference. Then, the stage of the free surface of the channel will be $z = h + e$.

To study the time evolution of the channel bed height, we discretize the problem in cells of length L_i and height z_i and discrete times $t^{(n)}$, with time step $\Delta t = t^{(n+1)} - t^{(n)}$. We also consider the additional simplification that the height z is almost constant for all the channel length.

Then, for the i-th cell the height of the bed and the water depth will be e_i and h_i, respectively (see Fig. 2). At time $t = n \cdot \Delta t$ the mean bed height in the i-th cell could be described by the erosion-sedimentation rate T_i. This rate involves the following process: $a)$ the suspension and sedimentation in the i-th cell, $b)$ the bed contribution balance of the solid inflow and outflow of the nearest neighbour cells, that is, the contributions of the $i - 1$ and $i + 1$-th cells. Then

$$T_i = \frac{R_i^{(n+1)}}{e_i^{(n)}} \, , \tag{1}$$

where $R_i^{(n+1)}$ and $e_i^{(n)}$ are the contribution of both process above mentioned to the bed height at time $(n + 1) \cdot \Delta t$, and the bed height at time $n \cdot \Delta t$ in the i-th cell respectively.

For the height $R_i^{(n+1)}$ we can write:

$$R_i^{(n+1)} = e_i^{(n+1)} - a_i^{(n)} \cdot e_i^{(n)} = e_i^{(n+1)} - (a_{i+1}^{(n)} - a_{i-1}^{(n)}) \cdot e_i^{(n)} \, , \tag{2}$$

where $e_i^{(n+1)}$ is the new height of the bed due to the erosion and sedimentation process in the i-th cell and the second term represents the transference of the bed mobile material in the i-th cell to the nearest neighbour cells at time $n \cdot \Delta t$ and $a_i^{(n)}$ is the transfer ratio ($-1 \le a_i^{(n)} \le 1$). Transfer ratio $a_i^{(n)} < 0$ means that the behavior of the i-th cell could be consider as source of bed material for the two nearest neighbours and $a_i^{(n)} > 0$ implies sink behavior.

Following Chang and Richards [2], the mean sediment concentration C_s is defined by the following functional relation

$$C_s = \frac{k \cdot V^m \cdot h^n}{w \cdot g} \, , \tag{3}$$

where k is the coefficient of sediment transport capacity, w is the mean fall velocity of the sediment, V is the mean flow velocity in the direction of flow discharge and g is the gravity aceleration. The values of constant exponents m and n are typically taken as 3 and -1 respectively.

Normalizing the height of the free surface of the channel z by a characteristic height H, ($z/H = 1$) and taking the usual values for the constants m and n, eq. (3) can be written as

$$C_s = \frac{k' \cdot V^3}{w \cdot g \cdot (1 - e')} \, , \tag{4}$$

where $k' = k \cdot H$ and $e' = e \cdot H$.

Making the Ansatz that for the i–th cell the erosion sedimentation rate, T_i, is directly proportional to the normalized coefficient of sediment transport capacity k_i',

$$T_i = \alpha \cdot k_i' \tag{5}$$

with α the proportionality constant, from eqs. (1) and (4) we can write for the height of the mobile bed material (in dimensionless form) at time $(n+1) \cdot \Delta t$ the following recurrence:

$$e_i^{(n+1)} = \xi_i \cdot e_i^{(n)} \cdot (1 - e_i^{(n)}) + a_i^{(n)} \cdot e_i^{(n)}, \tag{6}$$

where $\xi_i = (\alpha \cdot g \cdot w_i \cdot C_s^i)/V_i^3$; without loss of generality we have chosen $H = 1$.

Note that the above expression is the well known generalized logistic model [3]. Making the change of variables

$$y^{(n)} = \frac{\xi_i \cdot e_i^{(n)}}{\xi_i + a_i^{(n)}} \tag{7}$$

where

$$\rho = \xi_i + a_i^{(n)}, \tag{8}$$

the above recurrence relation can be written as

$$y^{(n+1)} = \rho \cdot y^{(n)} \cdot (1 - y^{(n)}) \tag{9}$$

which is the logistic parabola. As usual the parameter ρ takes values in the interval $[0, 4]$.

3 Results and Discussion

Our model is based on the Ansatz that T_i, the erosion sedimentation rate, is proportional to $C_s(k, V, z)$, the mean sediment concentration. In particular for $C_s(k, V, z)$ we adopted the functional dependence given by Chang and Richards [2] (see equation (4)). To validate this functional dependence for C_s, we solved the problem of mobile bed for one-dimensional shallow water equations by the method of the characteristics, with a time step of 60 sec and space grid of 450 m. The total reach length was 20 km, and the total simulation time was 30 min. The obtained results are showed in the Fig. 3. From this figure, we can conclude that the predicted and experimental values are in very good agreement.

Numerical simulations for our model are presented in Fig. 4. In all the cases, the initial bed height was $e_i^{(0)} = 0.5$. The time step Δt is given by $\Delta t = L_i/V$, where L_i is width of the i-th cell, and V is the fluid velocity.

For $\rho = 1$ we are in presence of an erosion process behavior. That is, the fluid drags the bed particles and the final height will be zero. This regime is characterised by a flow velocity higher than the minimum shear velocity necessary to give away bed material.

For $\rho = 2$ we have one state characterised by a constant height. This behavior corresponds to a regime where the shear stress is not strong enough to remove particles of the bed.

For $\rho = 3$ an oscillatory behavior is observed. At the beginning the height is increased up to 0.75 and then it presents regular fluctuation which could be associated with a weak turbulence effect. Finally for $\rho = 3.75$ we have a chaotic behavior with fast resuspension and sedimentation process due to the complete developed turbulence regime.

In conclusion, although our model is very simple and can be thought of as a toy model, it reproduces all the observed behaviors of mobile bed in natural channels. Moreover the involved parameters have a clear physical meaning.

Acknowledgements

This work was partially supported by CONICET and UBA, Argentina. Two of the authors (W.L. and O.A.R.) are very grateful to the organiser of the workshop for their very kind hospitality during their stay in Chile.

References

[1] A. J. Raudkivi, *Loose Boundary Hydraulics*, Pergamon Press, New York, 1990.

[2] F. F. Chang, D. L. Richards, Deposition of Sediment in Transient Flow, *Journal of Hydraulics Division*, ASCE 97, HY-6, 1971, 837-849.

[3] H. G. Schuster, *Deterministic Chaos*, 2nd. edition, VCH Verlagsgesellschaft, Weinheim, 1988.

E-mail:

jacovkis@ulises.ic.fcen.uba.ar
walter@ulises.ic.fcen.uba.ar
rosso@ulises.ic.fcen.uba.ar

a)   b)

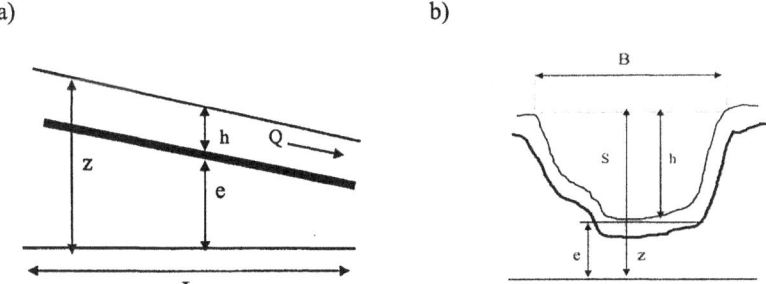

Fig. 1: a) Longitudinal and b) cross section of a natural channel.

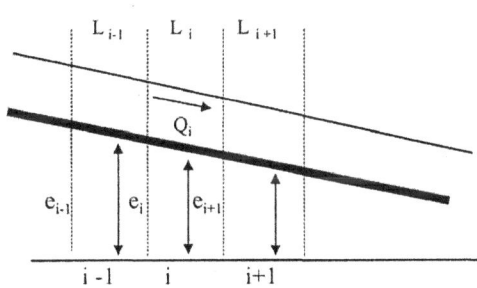

Fig. 2: Cells of a one-dimensional channel.

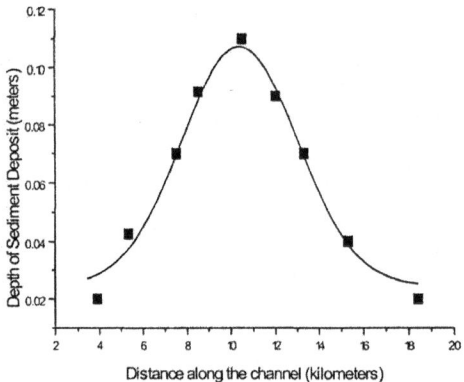

Fig. 3: Sediment load function C_s(k,V,h,e). Simulation (line). Experiment (bullet).

a)   b)

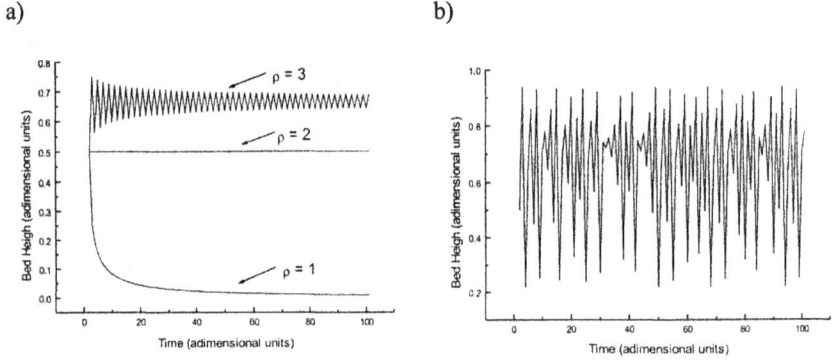

Fig. 4: Bed high time evolution for a) rho=1; rho=2; rho=3 and b) rho= 3.75.

Chaotic motion and the classical-quantum border

A.M. Kowalski, and A.N. Proto

Dpto. de

Computación, Facultad de Ingeniería,

Universidad de Buenos

Buenos Aires Scientific Research Commission (CIC)

Abstract

Based on a two quantum dynaminal invariants of motion, I, related with the Uncertaity Principle and E_r, adimensional and associated to the energy of the system, we study the classical-quantum transit of a semiclassical hamiltonian whose purely classical counterpart exhibits chaotic motion. The transit (no assumption concerning sizes or masses are done) between quantum non-chaotic to the classical chaotic regime is shown. Particularly, through E_r we define the threshold above which quantum chaos appears, and the interval during which both regimes co-exist.

(PACS: 03.65.Sq;05.45.Mt;89.70.+c) KEYWORDS: Semiquantum Chaos, Uncertainty Principle

I. INTRODUCTION

The definition of classical chaos is normally done in terms of the complexity of the trajectories in phase space, and its sensitivity to small changes in the initial conditions. Poincare´ surfaces give a quick look of the behaviour of the system. For quantum systems it is well known that chaotic motion can not be seen in Hilbert space. However, when a quantum and a classical systems are coupled chaotic motion of the the quantum state can

O. Descalzi et al. (eds.), Instabilities and Nonequilibrium Structures VII & VIII, 219–235.

220

be expected. These kind of systems give rise to a great variety of problems englobed in the common field of semi quantum chaos.

Although it is easy to show, from the mathematical point of view, that these semi-quantum (or semiclassical) systems should exhibit chaotic motion, there is not a clear and definite answer about how they are to be treated, considering that in general all the proposed calculations imply an approximation in one of both counterparts: or the approximation is made on the quantum system or in the classical one.

More fundamental are the questions about the emergence of classical chaos from a quantum system, as so as the study of the quantum to classical transition. The normal procedures to address these problems use one of the limits: or $\hbar \to 0$ (the semiclassical limit), or $t \to \infty$ (the late-time limit necessary to describe chaos). Unfortunately, the noncommutativity of these twin limits prevents the elucidation of the fundamental mechanism that makes classical chaos to emerge from a quantum system. Even when a great ammount of work have been done, a quantifier of "quantum chaos" has not been found. On the basis of the above open problems, it should be clear that the analysis of semi-classical models is a task which deserves renewed interest (see, also, for instance, and references therein).

It is the aim of this contribution to present a method to deal with these kind of systems, where a quantum and a classical systems are coupled. Our procedure is based on previous works related with an information theoretical treatment of hamiltonian systems [1–4] (and references therein). Under our formalism, [5–7], which we briefly outlined below, we do not make any approximation on any of the counterparts (the classical or the quantum ones) neither make use of the $\hbar \to 0$, or $t \to \infty$ limits. We concentrate our efforts upon the following Hamiltonian [8]

$$\hat{H} = \frac{1}{2}[\frac{\hat{p}^2}{m_q} + \frac{P_A{}^2}{m_{cl}} + m_q\omega^2\hat{x}^2],\tag{1}$$

where $\hat{x}$ and $\hat{p}$ are quantum operators, A y P_A *classical canonical conjugate variables* and

$$\omega^2 = \omega_q{}^2 + e^2 A^2,\tag{2}$$

where ω_q is a frequency, and m_q and m_{cl} the masses corresponding to the quantum and classical systems, respectively. Let us set (in convenient units)

$$m_q = 1, \tag{3a}$$

$$m_{cl} = 1, \tag{3b}$$

$$\omega_q = m, \tag{3c}$$

We will deal with a non conventional invariant of motion, [7], closely related to the Uncertainty Principle, in terms of which *we will show that it is possible to go continuously from a chaotic-classical regime to a "quantal" one.* In other words, it will be seen that the classical-quantum border is smoothly traversed.

II. THE FORMALISM

Following previous works, [1–3] we focus our attention upon q relevant operators $\hat{O}_j$ that are able to characterize the dynamics generated by the hamiltonian (1) in sufficient detail if we know their expectation values as a function of the time [5]. The trick is to select our quantum observables so that they satisfy the closure relations [1–3]

$$[\hat{H}(t), \hat{O}_j] = i\hbar \sum_{i=1}^{q} g_{ij}(t)\, \hat{O}_i; \;\; j = 1, 2, \ldots, q, \tag{4}$$

so that their evolution can be followed in exact fashion. The g_{ij} are the elements of a q x q matrix G with whose help we will find invariants of the motion (see [6,7] for more details). On account of (4) the temporal evolution of the expectation values (EVs) is given by the generalized Ehrenfest relationships [3]

$$\frac{d\langle \hat{O}_j \rangle}{dt} = -\sum_{i=1}^{q} g_{ij}(t)\langle \hat{O}_i \rangle, \; j = 1, 2, \ldots, q, \tag{5}$$

The generalized Ehrenfest theorem [1–3,12] yields in this case a set of first-order differential equations for the temporal evolution of the EVs of our q relevant operators, which, in turn, will, *for our Hamiltonian (1)*, depend on the classical ones through the g_{ji} elements of the matrix G [1,2,4]. The time evolution being canonical, all commutation-relations are trivially conserved for all times [5].Following standard procedures (i.e. [11]) the energy is taken to coincide with the quantum expectation value of the Hamiltonian, that in turn generates the temporal evolution of the classical variables.

The equations of motion for the classical variables, the position A and the momentum P_A, are given by the Poisson parenthesis (see for more details for instance [1,2])

$$\frac{dA}{dt} = \{\langle \hat{H} \rangle, P_A\}, \tag{6a}$$

$$\frac{dP_A}{dt} = \{\langle \hat{H} \rangle, A\}. \tag{6b}$$

The systems of equations given by (6-5), and also by (9) (to be presented below) configure autonomous sets of first-order coupled differential equations of the form

$$\frac{d\vec{u}}{dt} = \vec{F}(\vec{u}), \tag{7}$$

where $\vec{u}$ is a generalized variable (a "vector" with both classical and quantum components). This system is obviously conservative [1–3]. Both types of dynamics, the classical and the quantal ones, are derived from a Hamiltonian.as the divergence of $\vec{F}$ is easily seen to vanish: i) Eqs. (6a) are divergenceless and ii) the matrix G of the set of equations derived from (5) is traceless (on account of the canonical nature of (5)). As a result, our system is a conservative one. Of course, the remaining (quantal) subsystem of the total system should be such that relations of the type (4) hold.

III. DYNAMICS OF THE MODEL AND INVARIANTS OF MOTION

The closure relationship (4) can be applied to the hamiltonian we are dealing with, (1) using $\{\hat{x}, \hat{p}\}$ or $\{\hat{x}^2, \hat{p}^2, \hat{L} = \hat{x}\hat{p} + \hat{p}\hat{x}\}$, or both, as the relevant sets. The first set leads to the Heisemberg group, and the second to a S(1,1) algebra. Both sets are not related through the dynamics, and taken into account that we are dealing with a semiclassical (or semiquantal) system, we select the second one as it carries out information about the Uncertaity Principle

(UP), as we have shown in [9]. The basic commutation relation $[\hat{x}, \hat{p}] = i\hbar$ holds for any time t. The generalized Ehrenfest theorem (5) applied to

$$\hat{H} = \frac{1}{2}[\frac{\hat{p}^2}{m_q} + \frac{P_A{}^2}{m_{cl}} + m_q(\omega_q{}^2 + e^2A^2)\,\hat{x}^2],\tag{8}$$

gives ·

$$\frac{d\langle\hat{x}^2\rangle}{dt} = \frac{\langle\hat{L}\rangle}{m_q},\tag{9a}$$

$$\frac{d\langle\hat{p}^2\rangle}{dt} = -m_q\,(\omega_q{}^2 + e^2A^2)\,\langle\hat{L}\rangle,\tag{9b}$$

$$\frac{d\langle\hat{L}\rangle}{dt} = 2\left(\frac{\langle\hat{p}^2\rangle}{m_q} - m_q(\omega_q{}^2 + e^2A^2)\langle\hat{x}^2\rangle\right),\tag{9c}$$

$$\frac{dA}{dt} = \frac{P_A}{m_{cl}},\tag{9d}$$

$$\frac{dP_A}{dt} = -e^2 m_q\,A\langle\hat{x}^2\rangle,\tag{9e}$$

with ω^2 given by Eq. (2). These equations give the time-evolution of our classical and quantal relevant variables. So, we have been *translated our "quantum-classical" coupled problem into the language of a classical dynamical system.* We did not appeal to any consideration about the wave function, and the quantum temporal evolution of our relevant variables are taken in mean value, making it easy to be compare with their classical counterpart. Also notice that the above equations are independent of $\hbar$. Of course, if the the variables A y P_A were of a quantal nature (operators), the concomitant problem would lead to a nonclosed semi-algebra. No exact solution would then be available. Observing the equations it is easy to see that

$$I = \langle\hat{x}^2\rangle\langle\hat{p}^2\rangle - \frac{\langle\hat{L}\rangle^2}{4},\tag{10a}$$

$$E = \frac{1}{2}[\frac{\langle\hat{p}^2\rangle}{m_q} + \frac{P_A{}^2}{m_{cl}} + m_q(\omega_q{}^2 + e^2A^2)\,\langle\hat{x}^2\rangle],\tag{10b}$$

are *invariants of the motion* [6]. The first one, I, is in fact a time-dependent invariant of motion, as there is no any quantal operator associate to it, althougth $dI/dt = 0$ for all t, as the result of using the Ehrenfest theorem. The other is the energy of the system, which

224

always conmutes with itself, and besides it is trivial to see that $dE/dt = 0$. In order to analyze a little bit the invariant I , let us write th UP in the complete form (see i.e. [12])

$$\left[\langle \hat{x}^2 \rangle - \langle \hat{x} \rangle^2\right]\left[\langle \hat{p}^2 \rangle - \langle \hat{p} \rangle^2\right] \geqq \frac{1}{4}\left(\langle [\hat{x},\,\hat{p}]_+\rangle^2 + \hbar^2\right) \tag{11}$$

where $[\hat{x},\,\hat{p}]_+,$ is the anticonmutator, and it is equal to $\hat{L}$. For the particular case that $\langle x \rangle = \langle p \rangle = 0$ for all t (e.g. a pure quantum H.O.) our time-dependent invariant of motion is obtained. So, I, is in fact the "basic or elementary" invariant for this kind of systems as it can be related with the U.P straightfowardly (see also [9]).

For the classical case, the hamiltonian is

$$H = \frac{1}{2}[\frac{p^2}{m_q} + \frac{P_A{}^2}{m_{cl}} + m_q(\omega_q^2 + e^2 A^2)x^2], \tag{12}$$

and applying the Poisson brackets, the pertinent and equivalent classical equations are obtained, and the invariant, I, is now identically to zero.

Concerning the energy, again it is easy to prove that $dE/dt = 0$ as before, provided the equations are formally the same. The classical Hamiltonian is a chaotic one and was previously exhaustively analyzed [8]. We are now in position to analyze the transition between the quantum and classical-chaotic regimes, which is the main idea of this contribution.

IV. I-ANALYSIS OF THE CLASSICAL-QUANTUM BORDER AND CHAOTIC MOTION

A. General features

As we have seen before, for the classical regime $I = 0$, but as the UP says (11), this value cannot be available for the quantum case. The minimum value of our invariant for the quantum case is $\hbar/4$. So, we can ascertain that

$$I = 0, \quad classical \ \ regime \tag{13}$$

$$I \geqq \hbar/4, \quad quantum \quad regime \qquad (14)$$

In order to study the transit between both regimes we have the following tools: The temporal evolution of the relevant observables, two invariants of motion and the initial conditions. Particularly, the initial conditions for the relevant operators for the quantum case should be chosen according with the UP. This is a central point for the quantum case: always a set a coherent initial conditions is required (see i.e. [10]).

I also has an (energy-dependent) upper bound as well, as it is easily seen from Eqs. (10b), by considering A, $\langle \hat{L} \rangle$, and $\langle \hat{x}^2 \rangle$ as independent variables

$$\frac{\hbar^2}{4} \leq I \leq \frac{E^2}{\omega^2} - \frac{\langle \hat{L} \rangle^2}{4}. \qquad (15)$$

so that I can varies within the above range. Particularly, the $I = \hbar^2/4$-value was previously analyzed [8]. This value corresponds to minimum uncertainty (coherent state [9]). A chaotic behaviour was found there [8], and so, one may wonder whether by *increasing I , which is equivalent to make UP more evident*, one may not be able to observe a gradual vanishing of the chaotic features. In fact, our answer to this question is positive. *It is possible to show a gradual vanishing of the chaotic regime, which in fact means that the classical-quantum border is crossed.* The border lies somewhere in the interval (15), and I tells us whether we face a classical-chaotic or a quantal regime.

In order to arrive to our results, we applied the following recipe:

a) We fixed an E-value and then I- values laying into (15). This step determines segment of interest in the phase space ;

b) For each of the I-values which fulfill (15) (and used in the numerical calulations) we selected an initial value for A (which fixes ω in (15)), and then a couple of initial values for $\langle \hat{L} \rangle$ and $\langle \hat{x}^2 \rangle$ in agrement with the selected I-value. With this step we are able to follow those trayectories whose initial conditions satisfy UP.

c) We let the system evolve and draw graphs (Poincaré's surfaces of section) $\langle \hat{L} \rangle$ versus $\langle \hat{x}^2 \rangle$, and $\langle \hat{p}^2 \rangle$ versus $\langle \hat{x}^2 \rangle$ for different E, I pairs.

Our numerical results show that classical system exhibits *always chaotic behavior*. As I grows, i.e., the uncertainty grows, we gradually leave the purely classical realm and enter a new regime. The point to be stressed in this connection is that chaoticity does not vanish in sudden fashion. Notice that for (numerical) simplicity's sake, we are using Eqs. (3a) and $\hbar = 1$. In Figs. 1 and 2 we show the Poincare's sections. The UP is clearly seen in Fig.1 as the hyperbolic limiting curve is that which correspond to a simple pure quantum H.O. Particularly, the hyperbola "disappears" as I goes to zero. In both figures the restrictions imposed by UP to the Poincare's maps can be seen as "forbidden areas". Slowly, as I goes to zero, these areas vanished.

In more detail, the solutions to (9a) and its classical counterparts, are located on the hyper-surfaces (10a) and (10b), which restricts the associated dynamics to a sub-manifold of dimension 3. The following restrictions apply in the case of the quantum variables

$$\frac{\langle \hat{p}^2 \rangle}{m_q} + m_q \omega_q^2 \langle \hat{x}^2 \rangle \leq 2E, \tag{16a}$$

$$\left(m_q \omega_q^2 \langle \hat{x}^2 \rangle - E \right)^2 + \omega_q^2 \left(\frac{\langle \hat{L} \rangle}{2} \right)^2 \leq E^2 - \omega_q^2 I. \tag{16b}$$

We see that harmonic oscillator-like solutions delimit a special region, let us call it $\mathcal{B}$, such that the quantum variables take values precisely in $\mathcal{B}$, which makes it convenient to effect the parameterization

$$\langle \hat{x}^2 \rangle = \frac{1}{m_q \omega_q^2} \left(E + r_q \cos \theta_q \right), \tag{17a}$$

$$\langle \hat{L} \rangle = \frac{2}{\omega_q} r_q \sin \theta_q, \tag{17b}$$

$$\langle \hat{p}^2 \rangle = m_q \left(E - r_q \cos \theta_q \right), \tag{17c}$$

with an analogous classical counterpart, since the classical variables $\{x^2, p^2, L\}$ obey

$$\frac{p^2}{m_q} + m_q \omega_q^2 x^2 \leq 2E, \tag{18a}$$

$$\left(m_q \omega_q^2 x^2 - E \right)^2 + \omega_q^2 \left(\frac{L}{2} \right)^2 \leq E^2, \tag{18b}$$

so that the association parameterization is here

$$x^2 = \frac{1}{m_q \omega_q^2} \left(E + r_{cl} \cos \theta_{cl} \right), \qquad (19a)$$

$$L = \frac{2}{\omega_q} r_{cl} \sin \theta_{cl}, \qquad (19b)$$

$$p^2 = m_q \left(E - r_{cl} \cos \theta_{cl} \right), \qquad (19c)$$

with r_q and r_{cl} limited by

$$0 \leq r_q \leq \sqrt{E^2 - I \omega_q^2}, \qquad (20a)$$

$$0 \leq r_{cl} \leq E. \qquad (20b)$$

It is apparent that one should associate classical and quantum initial conditions by taking $\theta_q = \theta_{cl} = \theta$, as above, which correctly links classical and quantum curves in the limit $I \to 0$ and $E \to \infty$. θ varies in $0 < \theta \leq 2\pi$ while r_q, r_{cl} do so according to (20) in (17) and (19).

It seems convenient to study the transition in terms of a variable relating E and I. Inspection of the pertinent equations [7] suggests the following dimensionless quantity

$$E_r = c \frac{E}{I^{1/2}}. \qquad (21)$$

¿From Eqs. (20) one finds that E_r obeys

$$E_r = \frac{E}{I^{1/2} \omega_q}, \qquad (22)$$

its range of variation being

$$1 \leq E_r \leq \frac{2E}{\hbar \omega_q}, \qquad (23)$$

which reminds one of the harmonic oscillator situation.

In Fig. 1 we depict some relevant Poincaré surfaces (sections' cuts with $A = 0$) for $E = 0.6$ and for i) the semi-quantum instance, with I-values such that $I \to 0$ (Figs. 1 (a-e)), and ii) the classical case (Fig. 1 (f)). All the quantum orbits are enclosed within the region circumscribed by the curves

$$\frac{\langle \hat{p}^2 \rangle}{m_q} + m_q \omega_q^2 \langle \hat{x}^2 \rangle = 2E, \qquad (24a)$$

$$\langle \hat{x}^2 \rangle \langle \hat{p}^2 \rangle = I. \qquad (24b)$$

The curve (24a) (with $A = 0$, and $P_A = 0$) represents a periodic, stable solution for the system of equations (9). Here we want to remark that the hyperbole represents the Uncertainty Principle and the straight line is the locus of the pertinent harmonic oscillator evolution. The classical orbits are enclosed within the region circumscribed by the classical counterpart of the curves (24) (the classical case $I = 0$ may be regarded as a *"degenerate"* *hyperbole, that overlaps the coordinate axis*). These graphs, clearly show that the classical regime becomes noticeable in Fig. 1 (c) for a special value of I, denoted by I_{cl}, and whose value is $I_{cl} = 6.1242 \ 10^{-4}$. The hyperbole (24b) approaches the coordinate axis (towards the classical "degenerate" hyperbole) more and more closely as I diminishes.

In Figs. 2 show the same relevant Poincaré surfaces as in Figs. 1 in the sub-manifold $\langle \hat{L} \rangle$ vs. $\langle \hat{x}^2 \rangle$, for i) the semi-quantum instance, with E_r-values such that $E_r \to \infty$ (Figs. 2 (a-e)), and ii) the classical case (Fig. 2 (f)). All the semi-classical (or semi-quantum) orbits are enclosed within the region (16b), while the the classical orbits are enclosed within the region (18b). We find regular orbits when E_r approaches its minimum allowed value ($E_r = 1$). Starting from the special value $E_r = E_r^{\mathcal{P}}$ (in our present example we have $E_r^{\mathcal{P}} = 2^{1/2}$) a chaotic regime is detected. Notice that the change of regime is again present (Fig. 3c) for a special value of E_r, to be henceforth denoted by E_r^{cl}. In the present example we have $E_r^{cl} = 24.24524$.

B. Following the transit between the quantum to chaotic-classical regimes

To follow the transition between both regimes, we can define some quantitative indicator through the definition of a norm, between a quantum vector u

$$u = (\langle \hat{x}^2 \rangle, \langle \hat{p}^2 \rangle, \langle \hat{L} \rangle), \tag{25}$$

that "lives" at the intersection of the two surfaces determined by, respectively, (10a) and (10b), and its to its classical counterpart u_{cl}, expressed as

$$u_{cl} = (x^2, p^2, L). \tag{26}$$

A measure of this convergence is given by the norm of the vector Δu^n

$$\Delta u^n = u^n - u_{cl}{}^n, \tag{27}$$

given by

$$N_{\Delta u^n} = |u^n - u_{cl}{}^n|, \tag{28}$$

where the supra-index "n" indicates the kind of normalization one has employed in this endeavor. In order to compute Δu^n we need to identify u_{cl} also as the result of a limiting process $E_r \to \infty$ of u. Both classical and quantum initial conditions are to be correctly identified to such an effect.

So as to compute $N_{\Delta u^n}$ we choose the u-normalization

$$u^n = (m_q \omega_q^2 \langle \hat{x}^2 \rangle, \frac{\langle \hat{p}^2 \rangle}{m_q}, \frac{1}{\sqrt{2}} \omega_q \langle \hat{L} \rangle). \tag{29}$$

Notice that $N_{\Delta u^n}$ depends upon t. A time average then becomes the appropriate tool in such circumstances, to be evaluated in numerical fashion. For both the quantum variables and their classical counterpart, the initial conditions $\{\langle \hat{x}^2 \rangle(0), \langle \hat{p}^2 \rangle(0), \langle \hat{L} \rangle(0)\}$, and $\{x^2(0), p^2(0), L(0)\}$, are determined according to

$$r_q(0) = a \sqrt{E^2 - I\omega_q^2}, \tag{30a}$$

$$r_{cl}(0) = a\, E, \tag{30b}$$

$$0 < \theta(0) \le 2\pi, \tag{30c}$$

with $0 < a \le 1$. With such values fixed, $A(0)$ and $P_A(0)$ are chosen so as to comply with Eqs. 19 and 21. Depicting this semi-quantum norm in Fig. 3 one notices the precise meaning to the points $E_r{}^{\mathcal{P}}$ and $E_r{}^{cl}$ and the different regions can be clearly distinguished.

V. CONCLUSIONS

We have here studied a "quantum-classical" coupled problem which has been expressed as a dynamical system, taking advantage (4), in the classical (Poincare parenthesis) and quantum version. The system described by Eqs. (9a) has two invariants of motion, I , and the energy, E, in terms of which the two extremes regimes of the complex system, the quantal and the classical one, can be distinguished. Particularly, it can be seen that there exists a gradual transition between them, and we have found the I-interval along which both regimes coexist. The uncertainty growth associated with larger I-values yields the signature of having reached the more quantal zone. Considering that as it shown in Fig. 3, the system "moves" from the chaotic-classical regime, to the quantum one, smoothly, the present method allows to study frontier regions associated to the quantum-classical border, as well as transitions between chaotic to nonchaotic regimes, depending on the selected Hamiltonian. The main condition to this extent is to arrive to a semi-Lie algebra ((4)) using conmutator relationships, or Poincare parenthesis or both, as in the present case. Notice that the semi-Lie algebra can be also of infinite nature as it is shown in [13] and references therein. The closure relationship allows to construct the "canonical" dynamical system upon which the method is based on.

REFERENCES

[1] Y. Alhassid and R. D. Levine, J. Chem. Phys. **67** (1977) 4321.

[2] Y. Alhassid and R. D. Levine, Phys. Rev. A **18** (1978) 89.

[3] A. Plastino, A. N. Proto, and C. Sarris, Physica A **241** (1997) 649; A. M. Kowalski, A. Plastino and A. N. Proto, Phys. Lett. A **187** (1994) 220;E. Duering, D. Otero, A. Plastino, and A.N. Proto. Phys. Rev. A35 (1987) 2314

[4] D. Otero, A. Plastino, A.N. Proto and G. Zannoli. Phys. Rev. A26 (1982).1209

[5] A. M. Kowalski, A. Plastino, and A. N. Proto, Phys. Rev. E **52** (1995) 165.

[6] A. M. Kowalski, A. Plastino, and A. N. Proto, Physica A , **236** (1997) 429.

[7] A. M. Kowalski, M. T. Martin, J. Nuñez, A. Plastino, and A. N. proto, Phys. Rev. A **58** (1998) 2596.

[8] F. Cooper, J. F. Dawson, D. Meredith and H. Shepard, Phys. Rev. Lett. **72** (1994) 1337.

[9] J. Aliaga and A.N. Proto. Phys. Lett. A**142** (1989) 63;Crespo, and A.N. Proto. Phys. Rev. A**42** (1990)4325

[10] J. Aliaga, D. Otero, A. Plastino, and A.N. Proto. Phys. Rev. A**37** (1988) 918

[11] A. Messiah, *Quantum mechanics* (North Holland, Amsterdam, 1961), Vol. I, Chap. 6.

[12] E. Merzbacher, *Quantum mechanics* (Wiley, New York, 1963).

[13] Gruver, J.L., Plastino, A. and Proto, A.N. ,Phys. Lett. A, **246** (1-2) (1998) pp. 61-65 , and references therein

FIGURES

FIG. 1. Poincaré surface of section : $\langle \hat{p}^2 \rangle$ versus $\langle \hat{x}^2 \rangle$, for $E = 0.6$, $A(t = 0) = 0$, $m_q = m_{cl} = \omega_q = e = 1$. I adopts the following values: a) 0.35 ("quantum-like" regime), b) 0.15 (semi-classical regime), c) $I_{cl} = 6.1242 \; 10^{-4}$ (convergence to the classical limit becomes noticeable), d) 10^{-5}, e) 10^{-7} (note that, "visually", this inset resembles the next one, corresponding to the classical regime), and f) $I = 0$ (classical instance). The different curves (inset Figs. (a)-(e)) are bounded by the curve $\langle \hat{x}^2 \rangle \langle \hat{p}^2 \rangle = I$, the "Uncertainty Principle orbit" (below), and by the energy curve $E = \frac{1}{2}(\langle \hat{x}^2 \rangle + \langle \hat{p}^2 \rangle)$ one (top), that corresponds to that of a quantum H. O. of unit frequency. Notice the coexistence of the Uncertainty Principle with chaos, specially in Fig. (b). All quantities are given in arbitrary units.

FIG. 2. Poincaré surfaces of section: $\langle \hat{L} \rangle$ vs. $\langle \hat{x}^2 \rangle$, for $E = 0.6$, $A(t = 0) = 0$, $e = m = 1$. E_r adopts the following values: a) $E_r = 1.0142$ (quantum-like regime), b) $E_r = 1.5492$ (semi-classical regime), c) $E_r = 24.2452$ (convergence to the classical limit becomes noticeable), d) $E_r = 189.7367$, e) $E_r = 1897.3666$ and, f) $I = 0$, classical situation.

FIG. 3. $N_2[\Delta_{cl}^2 x]$ and $N_{max}[\Delta_{cl}^2 x]$ vs. E_r, for the $a = 0.98$ orbit, drawn with the data of Fig. 1 for 100 different values of E_r. Two different regions can be clearly distinguished. The semi-quantal zone lies at the left of E_r^{cl}, and its classical counterpart at the right. These quantities are dimensionless.

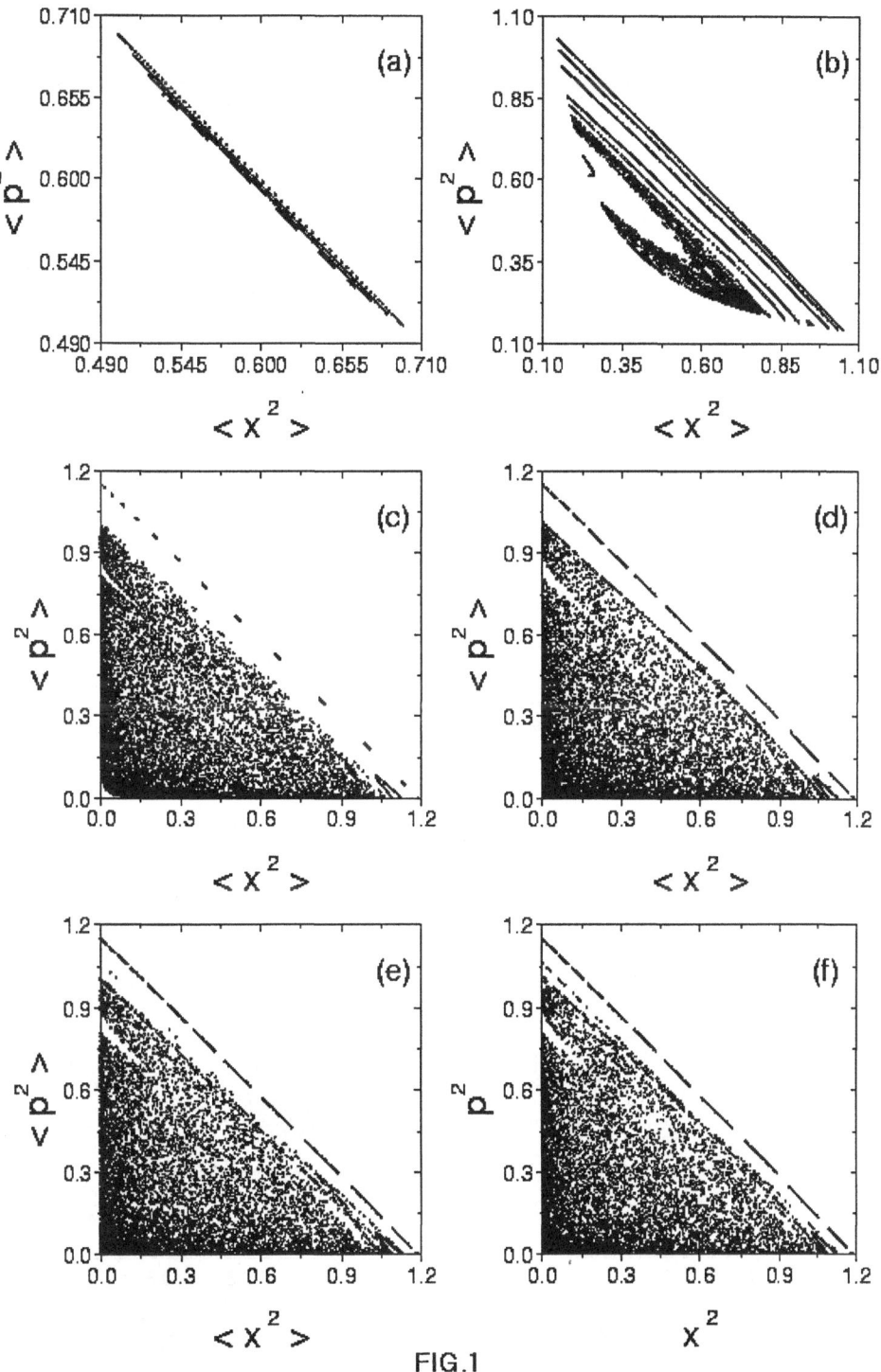

FIG.1

234

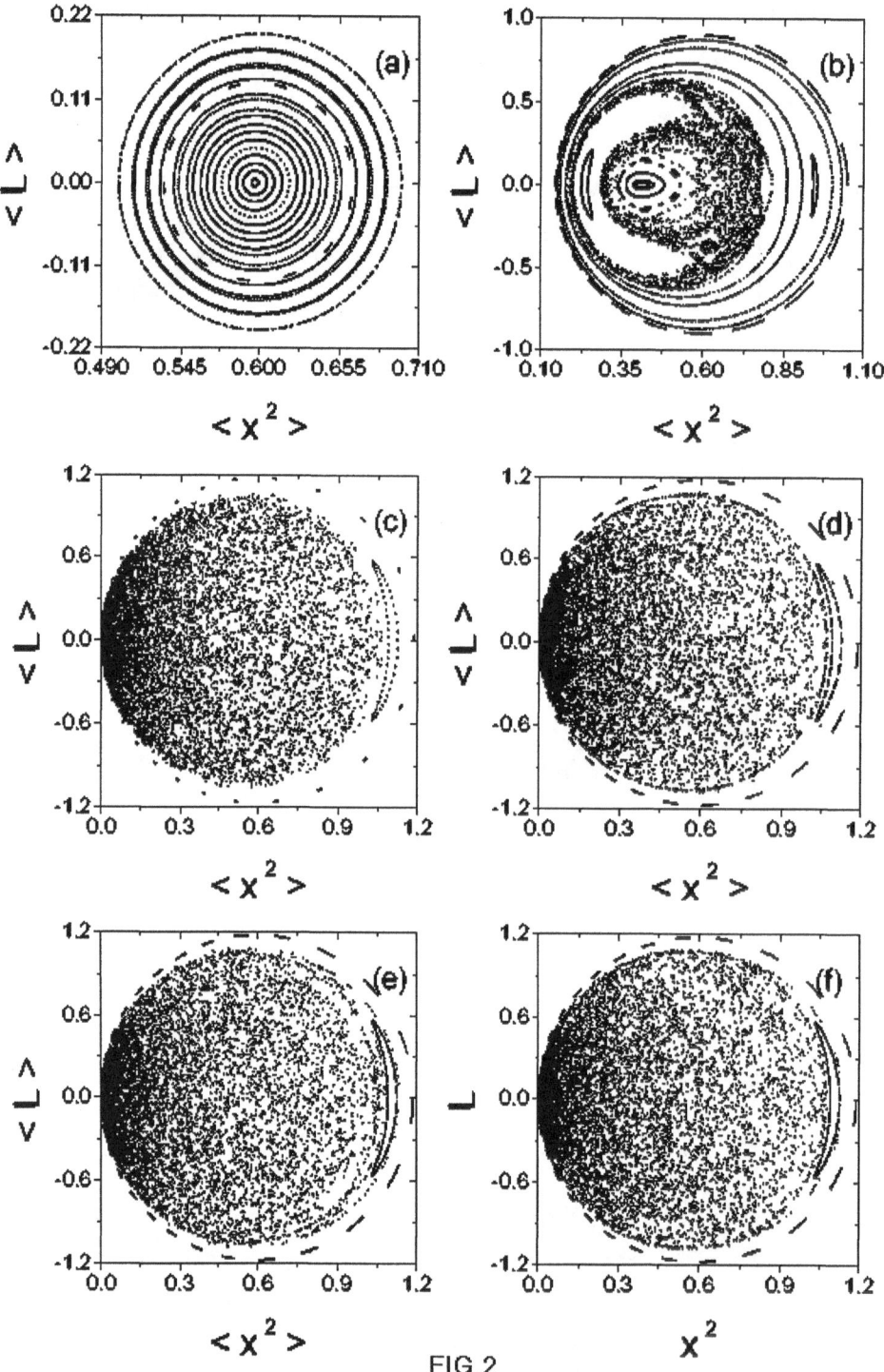

FIG.2

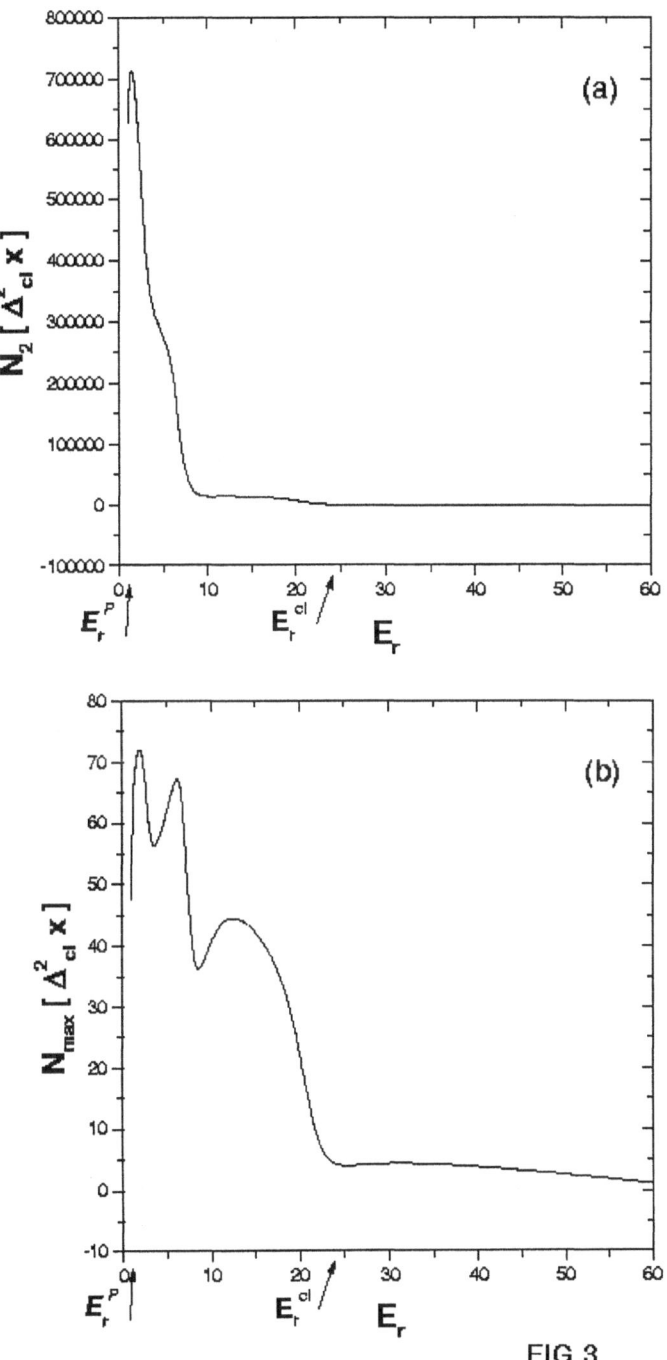

FIG.3

Thermodynamic behavior of a stain

Víctor A. Kuz

IFLYSIB,C.C.565, (1900) La Plata, Argentina.

A thermodynamic model is developed to explain the formation of a solid ridge of polystyrene spheres on the periphery of a sessile drop. The drop initially contains a uniform solution of water and polystyrene spheres. Evaporation creates a crater. The spheres leave the center and a ridge, whose width increases with increasing initial concentration, forms. For each equilibrium state, three contributions to the Gibbs free energy mainly control this process. One is related with the mechanical work of formation, the other with the spreading of the diameter and the last with the variation of the number of particles in the ridge. The model predicts, in agreement with the experiment, the initial and final value of the contact angles and the exponential diminution of concentration of particles with the ratio of the diameters of the ridge.

O. Descalzi et al. (eds.), Instabilities and Nonequilibrium Structures VII & VIII, 237–253.
© 2004 *Kluwer Academic Publishers. Printed in the Netherlands.*

1   Introduction

Complex fluids e.g. liquids containing impurities "paints, inks, liquids nutrients,..." are important in technology and applied sciences. Evaporation of polystyrene sphere-H_2O suspensions placed on an horizontal non-wetting solid surface promotes the formation of a (2D) two-dimensional [1],[2] or (3D) three-dimensional array of particles on a solid substrate. These interesting problems have been experimentally [3] and theoretically studied [4].

A colloidal suspension placed on a solid (a non-wetting surface) free of a surrounding (sessile drop), spreads over a certain area of a substrate [5], [6], [7]. In general it is found that, when a drop of complex liquid dries, it leaves a dense, ring-like deposit along the perimeter. The particles, initially dispersed over the entire drop, becomes concentrated into a small fraction of it. The pinning of the contact line of the drying drop ensures that liquid evaporating from the edge is replenished by liquid from the interior. The resulting outward flow can carry almost all the dispersed material to the edge. Different experimental observations show that a ring or a solid ridge forms for a wide variety of substrates, dispersed materials (solutes), and carrier liquids(solvents), as long as the solvent meets the surface at a non-zero contact angle, the contact line is pinned to its initial position and the solvents evaporates. These are the general condition which must be fulfilled to have a ridge from the spreading of a complex liquid suspension.

Of this general problem described above, some specific aspects have been considered recently. Parisse *et al.* [7], investigated experimentally the drop changes produced during drying and developed calculations of the shape changes under solvent evaporation. Deegan *et al.* [6], studied experimentally and predicted theoretically the power-law growth of the ring mass with time. Conway *et al.* [5], observed experimentally height, diameter, contact angle and mass as a function of time for a range of sessile drop sizes, polystyrene sphere diameters and suspension concentrations. Height and mass data were predicted by a theoretical model proposed by the authors. They also found experimentally a quantitative relation between the inner diameter of a ridge and the concentration of particles. The particles (polystyrene spheres) had diameters varied from 110 nm to 1000 nm and they were suspended in deionized water, surfactant-free, and at concentrations of 8 %. Higher and lower concentrations were also used. The initial drop volumes (spherical caps) varies from 30 μL to 100μL . During ridge formation the contact angle, as well as the external drop diameter remain unchanged. The ridge follows from the directional motion of the water and particles from the center toward the drop's boundary; its formation is independent of the drop's initial volume. This phenomena occurs in solutions [6], such as salt water or coffee, in mixtures, in suspensions of polystyrene beads [5] or in the process of wiping out a stain from a fabric with a solvent.

It is a well accepted fact that evaporation is the main driving force in the process of formation of the ridge. Particles close to the wedge of the drop attract each other by lateral capillary forces. Two or more of these particles close together create a capillary tube like the Plateau border in an emulsion. The evaporation and restitution of water in these capillary structures formed at the edge of the drop, induced a convective flow of water which drags the solid particles from the center to the periphery. The effect of this convective flow is to restore the equilibrium thickness and shape of the wedge which is also constantly reduced by evaporation and diffusion into the gas phase. This is the basic mechanism of the ridge formation. The present work is a model of the ridge formation, and its relation with the quantitative experiment done by Conway *et al.*[5]. These authors found that the inner diameter of the ridge decreases with increasing initial concentration of polystyrene spheres. Below we present a thermodynamical model to explain this experimental fact.

2 The ridge width and the concentration of particles.

Let us consider an enclosed system in thermal equilibrium composed by a sessile drop of polystyrene sphere-H_2O (a suspension) placed on a solid substrate (see Figure 1). Initially, the drop spreads and reaches a maximum external diameter with a spherical-cap profile. The contact line is pinned. Evaporation of the water reduces the height and changes the profile of the drop with time and the external diameter persists constant. Each evaporation results in the formation of a solid ridge on the perimeter of the drop, the polystyrene particles leave the center and move towards the periphery. As evaporation goes on, the initially spherical-cap shaped drop evolves and finally a crater is developed in its center. The final center height is of the order of the polystyrene bead diameter [5].

The ridge starts to be formed with a fixe external diameter. The inner diameter, developed in the drop, decreases with the concentration of polystyrene particles.

After these considerations let's use a thermodynamic approach to explain this phenomena. At any equilibrium state of the ridge formation, the variation of the Gibbs free energy G' of the liquid ridge suspension (polystyrene particles plus water) including the surface is:

$$dG' = V^s dp^s - A d\sigma + \mu_r^p dn_r^p + \mu^l dn^l \tag{1}$$

The first two terms of this equation denote the bulk and the surface work required to form the ridge while the third is linked with the variation of particles in the ridge. V^s, p^s, A and σ indicate the volume, the pressure, the area and the surface tension of the suspension; μ_r^p, μ^l and n_r^p, n^l are the chemical potentials and the number of polystyrene particles and molecules in the ridge respectively.

The variation of the Gibbs free energy G'' of the gas phase is:

$$dG'' = V^g dp^g + \mu^g dn^g \tag{2}$$

where V^g, p^g, μ^g and n^g indicate the volume, the pressure, the chemical potential and the number of molecules of the gas phase respectively.

Close to any equilibrium state, a transformation of liquid into gas is developed and we have that

$$dG = 0 \tag{3}$$

where G, is the Gibbs free energy of the system (gas phase plus ridge liquid suspension). Temperature, liquid solution-gas, solid-liquid solution surface tensions and the energy of substrate are assumed to remain constant. In any state of equilibrium during the ridge formation, the variation of the number of particles in the system (gas phase plus liquid suspension) of the ridge is zero, then

$$dn^t = dn^g + dn^l + dn_r^p = 0 \; ; \; dn_r^p = -\left(dn^g + dn^l\right) \tag{4}$$

It must be mentioned here that the center of the ridge acts as a source of polystyrene particles and water to the ridge. The transformation of liquid into gas and the simultaneous flow of suspension (polystyrene particles plus water) from the center to the border induces the ridge formation. Equation (4), is valid for each equilibrium state of the ridge formation. The total number of polystyrene particles is constant for each equilibrium state, but it increases in each successive equilibrium state. An equilibrium state may be reached by only stopping evaporation. The last state of equilibrium of the system is attained when thickness of the crater formed at the center of the ridge is of the order of the polystyrene particles. Here the flow ceases and the excess of solid particles remain filling the crater.

From the substitution of equations (1), (2), and (4) into (3) we have

$$V^s dp^s + V^g dp^g - A d\sigma + \mu_r^p dn_r^p - \mu^g dn_r^p = 0 \tag{5}$$

we have used the fact that $\mu^l = \mu^g$. The gas volume is $V^g >> V^s$ and the pressure p^g remains nearly constant, then equation (5) become

$$V^s dp^s - A d\sigma + \Delta\mu dn_r^p = 0 \tag{6}$$

where $\Delta\mu = (\mu_r^p - \mu^g)$.

Let us analyze some geometrical aspect of the ridge. The ridge volume changes due to evaporation and accumulation of polystyrene particles and is given by:

$$V^s = A_t R \tag{7}$$

$A_t = \pi(\pi r^2)$ is proportional to the transversal area and $A = 2\pi^2 r R$ is the liquid-gas area, r and R are the two radii (see figure 3a).

The ridge has two curvatures; the pressure is given by the Laplace equation

$$p^{in} - p^{ext} = \frac{\sigma^{l,g}}{R} + \frac{\sigma^{l,g}}{r} = \frac{\sigma^{l,g}}{R}\frac{(a+1)}{a} \tag{8}$$

where $\sigma^{l,g}$ is the liquid-gas surface tension; p^{in} and p^{ext} are the internal and the external pressure in the ridge. The radius r, is proportional to R, $(r = aR)$, where a

is a number that depends on the polystyrene particles diameter [5]. The substitution of equations (7) and (8) into (6) results in

$$-A_t \sigma^{l,g} \frac{(a+1)}{a} \frac{dR}{R} - A d\sigma + \Delta\mu dn_r^p = 0 \tag{9}$$

integrating between the outer and the inner radius, equation (9) becomes:

$$\ln\left(\frac{R_1}{R_2}\right) - \frac{a}{(a+1)} \frac{A}{A_t} \frac{[\sigma(R_2) - \sigma(R_1)]}{\sigma^{l,g}} + \frac{a}{(a+1)} \frac{\Delta\mu}{A_t \sigma^{l,g}} [n^p(R_2) - n^p(R_1)] = 0 \tag{10}$$

The theory of contact angle allows us to resolve horizontally the equilibrium tensions. For any finite contact angle, we have

$$\cos\theta = -\frac{\left[\sigma^{s,g} - \sigma^{s,l}\right]}{\sigma^{l,g}} \tag{11}$$

where $\sigma(R_2) = \sigma^{s,l} + \sigma^{l,g} \cos\theta - \sigma^{s,g}$, and $\sigma(R_1) = \sigma^{l,g} \cos\theta$, are the horizontal tensions acting at R_2 and R_1, respectively (see figure 3c). In the equation for the tension at the position R_1, we have disregarded the tension of the film which is in equilibrium with the ridge.

Equation (10) can be written as follows:

$$\frac{R_1}{R_2} = \exp(-\frac{a}{(a+1)} \frac{A}{A_t} \cos\theta) \exp\left[-\alpha \left(c(R_2) - c(R_1)\right)\right] \tag{12}$$

where $c(R_2) - c(R_1) = \left[\frac{n_r^p(R_2)}{V^s} - \frac{n_r^p(R_1)}{V^s}\right]$ is the concentration of particles. This equation relates the ratio of the ridge radii with the concentration of solid particles, the contact angle and the size of the solid particles. The equation is valid for any of the equilibrium state reached by the system.

The coefficient α, in equation (12), is obtained by replacing equations (7) and (8), into the third term of equation (10). Then using the Kelvin equation together with the expression of the chemical potential of an ideal gas, α reduces to:

$$\alpha = 2v^l \left(\frac{\mu_r^p}{\mu^g} - 1\right) \frac{a}{(a+1)} \tag{13}$$

244

where v^l indicates the specific volume of the liquid.

At the final state of equilibrium, the concentration of polystyrene spheres is more or less constant in the ridge, $c(R_2) \simeq c(R_1)$, then equation (12) reduces to

$$\left(\frac{R_1}{R_2}\right)_{final} = A_o = \exp(-2/\left(1+\frac{r}{R}\right)_{final} \cos\theta_{final}) \qquad (14)$$

The quantity $\frac{a}{(a+1)}\frac{A}{A_t} = 2/\left(1+\frac{r}{R}\right)_{final}$. θ_{final} indicates the final contact angle. A_o, is a constant which depends on the final contact angle and on the quotient $(r/R)_{final}$.

In the earlier beginning of the ridge formation, the pressure is given by $p^{in} - p^{ext} \simeq \frac{\sigma^{l,g}}{R}$, see equation (8). By using this approximation and assuming that the concentration at the pinned line is bigger than the concentration at the inner region, $c(R_2) - c(R_1) \simeq c$, equation (12) reduces to:

$$\frac{R_1}{R_2} = \exp(-2\left(\frac{R}{r}\right)\cos\theta_{initial})\exp\left[-\alpha\left(c(R_2)-c(R_1)\right)\right] = A_1 \exp\left[-\alpha\,c\right] \qquad (15)$$

where A_1 is another constant which depends on the initial contact angle $\theta_{initial}$.

In general, for any equilibrium situation, we have

$$\frac{D_1}{D_2} = A_o + A_1 e^{-\alpha c} \qquad (16)$$

where $D_1 = 2R_1$ and $D_2 = 2R_2$; A_o and A_1 are given by equations (14) and (15).

Let us compare this equation with the corresponding experimental expression presented recently by Conway et al. [5]

$$\frac{D_1}{D_2} = 0.3 + 0.6\,e^{-0.1\,c} \qquad (17)$$

Here D_1 and D_2 also indicate the inner and the outer diameter respectively. By equating (16) and (17), we see that $A_o = 0.3$, $A_1 = 0.6$, $\alpha = 0.1$. From these data we will estimate the initial and the final contact angle.

The final state of equilibrium is given by equation (14). In the experiment of Conway et al. is shown that for sphere diameter of 540 (nm), the relation between the inner and the outer diameter leads to $R_1/r = 4.4$ (see ref. [5]), then $2/\left(1 + \frac{r}{R}\right)_{final} \simeq 1.68$. With this data and the value of A_o, the contact angle predicted by equation (14) is

$$\theta_{final} \simeq 44.3° \tag{18}$$

To calculate $\theta_{initial}$, see equation (15), we assume that $R >> r$. Then $\theta_{initial}$ is

$$\theta_{initial} \simeq 90° \tag{19}$$

These values of the contact angles indicate that the polystyrene sphere-H_2O suspension was placed on a non-wetting surface as was mentioned in the experiment [5], and suggests that θ decreases with concentration of particles.

Expression (13) shows that the constant α, depends on the specific volume (water) v^l, on the quotient of the chemical potential $\frac{\mu_r^p}{\mu^g}$, and on $a = r/R$. The quantity $\frac{a}{(1+a)}$, for polystyrene spheres of 540 (nm) of diameter is $\frac{a}{(1+a)} \simeq 0.16$. The chemical potential of the latex particles μ^p is in general bigger than μ^g, but ($\mu_r^p \simeq \mu^g$) [8]. Then it is easy to see that $\alpha < 1$.

Finally, it is important to remark that expression (16) shows how the quotient $\frac{D_1}{D_2}$ depends on c (the concentration of particles). It also reveals the dependence of the coefficients A_o, A_1, and α, on the contact angles, chemical potential and particles size.

3 Comments

The present model explains the formation of a solid ridge of polystyrene spheres on the perimeter of a sessile drop. The formation of the ridge is an irreversible process caused by the evaporation of solvent (water). Each state of this evolution may be considered an equilibrium state that could be reach by stopping the evaporation of solvent. Therefore equilibrium thermodynamics is an appropriate tool to model this problem.

We write the Gibbs free energy of the system, and from its variation around equilibrium we obtain equation (12) and then (16). These equations show that in each equilibrium state there is a definite relation between the outer and the inner radii of the ridge, the concentration of polystyrene particles, the diameter of the solid particles and the surface tensions of the solid surface-gas phase, solid surface - liquid suspension and liquid suspension-gas phase; in short it depends on the contact angle (see equation (11)) formed between the solid substrate and suspension of particles. These are the relevant variables that control the evolution of the process. By changing the solid surface, it will be modified the common surface energies or tensions and therefore the contact angles and the relation between the outer and the inner radii. The radii relationship could be also changed by modifying the concentration of solid particles in the suspension or their size via the value of a, see equation (8).

On the basis of the above analysis let us compare our results with those found by experiment. In the experiment done by Conway et. al. [5], a complex liquid (a suspension of polystyrene particles and water) was placed on a glass plate, free of a surrounding (sessile drop). The suspension spreads over a certain area of a solid substrate with low free surface energy (a non-wetting surface).

The balance between spreading pressure and gravity allows the spherical cap approximation for the drop profile. In this situation, the external drop diameter remain unchanged as well as the corresponding contact angle. Then evaporation of water promotes the formation of the ridge. From the analysis of the experimental results Conway et al. [5], found a very definite relation between the inner and the outer diameters of the ridge with the concentration of polystyrene particles (see equation (17)). Three constants are involved in this relation. Their values are 0.3 , 0.6, and -0.1. From the experiment is not possible any further analysis of the origin of these constant or the relation that they have with others relevant parameters of the process. Also in the experiment they measured the final and the initial contact angle between the ridge and the solid surface and they found the following experimental values [5] : $\theta_{final} \simeq 45°$ and $\theta_{initial} \simeq 90°$, respectively. The liquid suspension was placed on a non-wetting surface [5].

Now if we compare equations (16) and (17), we see that the constants are: $A_o = 0.3$, $A_1 = 0.6$, $\alpha = 0.1$. A_o and A_1 are given, in our model, by equation (14) and (15) respectively. Both depend on the contact angle and on the relation between r/R. Then if we use the experimental data for the constants given by equation (17), we should be able to predict the values of the experimental contact angle. The values found in this case by our model are, $\theta_{final} \simeq 44.3°$ and $\theta_{initial} \simeq 90°$, see equations (18) and (19). Then, if it is not accidental these agreement, we could affirm that this two constants are basically related with the initial and final contact angles of the system (see equations (16) and (17)).

With the present thermodynamic model we are not able to predict the value of the constant α. But we are able to estimate it. The constant α basically depends on the relative value of the chemical potentials of the polystyrene particles and the chemical potential of the gas phase (see equation (13)). An accurate calculation of the molecule's free energy is in general a difficult problem. But the chemical potential [8] of the polystyrene particles μ_r^p is slightly bigger than μ^g, then α (see equation (13)), should be a quantity smaller than one.

Finally I want to remark the this simple model brings into light the relevant variables which govern this problem and the roles they play in the process of the ridge formation.

References

[1] Denkov, N.D.; Velev, O.D.; Kalchevsky, P.A.; and Ivanov, I.B.; Nature 26,361,(1993)

[2] Denkov, N.D.; Velev, O.D.; Kalchevsky, P.A.; Ivanov, I.B.; Yoshimura,H.; and Nagayama, K.; Langmuir 3183,8,(1992)

[3] Dushkin, C.D., Yoshimura,H., Nagayama,K.; Chem.Phys.Lett. 455,204,(1993)

[4] Kuz,V.A.; Langmuir, 13, 3900, (1997).

[5] Conway, J.; Korns,H.; Fisch,M.R.; Langmuir, 426, 13, (1997)

[6] Deegan, R.D.; Bakajin, O.; Dupont, T.F.; Huber,G.; Nagel, S.R.;Witten,T.A.; Nature, 827, 389,(1997).

[7] Parisse, F. and AllainC.; J.Phys.II France 6, 1111,(1996); Langmuir13,3598,(1997)

[8] Israelachvili, J.; Intermolecular & Surface Forces, Academic Press, London (1992).

Figure 1: A view of a drop of water with polystyrene beads on a solid substrate.

Figure 2: Polystyrene particles at the perimeter of the drop.

Figure 3: a) Coordinates of the ridge. b) View of the drop with polystyrene particles in suspension after evaporation. c) Profile of the ridge resting on a solid surface.

250

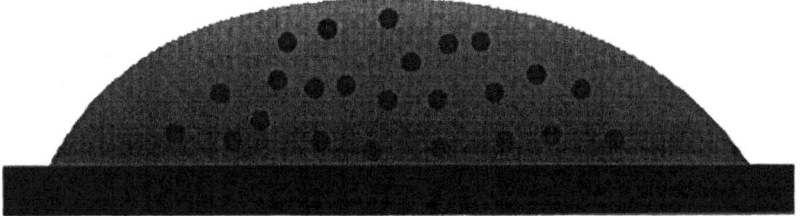

fig.1

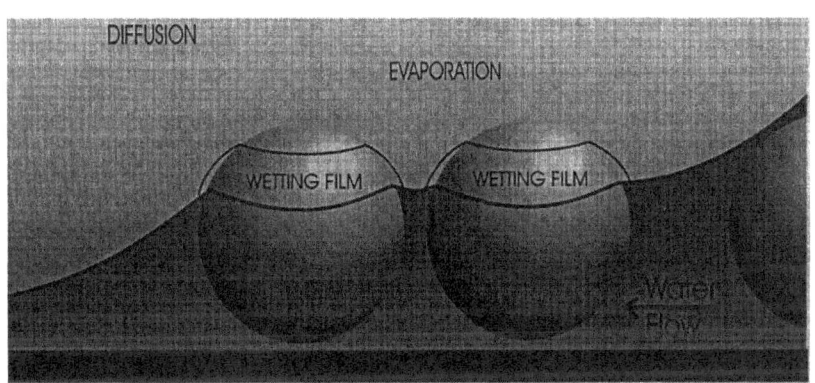

fig. 2

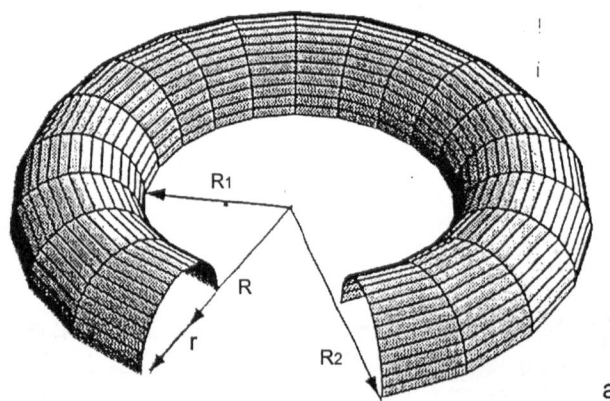

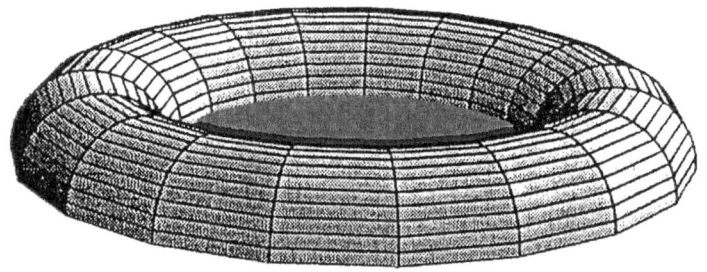

b

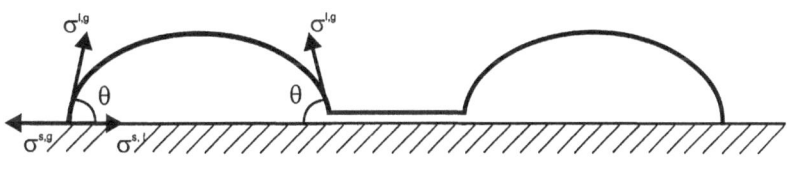

c

fig. 3 (a, b , c)

APPROACH TO EQUILIBRIUM FOR QUANTUM SYSTEMS WITH CONTINUOUS SPECTRUM

by

Roberto Laura

Departamento de Física, F.C.E.I.A., Universidad Nacional De Rosario
Instituto de Física Rosario; CONICET-UNR
Av. Pellegrini 250, 2000 Rosario, Argentina.
e-mail: laura@ifir.ifir.edu.ar

Abstract

Considering quantum states as functionals acting on observables to give their mean values, it is possible to deal with quantum systems with continuous spectrum, generalizing the concept of trace. Generalized observables and states are defined for a quantum oscillator linearly coupled to a scalar field, and the analytic expression for time evolution is obtained. The "final" state $(t \rightarrow \infty)$ is presented as a weak limit. Finite and infinite number of exited modes of the field are considered.

1   Introduction.

The search of a physical explanation for the evolution towards equilibrium of quantum systems has a great interest for quantum statistical mechanics.

In his pioneering work, E. Fermi et al. [1] showed that non linear couplings in general do not determine approach to equilibrium. I. Prigogine [2] showed that for a very big set of classical coupled oscillators, a weak non linear part in the interaction produce the approach to equilibrium.

The problem of a chain of independent quantum oscillators with a linear coupling between first neighbors was analyzed by P.Blaise et al. [3]. Numerical experiments show that for the initial condition of a single excited oscillator the system evolves first towards equilibrium (equipartition of energy), and later a recurrence time appears for which the energy return to the oscillator of the initial condition. The recurrence time grows proportionally to the number of oscillators.

This problem was also analyzed by J. L. Gruver et al. [4] by numerical methods, showing that a quantum "big" oscillator linearly coupled with a great number of identical "small" oscillators with a thermal distribution, evolve in such a way that it is possible to adjust its time evolution with an exponential that leads the "big" oscillator subsystem to the thermal equilibrium with the bath.

The works of references [3] and [4] deal with linear couplings for which the total Hamiltonian can be reduced to non interacting normal modes.

O. Descalzi et al. (eds.), Instabilities and Nonequilibrium Structures VII & VIII, 255–266.

When it is necessary to deal with systems with a huge number of particles, the standard procedure is to start with N particles in a box of volume V, making the limit $N \to \infty$, $V \to \infty$ with $\frac{N}{V} = c < \infty$ in the last step of the calculations. Even for a finite number of particles, the limit $V \to \infty$ produces a continuum spectrum in the unperturbed Hamiltonian. The usual formalism of quantum mechanics can not be used in this case, due to the appearance of diagonal singularities in states and observables [5][6][7].

Based in the pioneering work of I. E. Segal [8], I. Antoniou et al. [5] [6] developed a formalism for quantum systems with diagonal singularity. The quantum states of this theory are *functionals* over certain space of observables $\mathcal{O}$. Mathematically this means that the space $\mathcal{S}$ of states is contained in $\mathcal{O}^\times$. Physically it means that the only thing we can really observe and measure are the mean values of the observables $O \in \mathcal{O}$ in states $\rho \in \mathcal{S} \subset \mathcal{O}^\times$: namely $\langle O \rangle_\rho = \rho[O] \equiv (\rho|O)$. This is the natural generalization of the usual trace $Tr(\widehat{\rho}\widehat{O})$ which is ill defined in systems with continuous spectrum. We used this approach to describe the scattering problem and the decay process in Friedrichs model [9][10].

In this work we apply this formalism to a quantum oscillator linearly coupled to a scalar field, for a finite number of excited modes (decay process) and for the thermodynamic limit (infinite number of excited modes of the field). We will be able to obtain the time evolution of a finite number of excited modes, and to give an analytic expression for the "final" state ($t \to \infty$) as a weak limit on the "test observables" $\mathcal{O}$. As we shall see, the formalism can also be adapted to obtain exact expressions for the time evolution in the thermodynamic limit, where an infinite number of excited modes of the field drive the quantum oscillator to a final excited state. These results will be obtained without using coarse graining nor complex generalized eigenvalues nor box normalization.

The model is presented in section II. In section III we discuss the characterization of states and observables with diagonal singularity, both for a finite and infinite number of excited modes. In section IV we obtain the exact solution of the problem using the diagonalized form of the Hamiltonian.

An extended version of this report, where also the relation of the exact solution with the results of the Pauli master equation is discussed, will appear in Physical Review E [11]

2   The model.

We consider a quantum oscillator with the Hamiltonian:

$$H_S = \Omega\, b^\dagger\, b, \qquad [b, b^\dagger] = 1, \qquad (\hbar = 1), \tag{1}$$

and a quantum field with the Hamiltonian:

$$H_F = \int d\mathbf{k}\, \omega_k\, a_\mathbf{k}^\dagger\, a_\mathbf{k}, \quad [a_\mathbf{k}, a_{\mathbf{k}'}^\dagger] = \delta^3(\mathbf{k} - \mathbf{k}'),$$
$$\omega_k = k \equiv |\mathbf{k}|, \quad (c = 1). \tag{2}$$

The interaction is given by:

$$H_{INT} = \int d\mathbf{k}\, V_k[a_\mathbf{k}^\dagger b + b^\dagger a_\mathbf{k}],$$
$$V_k^* = V_k, \quad [a_\mathbf{k}^\dagger, b^\dagger] = [a_\mathbf{k}^\dagger, b] = 0. \tag{3}$$

If the function V_k is chosen in such a way that:

$$\eta_\pm(\omega_k) \equiv \omega_k - \Omega - \int \frac{d\mathbf{k}'\, V_{k'}^2}{\omega_k - \omega_{k'} \pm i0}, \tag{4}$$

do not vanish for any $k \in \mathbb{R}^+$, and the analytic extension $\eta_+(z)$ from the upper to the lower complex half plane of $\eta_+(k)$ has a simple zero at $z = z_0 \in \mathbb{C}^-$ ($\mathbb{C}^-$ is the lower part of the complex plane), The total Hamiltonian $H = H_S + H_F + H_{INT}$ can be diagonalized in terms of the creation (anhiquilation) operators $A_\mathbf{k}^\dagger$ ($A_\mathbf{k}$):

$$H = \int d\mathbf{k}\, \omega_k A_\mathbf{k}^\dagger A_\mathbf{k}, \qquad [A_\mathbf{k}, A_{\mathbf{k}'}^\dagger] = \delta^3(\mathbf{k} - \mathbf{k}'), \tag{5}$$

where:

$$A_\mathbf{k}^\dagger \equiv a_\mathbf{k}^\dagger + \frac{V_k}{\eta_+(\omega_k)}\left[b^\dagger + \int \frac{d\mathbf{k}'\, V_{k'}\, a_{\mathbf{k}'}^\dagger}{\omega_k - \omega_{k'} + i0} \right]$$
$$A_\mathbf{k} \equiv a_\mathbf{k} + \frac{V_k}{\eta_-(\omega_k)}\left[b + \int \frac{d\mathbf{k}'\, V_{k'}\, a_{\mathbf{k}'}}{\omega_k - \omega_{k'} - i0} \right] \tag{6}$$

The deduction of these expressions can be found in reference [11], and it follows essentially the same arguments of reference [12].

The operators $b^\dagger$, b, $a_\mathbf{k}^\dagger$ and $a_\mathbf{k}$ can be written in terms of $A_\mathbf{k}^\dagger$ and $A_\mathbf{k}$:

$$b^\dagger = \int \frac{d\mathbf{k}\, V_k}{\eta_-(\omega_k)} A_\mathbf{k}^\dagger, \qquad b = \int \frac{d\mathbf{k}\, V_k}{\eta_+(\omega_k)} A_\mathbf{k}$$
$$a_\mathbf{p}^\dagger = \int d\mathbf{k} \left[\delta^3(\mathbf{k} - \mathbf{p}) + \frac{V_k V_p}{\eta_-(\omega_k)(\omega_k - \omega_p - i0)} \right] A_\mathbf{k}^\dagger$$
$$a_\mathbf{p} = \int d\mathbf{k} \left[\delta^3(\mathbf{k} - \mathbf{p}) + \frac{V_k V_p}{\eta_+(\omega_k)(\omega_k - \omega_p + i0)} \right] A_\mathbf{k} \tag{7}$$

3 States and observables.

As it is discussed in references [5][6][7], for quantum systems with continuous spectrum the usual approach of density operators is not applicable. For a density operator $\hat{\rho}$ representing a state and an operator $\hat{O}$ representing an observable, expressions like $Tr(\hat{\rho}\hat{O})$ are meaningless due to the presence of singular diagonal terms. The usual way to avoid these problems is to "discretize" the spectrum by enclosing the system in a box with periodic boundary conditions. The size of the box is considered to be infinite only at the last step of the computation of relevant quantities.

A way to consider the continuous spectrum from the beginning was introduced in references [5][6][7]. For a given set $\mathcal{O}$ of operators representing physical observables, the states of the system can be represented by a set $\mathcal{S}$ of functionals acting on $\mathcal{O}$ ($\mathcal{S} \subset \mathcal{O}^\times$). The mean value of an observable $O \in \mathcal{O}$ in a state $\rho \in \mathcal{S}$ is given by the value of the functional ρ on O, denoted by $\langle O \rangle_\rho = (\rho|O)$. This expression generalizes the usual $Tr(\widehat{\rho}\widehat{O})$. As usual, the time evolution of the observables in Heisemberg representation is given by:

$$O_t = e^{i\mathbb{L}^\dagger t} O, \quad where \quad \mathbb{L}^\dagger O \equiv [H, O], \qquad O \in \mathcal{O}. \tag{8}$$

For the model presented in section II, let us consider a special set $\mathcal{O}$ of observables given by a generalized linear combination of products of one creation and one anhiquilation operators:

$$\begin{aligned} O &= O_1 b^\dagger b + \int d\mathbf{k} \int d\mathbf{k}' \, \widetilde{O}_{\mathbf{k}\mathbf{k}'} \, a^\dagger_\mathbf{k} a_{\mathbf{k}'} + \\ &\quad + \int d\mathbf{k} \, O_{\mathbf{k}1} a^\dagger_\mathbf{k} b + \int d\mathbf{k}' \, O_{1\mathbf{k}'} b^\dagger a_{\mathbf{k}'} \\ O_1^* &= O_1, \quad \widetilde{O}^*_{\mathbf{k}\mathbf{k}'} = \widetilde{O}_{\mathbf{k}'\mathbf{k}}, \quad O^*_{\mathbf{k}1} = O_{1\mathbf{k}} \end{aligned} \tag{9}$$

Due to the form of the total Hamiltonian given by equations (1), this form is preserved by time evolution ($e^{i\mathbb{L}^\dagger t}\mathcal{O} \subset \mathcal{O}$).

Some observables in $\mathcal{O}$ are specially important for our purposes:

$$\begin{aligned} n &\equiv b^\dagger b & &\text{number of discrete modes} \\ N &\equiv \int d\mathbf{k} \, a^\dagger_\mathbf{k} a_\mathbf{k} & &\text{number of continuous modes} \\ \sigma(\mathbf{r}) &\equiv \psi^\dagger(\mathbf{r})\,\psi(\mathbf{r}) & &\text{density of continuous modes at the point } \mathbf{r}, \end{aligned} \tag{10}$$

where in the last expression $\psi^\dagger(\mathbf{r}) \equiv \frac{1}{\sqrt{8\pi^3}} \int d\mathbf{k} \, e^{i\mathbf{k}\cdot\mathbf{r}} a^\dagger_\mathbf{k}$ is the creation operator for one continuous mode at the position $\mathbf{r}$. Notice that to include observables like N or H in the set $\mathcal{O}$ it is necessary to allow $\widetilde{O}_{\mathbf{k}\mathbf{k}'}$ to have a singular part proportional to $\delta^3(\mathbf{k} - \mathbf{k}')$.

For the cases in which the total number of excited modes is finite ($\langle N_T \rangle = \langle n \rangle + \langle N \rangle < \infty$), observables like H and N are well defined and should be included in $\mathcal{O}$. Therefore, in this case we denote by $\mathcal{O}_D$ (the D corresponds to "decaying" processes) the set of observables of the form (9) for which:

$$\widetilde{O}_{\mathbf{k}\mathbf{k}'} = O_\mathbf{k}\,\delta^3(\mathbf{k} - \mathbf{k}') + O_{\mathbf{k}\mathbf{k}'} \tag{11}$$

being $O_\mathbf{k}, O_{\mathbf{k}\mathbf{k}'}, O_{\mathbf{k}1}$ and $O_{1\mathbf{k}'}$ regular functions of the variables $\mathbf{k}$ and $\mathbf{k}'$.

For all $O \in \mathcal{O}_D$ we obtain:

$$\langle O \rangle_\rho = (\rho|O) = \rho_1^* O_1 + \int d\mathbf{k} \, \rho_\mathbf{k}^* O_\mathbf{k} +$$

$$+ \int\int d\mathbf{k}\, d\mathbf{k}' \, \rho^*_{\mathbf{k}\mathbf{k}'} O_{\mathbf{k}\mathbf{k}'} + \int d\mathbf{k} \, \rho^*_{\mathbf{k}1} O_{\mathbf{k}1} + \int d\mathbf{k}' \, \rho^*_{1\mathbf{k}'} O_{1\mathbf{k}'}, \tag{12}$$

where:

$$\rho_1^* \equiv (\rho|b^\dagger b), \quad \rho_{\mathbf{k}}^* \equiv (\rho|a_{\mathbf{k}}^\dagger a_{\mathbf{k}}),$$
$$\rho_{\mathbf{k}\mathbf{k}'}^* \equiv (\rho|a_{\mathbf{k}}^\dagger a_{\mathbf{k}'}), \quad \rho_{\mathbf{k}1}^* \equiv (\rho|a_{\mathbf{k}}^\dagger b), \quad \rho_{1\mathbf{k}'}^* \equiv (\rho|b^\dagger a_{\mathbf{k}'}) \tag{13}$$

Therefore, the states $\rho \in \mathcal{S} \subset \mathcal{O}_D^\times$ are represented by its "components" ρ_1^*, $\rho_{\mathbf{k}}^*$, $\rho_{\mathbf{k}\mathbf{k}'}^*$, $\rho_{\mathbf{k}1}^*$ and $\rho_{1\mathbf{k}'}^*$, while the observables $O \in \mathcal{O}_D$ are represented by the "components" O_1, $O_{\mathbf{k}}$, $O_{\mathbf{k}\mathbf{k}'}$, $O_{\mathbf{k}1}$ and $O_{1\mathbf{k}'}$. We stress the need of "singular components" $\rho_{\mathbf{k}}^*$ and $O_{\mathbf{k}}$ in both states and observables.

The condition on the states representing a finite number of excited modes is given by $\rho_1^* + \int d\mathbf{k}\, \rho_{\mathbf{k}}^* = \langle n + N \rangle_\rho < \infty$. As the total number of modes operator $N_T \equiv n + N$ commutes with the total Hamiltonian, the value of $\langle n+N \rangle_\rho$ is time independent. ρ_1^* gives the mean number of excited modes for the oscillator, while $\rho_{\mathbf{k}}^*\, d\mathbf{k}$ is the mean number of continuous modes having momentum between $\mathbf{k}$ and $\mathbf{k} + d\mathbf{k}$.

If we wish to describe a situation in which there is an infinite number of excited modes in the field ($\langle N \rangle = \infty$), as it is the case when the oscillator is interacting with a bath of radiation with uniform concentration of continuous modes, we need a different characterization of observables and states. In this case, extensive observables like N or H are not well defined and should be excluded from the set of observables $\mathcal{O}$. Only *intensive observables* of the field are accessible for measurement, and we define the class $\mathcal{O}_{TD}$ (the TD corresponds to 'thermodynamic limit') of observables of the form (9) for which $\widetilde{O}_{\mathbf{k}\mathbf{k}'} = O_{\mathbf{k}\mathbf{k}'}$, being $O_{\mathbf{k}\mathbf{k}'}$, $O_{\mathbf{k}1}$ and $O_{1\mathbf{k}'}$ regular functions of the variables $\mathbf{k}$ and $\mathbf{k}'$. For the mean value of any observable $O \in \mathcal{O}_{TD}$ we have:

$$\langle O \rangle_\rho = (\rho|O) = \rho_1^* O_1 + \int\int d\mathbf{k}\, d\mathbf{k}'\, \widetilde{\rho}_{\mathbf{k}\mathbf{k}'}^* O_{\mathbf{k}\mathbf{k}'} +$$
$$+ \int d\mathbf{k}\, \rho_{\mathbf{k}1}^* O_{\mathbf{k}1} + \int d\mathbf{k}'\, \rho_{1\mathbf{k}'}^* O_{1\mathbf{k}'} \tag{14}$$

where $\rho_1^* \equiv (\rho|b^\dagger b)$, $\widetilde{\rho}_{\mathbf{k}\mathbf{k}'}^* \equiv (\rho|a_{\mathbf{k}}^\dagger a_{\mathbf{k}'})$, $\rho_{\mathbf{k}1}^* \equiv (\rho|a_{\mathbf{k}}^\dagger b)$ and $\rho_{1\mathbf{k}'}^* \equiv (\rho|b^\dagger a_{\mathbf{k}'})$. A singular part should be included in $\widetilde{\rho}_{\mathbf{k}\mathbf{k}'}^*$, if we want to consider the possibility of states having a uniform concentration of continuous modes. This fact can be understood by computing the mean value of the density given by (10):

$$\langle \sigma(\mathbf{r}) \rangle = \frac{1}{8\pi^3} \int d\mathbf{k}\, d\mathbf{k}'\, e^{i(\mathbf{k}-\mathbf{k}')\mathbf{r}}\, \widetilde{\rho}_{\mathbf{k}\mathbf{k}'}^*$$

For a state with uniform concentration c we should have $\widetilde{\rho}_{\mathbf{k}\mathbf{k}'}^*$ proportional to $\delta^3(\mathbf{k} - \mathbf{k}')$. Therefore we assume the general form:

$$\widetilde{\rho}_{\mathbf{k}\mathbf{k}'}^* = \rho_{\mathbf{k}}^* \delta^3(\mathbf{k} - \mathbf{k}') + \rho_{\mathbf{k}\mathbf{k}'}^*, \tag{15}$$

being $\rho_{\mathbf{k}}^*$ and $\rho_{\mathbf{k}\mathbf{k}'}^*$ regular functions of the variables $\mathbf{k}$ and $\mathbf{k}'$. Combining (14) and (15) we obtain the general expression:

$$\langle O \rangle_\rho = (\rho|O) = \rho_1^* O_1 + \int d\mathbf{k}\, \rho_{\mathbf{k}}^* O_{\mathbf{k}\mathbf{k}} + \int\int d\mathbf{k}\, d\mathbf{k}'\, \rho_{\mathbf{k}\mathbf{k}'}^* O_{\mathbf{k}\mathbf{k}'} +$$
$$+ \int d\mathbf{k}\, \rho_{\mathbf{k}1}^* O_{\mathbf{k}1} + \int d\mathbf{k}'\, \rho_{1\mathbf{k}'}^* O_{1\mathbf{k}'}, \tag{16}$$

for computing the mean value of any observable $O \in \mathcal{O}_{TD}$ in a state $\rho \in \mathcal{S} \subset \mathcal{O}_{TD}^\times$. The states ρ are represented by their "components" ρ_1^*, $\rho_\mathbf{k}^*$, $\rho_{\mathbf{k}\mathbf{k}'}^*$, $\rho_{\mathbf{k}1}^*$ and $\rho_{1\mathbf{k}'}^*$, where in general $\rho_\mathbf{k}^*$ and $\rho_{\mathbf{k}\mathbf{k}}^*$ are independent objects ($\rho_\mathbf{k}^* \neq \rho_{\mathbf{k}\mathbf{k}}^*$), while O is represented by the "components" O_1, $O_{\mathbf{k}\mathbf{k}'}$, $O_{\mathbf{k}1}$ and $O_{1\mathbf{k}'}$. Notice that in this case there is no singular part for the observables.

4   Time evolution.

In the Heisenberg picture, the functionals ρ representing states are time independent, while the observables evolve in time according to $-i\frac{d}{dt}O = \mathbb{L}^\dagger O \equiv [H, O]$. For the creation and anhiquilation operators $A_\mathbf{k}^\dagger$ and $A_\mathbf{k}$ given in (6) we obtain

$$A_\mathbf{k}^\dagger(t) = e^{i\omega_k t} A_\mathbf{k}^\dagger(0), \qquad A_\mathbf{k}(t) = e^{-i\omega_k t} A_\mathbf{k}(0). \tag{17}$$

The time evolution of $a_\mathbf{k}^\dagger$, $a_\mathbf{k}$, $b^\dagger$ and b can be easily obtained replacing (17) in (7):

$$
b^\dagger(t) = \int \frac{d\mathbf{k}\, V_k}{\eta_-(\omega_k)} e^{i\omega_k t} A_\mathbf{k}^\dagger(0), \quad b(t) = \int \frac{d\mathbf{k}\, V_k}{\eta_+(\omega_k)} e^{-i\omega_k t} A_\mathbf{k}(0) \tag{18}
$$

$$
a_\mathbf{p}^\dagger(t) = e^{i\omega_p t} A_\mathbf{p}^\dagger(0) + \int \frac{d\mathbf{k}\, V_k V_p}{\eta_-(\omega_k)\,(\omega_k - \omega_p - i0)} e^{i\omega_k t} A_\mathbf{k}^\dagger(0)
$$

$$
a_\mathbf{p}(t) = e^{-i\omega_p t} A_\mathbf{p}(0) + \int \frac{d\mathbf{k}\, V_k V_p}{\eta_+(\omega_k)\,(\omega_k - \omega_p + i0)} e^{-i\omega_k t} A_\mathbf{k}(0) \tag{19}
$$

Using (18) and (19) we obtain:

$$
(\rho|b^\dagger b)_t = \int \frac{d\mathbf{k} d\mathbf{k}'\, V_k V_{k'} (\rho|A_\mathbf{k}^\dagger A_{\mathbf{k}'})_{t=0}\, e^{i(\omega_k - \omega_{k'})t}}{\eta_-(\omega_k)\,\eta_+(\omega_{k'})} \tag{20}
$$

$$
(\rho|a_\mathbf{p}^\dagger a_{\mathbf{p}'})_t = (\rho|A_\mathbf{p}^\dagger A_{\mathbf{p}'})_{t=0}\, e^{i(\omega_p - \omega_{p'})t} +
$$

$$
+ \int \frac{d\mathbf{k}'\, V_{k'} V_{p'} (\rho|A_\mathbf{p}^\dagger A_{\mathbf{k}'})_{t=0}\, e^{i(\omega_p - \omega_{k'})t}}{\eta_+(\omega_{k'})\,(\omega_{k'} - \omega_{p'} + i0)} + \int \frac{d\mathbf{k}\, V_k V_p (\rho|A_\mathbf{k}^\dagger A_{\mathbf{p}'})_{t=0}\, e^{i(\omega_k - \omega_{p'})t}}{\eta_-(\omega_k)\,(\omega_k - \omega_p - i0)} +
$$

$$
+ \int \frac{d\mathbf{k} d\mathbf{k}'\, V_k V_{k'} (\rho|A_\mathbf{k}^\dagger A_{\mathbf{k}'})_{t=0}\, e^{i(\omega_k - \omega_{k'})t}}{\eta_+(\omega_{k'})\,(\omega_{k'} - \omega_{p'} + i0)\eta_-(\omega_k)\,(\omega_k - \omega_p - i0)}. \tag{21}
$$

In the previous expressions $A_\mathbf{k}^\dagger(0)$ and $A_\mathbf{k}(0)$ can be expressed in terms of $a_\mathbf{k}^\dagger(0)$, $a_\mathbf{k}(0)$, $b^\dagger(0)$ and $b(0)$ using (6).

The total number of modes, given by the operator $N_T \equiv n + N = b^\dagger b + \int d\mathbf{k}\, a_\mathbf{k}^\dagger a_\mathbf{k}$ satisfies $[H, N_T] = 0$ and therefore N_T is conserved during time evolution. Two different physical situations will be discussed in the following subsections: first we will analyze the case with a finite number of modes, and then the infinite number of modes (thermodynamic limit).

4.1 Finite number of excited modes (decaying process).

In this case, extensive observables of the field like the number of modes $\langle N \rangle$ or the energy $\langle H_F \rangle$ are finite. In equations (9), (11), (12) and (13) we obtained the general form for states and observables.

Let us consider an initial state for which the mean number of excited modes in the oscillators is $\langle n \rangle_0$, and the mean number of modes of the field is $\langle N \rangle_0$, with a momentum distribution $f(\mathbf{k})$, i. e.:

$$\rho_1^* = \langle b^\dagger b \rangle_0 = \langle n \rangle_0, \quad \rho_{\mathbf{k}}^* = \langle a_{\mathbf{k}}^\dagger a_{\mathbf{k}} \rangle_0 = f(\mathbf{k}),$$

$$\int d\mathbf{k}\, f(\mathbf{k}) = \langle N \rangle_0, \quad \rho_{\mathbf{k}\mathbf{k}'}^* = \rho_{\mathbf{k}1}^* = \rho_{1\mathbf{k}'}^* = 0. \tag{22}$$

The density $\sigma(\mathbf{r})$ of continuous modes at the point $\mathbf{r}$, defined in (10) is of the form given in (9) and (11), with:

$$[\sigma(\mathbf{r})]_1 = [\sigma(\mathbf{r})]_{\mathbf{k}} = [\sigma(\mathbf{r})]_{\mathbf{k}1} = [\sigma(\mathbf{r})]_{1\mathbf{k}} = 0, \quad [\sigma(\mathbf{r})]_{\mathbf{k}\mathbf{k}'} = \frac{1}{8\pi^3} e^{i(\mathbf{k}-\mathbf{k}')\mathbf{r}}$$

The mean value of the density $\sigma(\mathbf{r})$ in the state defined by (22) is obtained using (12): $\langle \sigma(\mathbf{r}) \rangle_\rho = (\rho|\sigma(\mathbf{r})) = 0$ which is a reasonable result, as we are dealing with a finite number $\langle N \rangle_0$ of continuous modes in an infinite volume ($\mathbb{R}^3$).

From (6) and (22) we obtain:

$$(\rho|A_{\mathbf{p}}^\dagger A_{\mathbf{p}})_0 = f(\mathbf{p}) + \frac{V_p^2 \langle n \rangle_0}{\eta_+(\omega_p)\eta_-(\omega_p)} \tag{23}$$

$$(\rho|A_{\mathbf{p}}^\dagger A_{\mathbf{p}'})_0 = \frac{V_p V_{p'} \langle n \rangle_0}{\eta_+(\omega_p)\eta_-(\omega_{p'})}. \tag{24}$$

Replacing (24) in (20) we obtain:

$$\langle b^\dagger b \rangle_t = \langle n \rangle_0 \left| \int \frac{d\mathbf{k}\, V_k^2 e^{i\omega_k t}}{\eta_+(\omega_k)\,\eta_-(\omega_k)} \right|^2 \tag{25}$$

Replacing (23) and (24) in (21) (with $\mathbf{p} = \mathbf{p}'$) we obtain:

$$\begin{aligned}
\langle a_{\mathbf{p}}^\dagger a_{\mathbf{p}} \rangle_t = {} & f(\mathbf{p}) + \frac{\langle n \rangle_0 V_p^2}{\eta_+(\omega_p)\,\eta_-(\omega_p)} + \\
& + \frac{\langle n \rangle_0 V_p^2 e^{i\omega_p t}}{\eta_+(\omega_p)} \int \frac{d\mathbf{k}'\, V_{k'}^2 e^{-i\omega_{k'} t}}{\eta_+(\omega_{k'})\,\eta_-(\omega_{k'})\,(\omega_{k'} - \omega_p + i0)} + \\
& + \frac{\langle n \rangle_0 V_p^2 e^{-i\omega_p t}}{\eta_-(\omega_p)} \int \frac{d\mathbf{k}\, V_k^2 e^{i\omega_k t}}{\eta_+(\omega_k)\,\eta_-(\omega_k)\,(\omega_k - \omega_p - i0)} + \\
& \int\int \frac{d\mathbf{k}\, d\mathbf{k}'\, V_{k'}^2 V_k^2 \langle n \rangle_0 V_p^2 e^{i(\omega_k - \omega_{k'})t}}{|\eta_+(\omega_k)|^2 |\eta_+(\omega_{k'})|^2 (\omega_k - \omega_p - i0)\,(\omega_{k'} - \omega_p + i0)}.
\end{aligned} \tag{26}$$

Riemann Lebesgue theorem can be used in (25) and (26) to obtain:

$$\lim_{t\to\infty} \langle b^\dagger b\rangle_t = \lim_{t\to\infty} \langle n\rangle_t = 0, \qquad \lim_{t\to\infty} \langle a_{\mathbf{p}}^\dagger a_{\mathbf{p}}\rangle_t = f(\mathbf{p}) + \frac{\langle n\rangle_0\, V_p^2}{\eta_+(\omega_p)\,\eta_-(\omega_p)} \tag{27}$$

From the previous equation we obtain:

$$\begin{aligned}
\lim_{t\to\infty} \langle N\rangle_t &= \lim_{t\to\infty} \int d\mathbf{p}\, \langle a_{\mathbf{p}}^\dagger a_{\mathbf{p}}\rangle_t = \\
&= \int d\mathbf{p}\, f(\mathbf{p}) + \langle n\rangle_0 \int \frac{d\mathbf{p}\, V_p^2}{\eta_+(\omega_p)\,\eta_-(\omega_p)} = \langle N\rangle_0 + \langle n\rangle_0.
\end{aligned}$$

Therefore the discrete system decays to the vacuum, transferring the initial number of modes to the field (As we pointed out the total number of modes is a constant of motion).

The momentum distribution for $t \to \infty$ is equal to the initial distribution $f(\mathbf{p})$ plus an isotropic term having a sharp peak around $\omega_p = \Omega$ (it is easy to prove that for small interactions $\frac{V_p^2}{\eta_+(\omega_p)\,\eta_-(\omega_p)} \simeq \frac{\delta(\omega_p-\Omega)}{4\pi\Omega^2}$). The energy of the discrete system is completely transferred to the field.

Expression (25) is a well known result for Friedrichs model [12], that we reobtain in the functional approach. From this expression it is possible to prove that $\frac{d}{dt}\langle b^\dagger b\rangle_{t=0} = 0$ (Zeno regime) and also that for $t \to \infty$, the asymptotic form of $\langle b^\dagger b\rangle_t$ is proportional to t^{-6} (Khalfin regime [13] corresponding to $\omega_k = |\mathbf{k}|$). Therefore, the functional approach applied to the model with a finite number of excited modes do not allow pure exponential decay.

4.2 Infinite number of excited modes (thermodynamic limit).

In this case extensive observables like energy or number of modes of the field are not well defined, since they are really infinite and therefore cannot be considered. Only local observables of the field like the density $\sigma(\mathbf{r})$ defined in (10) are available. In equations (14) to (16) we obtained the general form of observables, states and mean values.

We are going to consider the initial condition for which the mean number of discrete modes in the system is finite ($\langle n\rangle_0 = (\rho|b^\dagger b)_{t=0} < \infty$), and the field has a uniform distribution in the space with concentration c:

$$(\rho|\sigma(\mathbf{r})) = \frac{1}{8\pi^3} \int d\mathbf{k} \int d\mathbf{k}'\, e^{i\,(\mathbf{k}-\mathbf{k}')\,\mathbf{r}} (\rho|a_{\mathbf{k}}^\dagger a_{\mathbf{k}'}) = c$$

From the previous equation we obtain:

$$\widetilde{\rho}_{\mathbf{k}\,\mathbf{k}'}^* = (\rho|a_{\mathbf{k}}^\dagger a_{\mathbf{k}'})_{t=0} = \rho(\mathbf{k})\,\delta^3(\mathbf{k}-\mathbf{k}'), \qquad \int d\mathbf{k}\,\rho(\mathbf{k}) = 8\pi^3 c. \tag{28}$$

We also assume for simplicity that there is no initial correlation between the discrete system and the field:

$$\rho_{\mathbf{k}1}^* = (\rho|\,a_{\mathbf{k}}^\dagger\, b)_{t=0} = 0, \qquad \rho_{1\mathbf{k}}^* = (\rho|b^\dagger\, a_{\mathbf{k}})_{t=0} = 0 \tag{29}$$

Using (28), (29) and the definition (6) for $A_{\mathbf{k}}^{\dagger}$ we obtain:

$$(\rho|A_{\mathbf{p}}^{\dagger}A_{\mathbf{p}'})_{t=0} = \rho(\mathbf{p})\,\delta^3(\mathbf{p}-\mathbf{p}') + \frac{\langle n\rangle_0\,V_p V_{p'}}{\eta_+(\omega_p)\,\eta_-(\omega_{p'})} +$$

$$+ \frac{V_p V_{p'}\rho(\mathbf{p})}{\eta_-(\omega_{p'})(\omega_{p'}-\omega_p-i0)} + \frac{V_p V_{p'}\rho(\mathbf{p}')}{\eta_+(\omega_p)(\omega_p-\omega_{p'}+i0)} +$$

$$+ \frac{V_p V_{p'}}{\eta_+(\omega_p)\,\eta_-(\omega_{p'})} \int \frac{d\mathbf{k}\,V_k^2\,\rho(\mathbf{k})}{(\omega_p-\omega_k+i0)(\omega_{p'}-\omega_k-i0)}. \tag{30}$$

The last expression can be replaced in (20) and (21) to obtain the explicit expressions for $\langle b^\dagger b\rangle_t$ and $\langle a_{\mathbf{k}}^\dagger a_{\mathbf{k}'}\rangle_t$. The time dependence is rather complicated, but the Riemann Lebesgue theorem can be used to eliminate the oscillating terms and to obtain the following asymptotic expressions for $t\to\infty$:

$$\lim_{t\to\infty}(\rho|b^\dagger b)_t = \int \frac{d\mathbf{p}\,V_p^2\,\rho(\mathbf{p})}{\eta_+(\omega_p)\,\eta_-(\omega_p)} \tag{31}$$

$$\mathrm{W}\lim_{t\to\infty}(\rho|a_{\mathbf{p}}^\dagger a_{\mathbf{p}'})_t = \rho(\mathbf{p})\,\delta^3(\mathbf{p}-\mathbf{p}') + \frac{V_p V_{p'}\rho(\mathbf{p}')}{\eta_-(\omega_{p'})(\omega_{p'}-\omega_p-i0)} + \tag{32}$$

$$+ \frac{V_p V_{p'}\rho(\mathbf{p})}{\eta_+(\omega_p)(\omega_p-\omega_{p'}+i0)} + \int \frac{d\mathbf{k}\,V_k^2\rho(\mathbf{k})V_p V_{p'}}{\eta_-(\omega_k)\,\eta_+(\omega_k)\,(\omega_k-\omega_p-i0)(\omega_k-\omega_{p'}+i0)}$$

The last limit should be understood in the weak sense, i.e.:

$$\lim_{t\to\infty}\int\int d\mathbf{p}\,d\mathbf{p}'\,(\rho|a_{\mathbf{p}}^\dagger a_{\mathbf{p}'})_t\,O_{\mathbf{p}\mathbf{p}'} = \int\int d\mathbf{p}\,d\mathbf{p}'\left[\mathrm{W}\lim_{t\to\infty}(\rho|a_{\mathbf{p}}^\dagger a_{\mathbf{p}'})_t\right]O_{\mathbf{p}\mathbf{p}'}.$$

The main difference with the results of subsection IV-A is that in this case equation (31) shows that the discrete system does not decay towards the vacuum. The infinite number of modes of the field surrounding the discrete system produces a final state of the system independent of $\langle n\rangle_0$, but dependent on the interaction and the initial momentum distribution of the field.

If the initial condition for the bosonic field is a canonical distribution with temperature $T=\frac{1}{\kappa\beta}$ (κ is the Boltzmann constant), we have:

$$\rho(\mathbf{k}) = \frac{1}{e^{\beta\omega_k}-1} \tag{33}$$

If, in addition, we assume a small interaction, we have:

$$\frac{V_p^2}{\eta_+(\omega_p)\,\eta_-(\omega_p)} \simeq \frac{\delta(\omega_p-\Omega)}{4\pi\Omega^2}. \tag{34}$$

Replacing (33) and (34) in (31) we obtain:

$$\langle b^\dagger b\rangle_{t\to\infty} \simeq \frac{1}{e^{\beta\Omega}-1},$$

which is the mean number of modes of the discrete system we would have obtained for a quantum oscillator of frequency Ω in a canonical distribution with temperature $T=\frac{1}{\kappa\beta}$. This result cannot be obtained without the assumption of weak interaction.

5 Conclusions.

Let us summarize our main results:

For a quantum oscillator ($H_S = \Omega b^\dagger b$) and a quantum field ($H_F = \int d\mathbf{k}\, \omega_k a_{\mathbf{k}}^\dagger a_{\mathbf{k}}$), we considered a linear coupling ($H_{INT} = \int d\mathbf{k}\, V_k[a_{\mathbf{k}}^\dagger b + b^\dagger a_{\mathbf{k}}]$). For a finite number of excited modes, we used the functional approach to obtain the time evolution of a state having a finite number of excited modes. The mean number of discrete modes is given by:

$$\langle b^\dagger b \rangle_t = \langle n \rangle_0 \left| \int \frac{d\mathbf{k}\, V_k^2 e^{i\omega_k t}}{\eta_+(\omega_k)\, \eta_-(\omega_k)} \right|^2, \qquad \eta_\pm(\omega_k) \equiv \omega_k - \Omega - \int \frac{d\mathbf{k}'\, V_{k'}^2}{\omega_k - \omega_{k'} \pm i0}.$$

This is a well known result [12], that we reobtain in the functional approach. From this expression it is possible to prove that $\frac{d}{dt}\langle b^\dagger b \rangle_{t=0} = 0$ (Zeno regime) and also that for $t \to \infty$, the asymptotic form of $\langle b^\dagger b \rangle_t$ is proportional to t^{-6}. Therefore, the functional approach applied to the model with a finite number of excited modes do not allow pure exponential decay.

The functional approach is a powerful tool to deal with the case of an infinite number of excited modes in the field. In this case, the mean number of discrete modes approach *in non exponential form* to:

$$\langle b^\dagger b \rangle_\infty = \lim_{t \to \infty} (\rho|b^\dagger b)_t = \int \frac{d\mathbf{p}\, V_p^2\, \rho(\mathbf{p})}{\eta_+(\omega_p)\, \eta_-(\omega_p)}.$$

This number depends on the initial momentum distribution $\rho(\mathbf{p})$ of the field and the form of the interaction V_p, but it do not depend on the initial condition of the quantum oscillator. If the interaction is very small, $\langle b^\dagger b \rangle_\infty$ is independent of the form of the interaction, precisely:

$$\langle b^\dagger b \rangle_\infty \simeq \frac{1}{4\pi\Omega^2} \int d\mathbf{p}\, \delta(\omega_p - \Omega)\rho(\mathbf{p}).$$

If in addition the initial momentum distribution of the field is the canonical distribution with temperature T, we obtained:

$$\langle b^\dagger b \rangle_\infty \simeq \frac{1}{e^{\frac{\Omega}{\kappa T}} - 1},$$

which is the mean number of modes for a single oscillator having temperature T. Therefore, for a small interaction, we proved the evolution of the oscillator to the thermal equilibrium with the field.

It is interesting to note that:

1.-The space of "test observables" $\mathcal{O}_D$ for a finite number of excited modes is different from the space $\mathcal{O}_{TD}$ for an infinite number of excited modes. In the later case only intensive observables of the field are accessible for measurement, and therefore we excluded the possibility of a singular part $\int d\mathbf{k}\, O_{\mathbf{k}} a_{\mathbf{k}}^\dagger a_{\mathbf{k}}$ from $\mathcal{O}_{TD}$. This difference between $\mathcal{O}_D$ and $\mathcal{O}_{TD}$ determinates different time evolution for the states, since they are functionals whose properties depend on the domain of definition.

2.-The strong limit of the state for infinite time does not exist, but a functional ρ_∞ exists such that $\lim_{t \to \infty}(\rho_t|O) = (\rho_\infty|O)$ for all O in the set of observables. We do not have, as it is the case in the "coarse graining" method, a preferred set of relevant components of the state, obtained for a set of preferred observables. In our case we really consider the set of "components" $(\rho|O)$ of the states ρ, labelled by all the observables O. The difficulty of defining a canonical coarse graining is avoided in this approach.

3.-As no analytic extensions have been involved to obtain the exact results, no special riggings of the spaces of states and observables have being used. Only mild conditions on the momentum distribution of the field are necessary to use Riemann Lebesgue theorem in order to obtain the states for $t \rightarrow \infty$.

Acknowledgments

This work was partially supported by Grant No. CI1-CT94-0004 of the European Community, Grant No. PID-0150 of CONICET (National Research Council of Argentina), Grant No. EX-198 of Buenos Aires University, Grant No. 12217/1 of Fundación Antorchas, and also a Grant from Foundation pour la Recherche Foundamentale OLAM.

The author is very grateful to the organizer of the workshop for their very kind hospitality during his stay in Chile.

References

[1] E.Fermi, J.Pasta and S.Ulam, in *"Collected papers of Enrico Fermi"*, Vol II, p.978, University of Chicago Press, (1965).

[2] I.Prigogine, *"Non Equilibrium Statistical Mechanics"*, Interscience Publishers, John Wiley and Sons Inc., (1962).

[3] P.Blaise, P.Durand, O.Henri-Rousseau, Physica A, **209**, 51-82, (1994).

[4] J.L.Gruver, J.Aliaga, H.Cerdeira, A.Proto, Phys.Rev.E, **51**, 6263, (1995).

[5] I.Antoniou, Z.Suchanecki, in *"Nonlinear, deformed and irreversible quantum systems"*.H.D.Doebner et al editors, World Scientific (1995).

[6] I.Antoniou, Z.Suchanecki, Found. of Physics, **24** 1439-1457 (1994)

[7] I.Antoniou, Z.Suchanecki, R.Laura, S.Tasaki, Physica A, **241**, 737-772 (1996).

[8] I.E.Segal, Annals Math., **48**, 930-948 (1947)

[9] R.Laura, Int.Jour.Theor.Phys., **36**, 2315-2334 (1997)

[10] R.Laura, M.Castagnino, Phys.Rev.A, **57**, 4140-4152 (1998)

[11] R.Laura, M.Castagnino, *"Functional approach for quantum systems with continuous spectrum"*, Phys.Rev.E (1998) (in press)

[12] E.C.G.Sudarshan, C.B.Chiu, V.Gorini, Phys. Rev. D, **18**, 2914-2929, (1978).

[13] L.A.Khalfin, Sov. Phys. JETP **6**, 1053-1063, (1958)

LINEAR MARANGONI PROBLEM IN A DILUTED POLYMER

J. Martínez-Mardones, M. Vega, and W. Zeller[*]
Instituto de Física
Universidad Católica de Valparaíso
Casilla 4059, Valparaíso-Chile

ABSTRACT

Linear stability analysis of the Bénard-Marangoni problem is a realistic polymeric fluid layer with a plane-free upper surface. Results of stationary and oscillatory instabilities obtained for models of diluted polymeric Jeffreys fluids and Maxwellian concentrated solutions were compared. It was shown that the oscillatory behavior may often occur before the onset of steady convection at physically meaningful parameter ranges.

I. INTRODUCTION

Extensive linear and non-linear analysis of the Rayleigh-Bénard convection in thin Newtonian fluid layers heated from below have been performed theoretically and experimentally [1]. In viscoelastic fluids we find similar questions [2-3]. Nevertheless, the behaviours registered in viscoelastic polymeric solutions are sometimes substantially different [4-5]. The introduction of realistic constitutive equations and capillarity effects due to surface tension has to be considered on a plane fee and nonconducting upper surface, whereas the bottom is a mechanically rigid and thermally conducting surface [6-9]. The present study considers diluted polymeric fluids in the context of linear stability theory, as a first step towards further inclusion of deflections in the upper limiting surface in the linear approach.

The convection of non-Newtonian fluids has been treated [6-8] specifically for some restrictive constitutive equations and, in general, it has been evaluated for unrealistic physical parameters.

II. FORMULATION

The dynamic of an infinite horizontal viscoelastic incompressible polymeric fluid layer of depth d, subject to Boussinesq approximation, is governed by the well-known mass, momentum and energy balance equations, and subject to boundary conditions. It is also assumed that we have to deal with viscoelastic fluid of flexible polymer molecules in diluted and concentrated solutions that obey the Jeffreys constitutive equation, which can be derived from a molecular theory,

---

[*] Deceased

O. Descalzi et al. (eds.), Instabilities and Nonequilibrium Structures VII & VIII, 267–272.
© 2004 *Kluwer Academic Publishers. Printed in the Netherlands.*

$$\hat{\sigma} + \lambda_1 \partial_t \hat{\sigma} = 2\mu\left(1 + \lambda_2 \partial_t\right)\mathbf{D}. \tag{1}$$

The rate strain tensor $\mathbf{D} = 1/2\left[\nabla \mathbf{v} + \left(\nabla \mathbf{v}\right)^T\right]$ is expressed in terms of the velocity field $\mathbf{v}$, and the fluid properties to be taken into account are the viscosity μ, the relaxation time λ_1, and the retardation time λ_2. Following the standard procedure, the set of governing relations and the boundary equations of the basic conductive state is perturbed. The dynamics of the linearized and adimensionalized system of perturbations may be reduced to a couple of equations of the form

$$\left(\partial_t - \nabla\right)^2 \theta = w \tag{2}$$

$$\left[\left(1 + \Gamma\Lambda\partial_t\right)\nabla^2 - \left(1 + \Gamma\partial_t\right)\mathrm{P}^{-1}\partial_t\right]\nabla^2 w = \left(1 + \Gamma\partial_t\right)R\nabla_H^2 \theta, \tag{3}$$

where the adimensional quantities w and θ are the vertical component of the velocity and the temperature perturbations, respectively; $R = \rho_0 g\alpha\Delta T d^3 / \kappa\mu$, the Rayleigh number; $\mathrm{Pr} = \mu/\rho_0\kappa$, the Prandtl number; $\Gamma = \lambda_1\kappa/d^2$, the Deborah number; $\Lambda = \lambda_2/\lambda_1$, the quotient between the relaxation and retardation times; and the horizontal Laplacian is $\nabla_H^2 = \nabla_{xx}^2 + \nabla_{yy}^2$. The thin fluid layer is limited by a mechanically rigid and heat-conducting bottom. The heat behavior at the top of the layer can be described by both the Biot law and the capillarity effect due to surface tension σ. With the introduction of normal modes perturbations, we obtain $\left[w(x,t), \theta(x,t)\right] = \left[W(z), \Theta(z)\right]\exp\left[i\left(k_x x + k_y y\right) + st\right]$, where $s = \alpha + i\omega$ is the complex growth rate; k_x and k_y are the components of the wave vector; and $\upsilon = \left(1 + \Gamma\Lambda s\right)/\left(1 + \Gamma s\right)$, we obtain the following pair of equations:

$$\left[D^2 - \left(k^2 + s/\upsilon P\right)\right]\left[D^2 - k^2\right]W = \left(k^2 R/\upsilon\right)\Theta \tag{4}$$

$$\left[D^2 - \left(k^2 + s\right)\right]\Theta = -W. \tag{5}$$

The boundary relations for the amplitudes $W(z)$ and $\Theta(z)$ on the bottom are

$$W = DW = \Theta = 0, \qquad\qquad \text{at } z = 0. \tag{6}$$

The upper plane is subject to Marangoni capillarity behavior from surface stress and, therefore, we find the following conditions:

$$W = 0, \quad D^2W + \left(k^2/\upsilon\right) = 0, \quad D\Theta + H\Theta = 0, \quad \text{at } z = 1, \tag{7}$$

where $M = \sigma' d\Delta T/\kappa\mu\rho$ is the Marangoni adimensional, number and $\sigma' = -d\sigma/d\theta$. Finally, by taking the whole set of equations (4-7) together, we obtain an eigenvalue problem expressed by the matrix of First order differential equations system:

$$DX = AX, \tag{8}$$

where the vector X is $\left(X^1 = W, \quad X^2 = DW, \quad X^3 = D^2W, \quad X^4 = D^3W, \quad X^5 = \Theta, \quad X^6 = D\Theta\right)^T$, and the matrix A is

$$A = \begin{bmatrix} 0 & 1 & 0 & 0 & 0 & 0 \\ 0 & 0 & 1 & 0 & 0 & 0 \\ 0 & 0 & 0 & 1 & 0 & 0 \\ -k^2\left(k^2 + \dfrac{s}{P\upsilon}\right) & 0 & 2k^2 + \dfrac{s}{P\upsilon} & 0 & \dfrac{k^2R}{\upsilon} & 0 \\ 0 & 0 & 0 & 0 & 0 & 1 \\ -1 & 0 & 0 & 0 & k^2 + s & 0 \end{bmatrix}. \tag{9}$$

The numerical solution, given via the shouting method, leads to an initial condition problem solved by straightforward Runge-Kutta integration.

III. RESULTS AND DISCUSSION

The onset of instability took place as the perturbation growth parameter vanished $(\alpha = 0)$. When the frequency ω was zero, the instability was stationary; otherwise, it was oscillatory. The Marangoni number M was presented as function of the wave number k in Fig. 1 a) for the marginal state (dot curve) and for oscillatory behaviors (full curves) for three characteristic relaxation times. The Taker-Bourdanoff two-codimensional and a high Deborah number solutions are found for $\Gamma = 0.083$ and $\Gamma = 0.143$, respectively. For a realistic fluid (P = 1.000, H = 0.1, and $\Gamma = 1.0$) (Fig. 1b), neutral Marangoni curves for five values of Λ on the range between the Maxwellian concentrated polymer ($\Lambda = 0.0$), passing through a diluted polymer ($\Lambda = 0.8$) and the pure Newtonian fluid ($\Lambda = 1.0$), were shown. The Deborah number ($\Gamma = 1.0$) was selected so that the oscillatory state became dominant. The critical values in capillary overstability for Marangoni, wave, and frequency numbers were given as functions of the relaxation time in Figs. 2 a) and c), respectively.

In Fig. 3, solid curves expressed the relation between the critical Marangoni and Rayleigh numbers for realistic diluted polymer. These curves normally ran below the Nield curve (dot) for stationary instability in a Newtonian fluid [7]. Therefore, oscillatory instabilities can appear for diluted polymers. Results were obtained just over the Nield curve at very low critical Rayleigh numbers, when the relaxation time had values below $\Gamma = 0.74$. In general, we expected oscillatory instabilities for diluted polymers, since both stationary and oscillatory instability only arose in a very narrow range of Γ.

In general, the presence of either travelling or standing waves could be found, as well as a stationary pattern formation. However, the lmited scope of linear theory of the present work could not give us insight into the persistence of eventual oscillatory pattern formation.

References

[1]   Drazin P.G. and Reld, W.H. (1981) Hydrodynamic Stability Cambridge University Press, Cambridge.

[2]   Martínez-Mardones J. and Pérez-García. C. (1990) "Linear Instability in in viscoelastic fluid convection", J. Phys.: Condens. Matter 2, 1281.

[3]   Joseph, D.D. (1990) Fluid Dynamics of Viscoelastic Liquids, Springer-Verlag, Berlin.

[4]   Tanner, R.I. (1985) Engineering Rheology, Claredon Press, Oxford.

[5]   Kolka, R.W. and Iarley, G. (1987) "On the convected linear stability of a viscoelastic Oldroyd B fluid heated from below", J. Non-Newtonian Mech. 25, 209.

[6]   Pérez-García C. and Carneiro, G. (1991) "Linear stability analysis of Bénard-Marangoni convection in fluids with a deformable free surface", Phys. Fluids A 3, 292.

[7]   Nield, D.A. (1964) "Surface tension and buoyancy effects in cellular convection", J. Fluid Mech. 19, 341.

[8]   Velarde, M.G. (ed.), (1993) "Proceedings of the Physicochemical Hydrodynamics Interfacial Phenomena", Plenum, New York.

[9]   Davis, S.H. (1987) "Thermocapillary Instabilities", Ann. Rev. Fluid Mech. 19, 403.

[10]   Benguria R.D. and Depassier, M.D. (1989) "On the linear stability theory Bénard-Marangoni convection", Phys. Fluids A 1, 1123.

[11]   Dauby. P.C., Parmentler. P., Lebon G., and Grmela. M. (1993) Coupled buoyancy and thermocapillary convection in a viscoelastic Maxwell fluid", J. Phys.: Condens. Matter 5, 4334.

[12]   Lebon G., and Cloot, A. (1988) "An extended thermodynamic approach to non-newtonian fluids and related results in Marangoni problem", J. Non-Newtonian Fluid Mech. 28, 61.

[13] Getachew, D., and Rosenblat, S. (1985) "Thermocapillary Instabillity of a viscoelastic liquid layer", Acta Mechanica 55, 137.

[14] Garazo A.N., and Velarde, M.G. (1991) "Dissipative Korteweg – de Vries description of Bénard-Marangoni convection", Phys. Fluids. A, 2295.

FIGURES

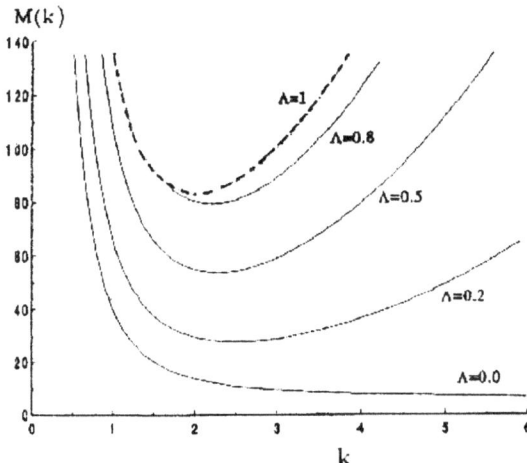

Figure 1a)   Marangoni neutral curves, with R = 0 corresponding to stationary (dot) and oscillatory (full) instability for P = 1.000, H = 0.1 and Γ = 1.

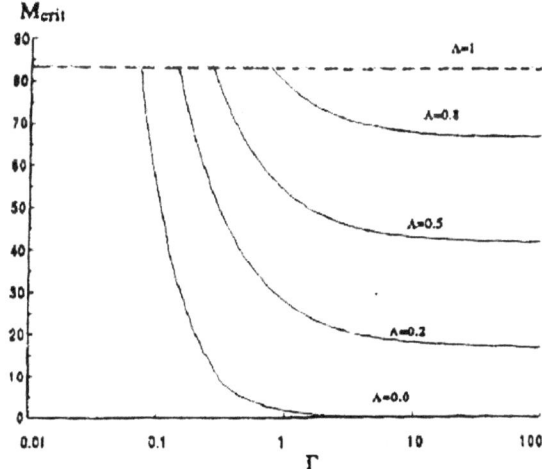

Figure 1b)   Critical Marangoni number for different values of Λ.

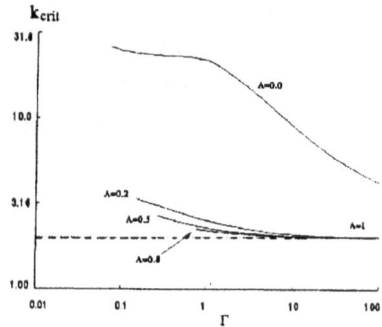

Figure 2a.- Critical wave number k_c for different values of Λ.

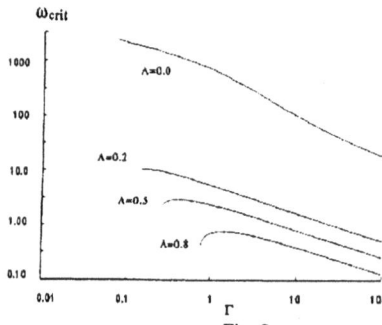

Figure 2b.- Critical frequency ω_c for different values of Λ.

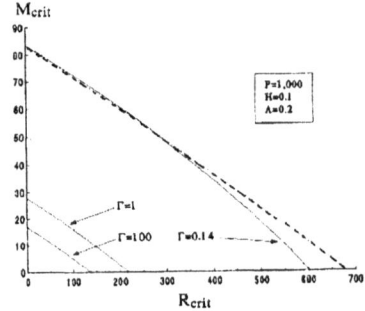

Figura 3.- Generalized Nield curves for stationary (dot) and some values of Γ for oscillatory (full) for critical Marangoni- Rayleigh instability.

On some localized solutions of coupled Ginzburg–Landau equations

by

R. Montagne[1][‡], E. Hernández-García [2][§]

[1]: Instituto de Física – Universidad de la República.
Iguá 4225, Montevideo 11400 –Uruguay.
[2]: Institut Mediterrani d'Estudis Avançats, IMEDEA[1].
CSIC-Universitat de les Illes Balears, E-07071 Palma de Mallorca –Spain.

Abstract

Coupled Complex Ginzburg-Landau equations describe generic features of the dynamics of coupled fields when they are close to instabilities leading to nonlinear oscillations. We study numerically this equation set within a particular range of parameters, and find uniformly propagating localized objects behaving as coherent structures. Some of the localized objects found are interpreted in terms of exact analytical solutions.

1 Introduction

When an extended system is close to a Hopf bifurcation leading to uniform oscillations, the amplitude of the oscillations can be generically described in terms of the complex Ginzburg-Landau (CGL) equation [1]. When there are two fields becoming unstable at the same bifurcation, coupled complex Ginzburg-Landau equations (CCGL) should be used instead. This model set of equations appears in a number of contexts including convection in binary mixtures and transverse instabilities in unpolarized lasers [2, 3].

Coherent structures such as fronts, shocks, pulses, and other localized objects play an important role in the dynamics of extended systems [4]. In particular, for the complex Ginzburg-Landau equation, they provide the *building blocks* from which some kinds of spatiotemporally chaotic behavior are built-up [5]. A systematic study of localized structures in CCGL equations in one spatial dimension was initiated in [6]. That paper focused on parameter ranges such that wave coexistence is unlikely so that the two oscillating fields are mutually excluding.

Here we present results on one dimensional CCGL equations in parameter ranges such that they can be written as

$$\partial_t A_\pm = \mu A_\pm + (1 + i\alpha)\partial_x^2 A_\pm - (1 + i\beta)\left(|A_\pm|^2 + \gamma|A_\mp|^2\right) A_\pm \,. \tag{1}$$

Group velocity terms of the form $\pm v_g \partial_x A_\pm$ are explicitly excluded, and γ is restricted to take real values (without loss of generality, α and β are also real parameters). In addition we just consider

---

[1]http://www.imedea.uib.es/Nonlinear

O. Descalzi et al. (eds.), Instabilities and Nonequilibrium Structures VII & VIII, 273–279.
© 2004 *Kluwer Academic Publishers. Printed in the Netherlands.*

$1 + \alpha\beta > 0$ (Benjamin-Feir stable range). These restrictions are the appropriate ones for the description of transverse laser instabilities [2]. In that case $A_\pm$ are related to the two orthogonal circularly polarized light components. We further restrict our study to the case $0 < \gamma < 1$ which is the range obtained when atomic properties in the laser medium favor linearly polarized emission. In terms of the wave amplitudes $A_\pm$, wave coexistence is preferred.

2   Localized objects

Many experiments on traveling wave systems or numerical simulations of Ginzburg–Landau–type equations [1, 7, 8] exhibit local structures that have an essentially time–independent shape and propagate with a constant velocity, at least during an interval of time where they appear to be coherent structures [9, 7, 6]. In order to analyze these structures it is common to reduce the initial partial differential equation into a set of ordinary differential equations by restricting the class of solutions to uniformly traveling ones. Localized structures are homoclinic or heteroclinic orbits in this reduced dynamical system, that is they approach simple solutions (typically plane waves) in opposite parts of the system, whereas they exhibit a distinct shape in between.

Instead of looking for solutions of the reduced dynamical system, we prefer here to resort to direct numerical solution of (1) under different initial conditions. A pseudo–spectral code [7] with periodic boundary conditions and a second–order accuracy in time is used. Spatial resolution was typically 512 modes. Time step was typically 0.05. The system size was always taken to be $L = 512$. Several kinds of localized objects which maintain coherence for a time appear and travel around the system. Different initial conditions give birth to different kinds of structures. Some of them decay shortly, and the qualitative dynamics at long times becomes determined by the remaining ones, and essentially independent of the initial conditions.

The upper part of Fig. 1 shows the spatiotemporal evolution of $|A_+(x,t)|$ and $|A_-(x,t)|$ at parameter values $\alpha = -0.35$, $\beta = -2.0$ and $\gamma = 0.2$. Time runs upwards and x is represented in the horizontal direction. Lighter grey corresponds to the maximum values of $|A_\pm(x,t)|$ and darker to the minima. This particular evolution was obtained starting from $A_+(x,0)$ equal to the *Nozaki-Bekki hole*, a known analytical solution of the single Ginzburg-Landau equation [10, 11], and for $A_-(x,0)$ a *Nozaki-Bekki pulse* [11]. These are not exact solutions of the set of equations (1) so that this initial condition decays and gives rise to complex spatiotemporal structures. After a transient that will be described below, the configuration of the system consists in portions with a modulus nearly constant (corresponding to plane wave states) interrupted by localized objects with particle-like behavior. Dark features in $|A_+|$ appear where $|A_-|$ has clear features, thus indicating that the localized object carries a kind of anticorrelation between the fields. The lower panels of Fig. 1 show the modulus of the two fields at $t = 399$ and $x \approx 300$, where one of such objects is present. One of the components shows a maximum in the modulus, whereas the other displays a deep minimum. We can call this object a "hole–maximum pair". It is the dissipative analog of the 'out-gap' solitons appearing in Kerr media with a grating [12], and here it is the characteristic object building-up the disordered intermittent dynamics seen at long times. It is clear that these objects connect the plane wave states (that is the constant modulus regions) filling most of the system.

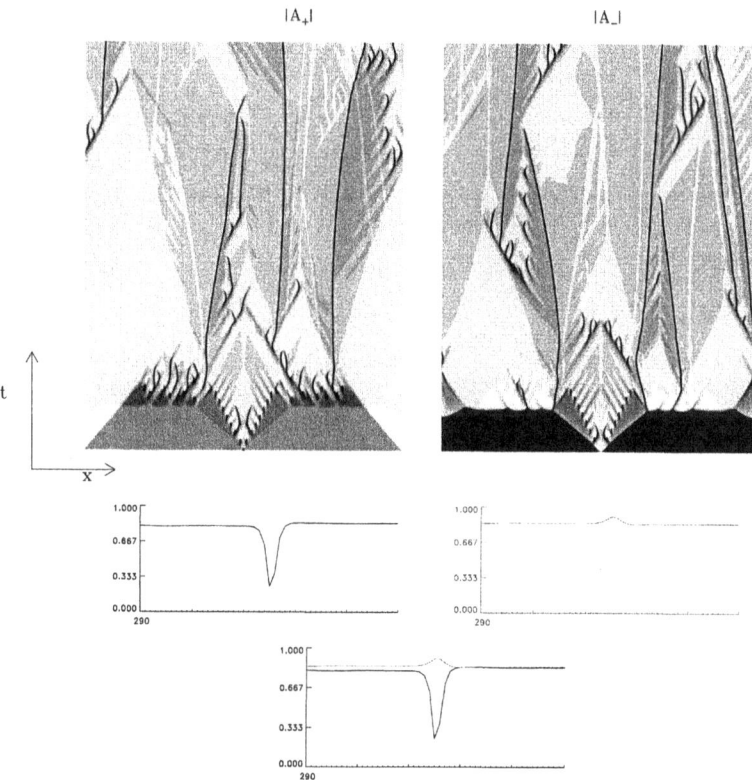

Figure 1 Upper panels: Spatiotemporal evolution of $|A_+(x,t)|$ and $|A_-(x,t)|$ with time running upwards from $t = 0$ to 400 and x in the horizontal direction, from $x = 0$ to $x = 512$. Lighter grey corresponds to the maximum value of $|A_\pm(x,t)|$ and darker to the minimum. Parameter values are $\alpha = -0.35$, $\beta = -2.0$, and $\gamma = 0.2$. Lower panels: A hole–maximum coupled pair at $t = 399$. This is the dominant coherent structure at long times. Left $|A_+(x,t)|$, right $|A_-(x,t)|$, and both graphs are superposed in the central bottom panel.

Before reaching this asymptotic state, the system evolves through configurations where additional kinds of localized objects are seen. The presence of the Nozaki-Bekki hole-pulse pair as initial condition in the central part of Fig. 1 gives birth to a pair of fronts which replace the initial lateral plane-waves by new ones. Interestingly, a different kind of localized object is seen to form just where the initial hole-pulse pair was placed. A close-up of it at $t = 90$ is displayed in Fig. 2. It is a kind of coupled maximum-maximum pair. The moduli of the two fields are superposed in the central panel showing the full object. The lateral small bumps are propagating waves that travel towards the maxima. Thus the center of the coherent structure acts as a wave sink [9].

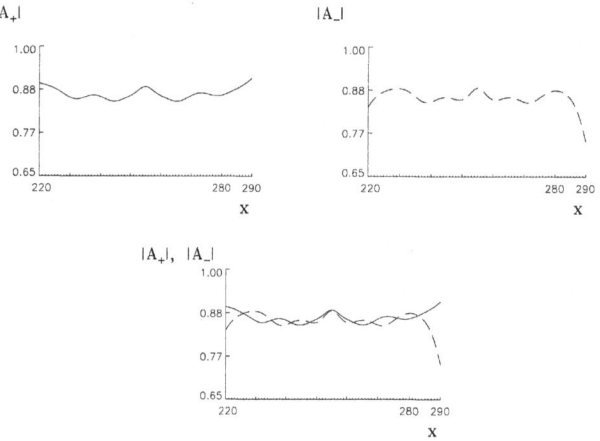

Figure 2 Snapshots of part of the system in Fig. 1 at $t = 90$ showing a localized maximum–maximum wave sink.

In Figure 3 the spatiotemporal evolution of $|A_+(x,t)|$ and $|A_-(x,t)|$ was obtained using as initial conditions a sharp phase jump at the center of the system, with small random white noise added. The parameter values are $\alpha = 0.6$, $\beta = -1.4$ and $\gamma = 0.7$. After a short time, the system reaches a state dominated by branching hole–hole pair structures. Lighter grey correspond to the maximum values of $|A_\pm(x,t)|$ and darker to the minima. The two big triangles correspond to regions of constant modulus, that is, plane waves. The bottom panels show $|A_+|$ and $|A_-|$ in a portion of the system at these early times. Both are superposed in the central panel to show the complete matching of the two solutions.

At longer times, all the hole-hole pairs disappear from the system, thus indicating that they are not stable objects at this value of the parameters. The system decays to the same state as at the end of Fig. 1: the dominant coherent structures are the maximum-hole pairs.

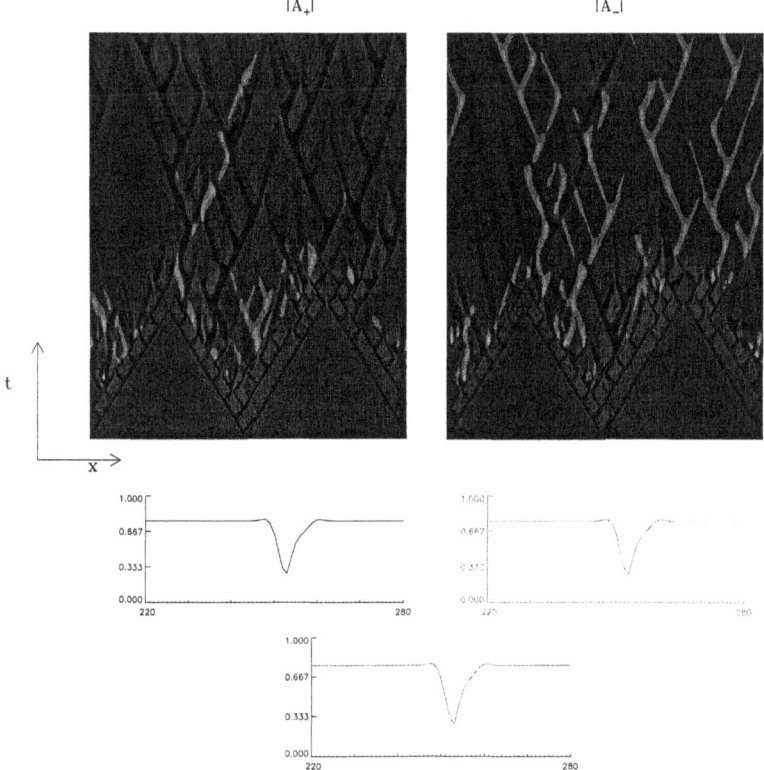

Figure 3 Spatiotemporal evolution of $|A_+(x,t)|$ and $|A_-(x,t)|$ with time running upwards from 0 to 400 and x in the horizontal direction from 0 to 512. Lighter red correspond to the maximum value of $|A_\pm(x,t)|$ and darker to the minimum. Bottom panels (left and right) show one of the localized hole-hole objects dominating the early dynamics. Central bottom panel superpose them, showing its perfect matching. Both fields, then, have exactly the same modulus around the core of the coherent structure, as in the ansatz (2).

3 Exact solutions

The different spatiotemporal evolutions shown in the previous figures (1) and (3) are themselves interesting enough for a detailed study. The localized objects appearing in the simulations are clearly responsible for most of the complex dynamics in the system. A systematic survey on these kind of solutions and a detailed analytical description for most of them is still lacking. Nevertheless, we can interpret some of the observed structures from a simple ansatz:

$$A_+(x,t) = e^{i\varphi} A_-(x,t) \tag{2}$$

where φ is constant, and $A_-(x,t)$ is any solution of the single CGL equation:

$$\partial_t A_- = A_- + b\,\partial_x^2 A_- - c \mid A_- \mid^2 A_- , \tag{3}$$

where $b = 1 + i\alpha$ and $c = (1 + \gamma) + i(1 + \gamma)\beta$.

This simple ansatz gives us a rather rich set of solutions: for each known analytical solution of the single CGL equation (3), there is a corresponding solution of the CCGL equation set, in which A_- and A_+ have essentially the same shape except for a constant global phase. In particular, hole, pulse, shock, and front solutions are localized solutions analytically known for the single equation [10, 11, 9, 13], so that hole-hole, pulse-pulse, shock-shock and front-front pairs are immediately found as analytical solutions of the CCGL set. Some of the localized structures seen in Figs. 1 and 3 are well described by the ansatz (2).

It is worthwhile to note that the studies of *instability* for these objects in the complex Ginzburg-Landau equation are immediately translated into instability results for the CCGL equations.

4 Summary

In summary, we have shown numerically the existence of different kinds of localized objects, responsible for the complex behavior or solutions of the CCGL equations. Some of these objects can be understood in terms of exact solutions arising from a simple ansatz. A more detailed analysis is still needed, however. In particular, the hole-maximum structure, which appears as the dominant coherent structure at long times, can not be described by our ansatz. In addition, much more work is needed in order to establish the stability properties of the different objects, and the nature of their interactions.

We acknowledge financial support from Programa de Desarrollo de Ciencias Básicas (PEDECIBA, Uruguay), Comisión Sectorial de Investigación Científica (CSIC, Uruguay), and MCyT (Spain) project CONOCE BFM2000-1108.

References

[‡] Email: montagne@fisica.edu.uy.

[§] Email: emilio@imedea.uib.es.

[1] M. Cross and P. Hohenberg, Rev. Mod. Phys. **65**, 851 (1993), and references therein.

[2] M. San Miguel, Phys. Rev. Lett. **75**, 425 (1995).

[3] L. Gil, Phys. Rev. Lett. **70**, 162 (1993).

[4] H. Riecke. *Localized structures in pattern-forming systems.* in *Pattern Formation in Continuous and Coupled Systems: a Survey Volume*, Ed. by M. Golubitsky, D. Luss, and S.H. Strogatz (Springer, Berlin, 1999).

[5] M. van Hecke, Phys. Rev. Lett. **80**, 1896 (1998).

[6] M. van Hecke, C. Storm, and W. van Saarloos. Physica D **134**, 1 (1999).

[7] R. Montagne, E. Hernández-García, A. Amengual, and M. San Miguel, Phys. Rev. E **56**, 151 (1997).

[8] E. Hernández-García, M. Hoyuelos, R. Montagne, M. San Miguel, P. Colet, Int. J. Bif. Chaos **9**, 2257 (1999).

[9] W. van Saarloos and P. Hohenberg, Physica D **56**, 303 (1992), and (Errata) Physica D **69**, 209 (1993).

[10] K. Nozaki and N. Bekki, Phys. Rev. Lett. **51**, 2171 (1983).

[11] K. Nozaki and N. Bekki, J. Phys. Soc. Japan **53**, 1581 (1984).

[12] K. Kneubühl and J. Feng, IEEE J. Quantum Electron. **29**, 590 (1993).

[13] R. Conte and M. Musette, Physica D **69**, 1 (1993).

STABLE-UNSTABLE TRANSITIONS IN VISCOUS FINGERING

S.Obernauer and M.Rosen

Grupo de Medios Porosos. Facultad de Ingeniería Universidad de Buenos Aires

Paseo Colón 850

(1063) Buenos Aires, Argentina

We have carried out experiences in a Hele-Shaw cell, with a radial injection at constant flow rate. A liquid, oil or glycerol solutions, pushes (or is pushed by) an aqueous polymeric solution of Xanthan. This solution is a shear thinning fluid. An oil and glycerol solution could be considering Newtonian fluids. In this type of displacement, when the mobility control parameter is greater than one, the characteristic Viscous Fingering appears. During the advance of the interface the crossover from a stable to an unstable displacement or vice versa is investigated. On the other hand, we have experimentally obtain the effective mobility of the shear thinning fluid as a function of the depth averaged velocity in order to predict the moment of transition. In the immiscible displacement case, the front becomes unstable and is finally stabilized. In this case, the non-Newtonian fluid displaces the Newtonian one. We observe that the transition occurs approximately at a critical mean velocity where the mobility control parameter reaches one. For miscible fluids, we have studied the displacement where the non-Newtonian fluid is the displacing one. In that case, the front becomes stable and finally instabilities appear. But contrary from we would expect, the critical velocity depends on the injection flow rate Q. As Q increases, the critical velocity increases too. Our experiments seem to show that the critical velocity is fairly constant and not so sensitive to changes in the Newtonian fluid mobility.

I.INTRODUCTION

When a fluid of viscosity μ_2 is injected into a fluid of higher one μ_1 the phenomenon of VF is unchained. In a horizontal displacement (despising gravitational effects), instability always results when a less viscous one displaces a more viscous fluid since the less viscous fluid has the greater mobility. The mobility control parameter μ_1/μ_2 measures the relation between viscosities and when the front is unstable it becomes higher than one. In this work, a Hele-Shaw (H-S) cell is used to visualize the phenomenon. It consists of two large transparent plates separated by a distance e in the order of half millimeter where the fluid is confined in a 2D geometry. Then, if the

281

O. Descalzi et al. (eds.), Instabilities and Nonequilibrium Structures VII & VIII, 281–296.

injected fluid enters through a hole at the center of the cell, the *VF* structure appears with a radial symmetry.

The classical displacement of a Newtonian (N) fluid in a H-S cell has received extensive theoretical and experimental investigation [1] [2] [3] [4] [5]. Numerous experimental studies have been conducted to collected data on the displacement, in an attempt to quantify the effects of different material and geometrical properties on the flow. However, the flow phenomenon with a non-Newtonian (n-N) fluid has not been very studied.

The displacement of a high-viscosity n-N fluid by a low viscosity N fluid in a H-S cell is capable of producing ramified *VF* patterns. Experiments using polymer solutions – typical n-N fluids- showed much richer fingering patterns than those produced by N fluids [6].

Nittmann et al. [7] have first performed the fingering experiments of aqueous polymer solutions in axial and radial H-S cells, and they observed the much more ramified fingering path. As an example, Sader et al [8] pointed out about n-N effects on immiscible viscous instability in a radial H-S. This work therefore settled the point of contention regarding the role of interfacial tensions. Makino et al. [9] Also show a morphological transition from a side branching pattern to a tip splitting one with aqueous hydroxypropyl methyl cellulose (HPMC) solution pushed by air. The observed morphology transition was correlated with the dimension of the fingering pattern. Van Damme in Ref [10] has shown the patterns obtained by displacing an aqueous colloidal clay suspension by air. It was pointed out there that a fractal behavior is obtained only by increasing the n-N character of the colloidal fluid. When that once sufficient n-N character has been obtained, lowering the interfacial tension by using miscible fluids he has indicated Thad does not introduce major changes in the geometry of the patterns. The boundary fractal dimension is basically unchanged, but the average finger width is smaller.

Zhao et al [11] [12] presented the VF problem for an N fluid that pushes an n-N fluid in the miscible case. In this work pattern formation is studied by injecting water into a radial H-S cell filled with an aqueous solution hydrophobically terminated polyxythhylene. A crossover form a fingering to a fracturing pattern is observed when the injection rate exceeds a threshold value. Despite the strong variation of patterns as four control parameters are independently varied (injection rate, molecular weight, gap of the cell, polymer concentration) the initial unstable characteristic length shows a smooth variation with a dimensionless number constructed from the ratio of the imposed shear time scale to the polymer relaxation time.

Kondic [13] has presented the Saffman Taylor instability of a gas bubble expanding into a shear thinning liquid in a radial cell. The numerical simulations show that the shear thinning significantly influences the developing interfacial patterns and can suppress tip splitting and produce dendrite fingers.

However, the investigation of a possible transition form from stable to unstable displacement or an inverse possibility one, when one of the fluids is a shear thinning, has been relatively unexplored.

In this work, we describe the experimental results of VF in the radial H-S cell using aqueous solutions of a shear thinning polymer called Xanthan, a n-N high molecular weight polysaccharide commonly used for rheological control. We have carried out experiments of immiscible and miscible displacements. Oil is used as the second fluid in the immiscible case, and glycerol solutions in the last type of experiments. We found experimental evidence that shows that the interface region between a fluids that pushes another, can be either unstable or stable depending on the mean velocity of the front and also on the initial conditions of the displacement. We have shown that the front could reach the unstable conditions whether the n-N fluid is the displacing fluid or the displaced fluid.

II. THE EXPERIMENTAL SYSTEM
A. Rheology of Xanthan solutions

Many n-N fluids can be exhibited both effects: shear thinning and elasticity. In particular experimental conditions, if one of these fluids was placed in a H-S cell and displaced at a rate which gives a very small Deborah number (the ratio of elastic relaxation time to the fluid flow time), the elastic properties of this fluid may not be exhibited. In the case of the more rigid road type molecule: the Xanthan structure gives a purely shear thinning behavior, and results in inelastic solutions at lower concentrations.

Figure 1 shows the characteristic behavior of the solution (1700ppm) as measured by us. This solution shows no yield value and the typical flow curve for these materials indicates that the ratio of shear stress τ to the shear rate $\dot{\gamma}$, which may be termed the apparent viscosity μ_a, falls progressively with shear rate and becomes linear only at high shear rates. This limiting slope is known as the viscosity at infinite shear and designated μ_∞. At low shear rates the viscosity becomes constant and in the limit to zero is defined μ_0.

An empirical functional relationship known as *power law* is widely used to characterize fluids of this type. This relation may be written as:

$$\tau = b\gamma^n$$

(1)

Where b and n are constants (n<1). The apparent viscosity μ_a can be written as a power law and may be expressed in terms of n:

$$i.e \mu_a = b\gamma^{n-1}$$
$$\mu_a = \frac{\tau}{\dot{\gamma}}$$

(2)

The shear flow measurements of Xhanthan solutions were carried out indirectly from observations of the flow rate Q and the correspondent pressure gradient ∇P [14] inside a rectangle HS cell. On the assumptions that flow is laminar and there is no slip at the walls we arrive at the following expression (Darcy law):

$$\overline{v} = \frac{Q}{S} = -(k/\mu)\nabla P$$

Where k is the permeability and S the section of the cell.

In the case of power law fluids it is possible to express a generalization of Darcy law as a function of the same exponent n. In the same way we obtained the apparent viscosity μ_a as the relation between $k\nabla P$ and the mean velocity v.

Additionally, $\dot{\gamma}$ could be expressed as a function of the mean velocity with an empirical relation [15].

Finally,

$$\dot{\gamma} = \frac{2n+1}{2n}\frac{4v}{e}$$

Applying this relationship, we obtained b=0.0319dyna.s^{n-1}/cm2 and n=0422 for our solutions in the n-N region of the rheological diagram.

B. The Hele-Shaw cell

The experiments were performed in a radial geometry, using two plane-parallel square glass plates (1x40x40) cm^3 with 16 rubber spacers of 0.05cm thickness clamped in between. The H-S cell was placed horizontally on a white board. Uniform lighting was provided from below and behind the cell. The resulting fingering structure was photographed sequentially and a record of the development patterns was taken by a video camera.

C. The fluids employed

Each experience was performed with a two-system fluid: one of them previously saturates the cell, and the other is injected at a constant flow rate. We have employed the polymeric solution described above. In the immiscible case, we have used cooking oil with an intermediate viscosity (52 cp) between η_0 and η_∞. (t=20^{a}c) of these polymeric solutions. In the miscible case, we have prepared a glycerol solution at different concentrations, where their viscosities vary from 10cp to 100cp.

As usual, it is necessary to put one colorant in one of the two fluids to allow us to observe the phenomenon. "Rodhamine" was used in the case of the glycerol solutions and bleu "Solofenil" for the polymeric solution. We have verified that the rheology of the solutions was not affected because the concentration of the colorants was low enough.

III. IMMISCIBLE FLUIDS

We have carried out experiments with flowing immiscible fluids because they have the advantage of having a sharp interface.

In this work, we show experimental results where an N fluid (oil) is pushed by an n-N one. We have performed experiments at three flow rate, Q=20mil/min, Q=10ml/min and Q=4mil/min respectively. At low flow rate it was shown that the displacement is always unstable and at high flow rate, the front is always stable. Figure 2 (a) shows the intermediate case: the evolution from an unstable front that is finally stabilized (at Q=10mil/min).

In order to compare the dynamics and morphologic differences of the structures, comparative experiments with N solutions were carried out (Fig. 2 (b)). In that way we have prepared glycerol solutions (43 %) corresponding to the viscosity of μ_∞ of the polymeric solution. These experiments were also made at Q=10ml/min., and in the case of the polymeric solution, a transition was observed in the course of the experiments. Comparing Figures 2(a) and 2(b), we were able to conclude that, there are some correspondences at short times but, thereafter, the amplitude grows in the case of the glycerin and decreases in the case of the polymeric solution. It indicates that at the beginning of the experiments the viscosity of the displaced fluids is the same for both cases, but then the viscosity of the n-N fluid increases and in consequence the mobility control changes.

As we pointed out, the *VF* experiment was performed with an N fluid with an intermediate viscosity value between the asymptotically one of the n-N fluid. At a constant flow rate, in a radial geometry, the velocity decreases as a function of the radius. Then, the mean velocity of the interface and the effective shear rate during the experiment vary between two values.

As there is a constant flow in each experiment, the meen front velocity (v_c) will be, in terms of the distance to the injection point:

$$v_c = \frac{Q}{2\pi e} \frac{1}{r}$$

where e represents the thickness of the cell.

The mean velocity, in its first approach, is proportional to $\dot{\gamma}$. Thus, the experiments have been carried out in the direction of the decreasing $\dot{\gamma}$.

Figure 3 shows three possible velocity intervals depending on the flow rate of the experiments. As μ_2 increased when the velocity decreased, the mobility control parameter μ_1/μ_2 between both fluids involved could change from over one to below one.

Then, the front that becomes unstable ends up stabilizing. The stability condition could be therefore varied with an external control parameter.

We think that the same analysis is also valid with an inversion of the fluids; that is, the N fluid pushing the n-N one. But in that case, we could expect a front that becomes stable and thereafter unstabilized.

In Fig. 3 we notice the cut of rheological curves of both fluids at 0.09 cm/s. In this point there is an inversion in the mobility control parameter. If we estimate for each flow rate, the radius that corresponds to this velocity we found the following results:

-At high flow rate (Q=20mil/min) we could expect the moment to shift from an unstable to a stable condition at 2.7cm. But this radius is not enough to develop instabilities because of the influence of the interfacial tension [16]. Therefore, and in agreement with this analysis we observe a stable displacement.

-At the intermediate flow rate (Q=10mil/min) we expect a transition at 6.8cm and we observe a displacement that began being clearly unstable but ended up being stable at 12cm approximately where we stop the injection.

- At low flow rate (Q=4mil/min) the position of the calculated transition is 14cm. Normally we do not reach this distance and in consequence we observe a front, which is always unstable.

IV. MISCIBLE FLUIDS

In Fig. 4 we observe two different fingering patterns. In the first case a glycerol solution pushes polymeric solution. And, in the other experiment, the fluids are inverted. In both cases instabilities appear.

The main objective of this work is to analyze experiences where the polymeric solution is displaced by glycerol solutions because this type of flood has a potential applicability in oil recovery. A traditional water flood, in turn displaces the slug of polymer that displaces the oil. It is necessary to prevent instabilities in the second interface so this is the reason for our interest.

The variables considered are the flow rate (Q) and the viscosity of the N fluid (μ_2). Contrary to the example in immiscible fluids here we expect fronts that begin being stable and finally become unstable. This is because an N fluid displaces an n-N one, that is the inverse case to the one presented for immiscible fluids.

We have estimated the moment of the transition from all the experiments and we have measured the radius of the structure at the moment that the instability appears. It is called r_c.

In Fig. 5 we compare three experiments with the same fluids but a different flow rates. A Newtonian fluid pushes the n-N one. At high flow rate it is possible to clearly observe an interface that begins being stable and then breaks in fingers. But at low flow rate this condition is reached nearest the injection point. These leds to study the transition stage more throughly. In Fig. 6 we observe the experimental results of the evolution of the front position as a function of injected volume at different flow rates (μ_2 remains constant). We compare this with a stable displacement. As the flow rate increases, the instability delay its arrival. There is a correspondence to the previous analysis.

In the case of a stable front, the traveled distance becomes:

$$r = \sqrt{\frac{3Qt}{2\pi e} + r_0^2}$$

We have calculated the mean velocity of the front in that moment as:

$$v_c = \frac{Q}{2\pi_c}$$

We have plotted v_c as a function of the viscosity of the N fluid (Fig. 7 (a)) and as a function of the flow rate (Fig. 7 (b)). Contrary to what we would expect v_c changes with the flow rate.

The tendencies shown by these experiences are that v_c grows linearly with the flow rate and apparently it is not too sensitive to changes in the viscosity of the N fluid. The critical velocity decreases smoothly while the Newtonian viscosity is increases [17].

When we compare the experimental values of v_c with the predicted ones, we do not find correspondence. Experimental values are always higher than predicted values (Fig. 8).

V. SUMMARY

We have found experimental evidence in order to demonstrate that the front of a fluid (nN) pushing another in a horizontal displacement, can go from an unstable initial situation to a stable one or vice versa if the mean velocity front is changed. This means that the mobility control parameter between both fluids is modified during the experiment because one of the fluids is shear thinning.

In this work, we have explored this possibility for the radial injection case. The decrease of the local velocity with time due to this radial displacement, leads to a decrease of the effective shear rate. At the same time this produced an increase in local viscosity of the n-N solution.

In the immiscible case we observe (displacement of a N fluid by a n-N one) an unstable to stable transition. The prediction that the transition occurs at the velocity when the viscosity of the polymeric solution reaches the same value as the viscosity of N fluid, agrees with the experimental observations. The viscosity of the polymeric solution was obtained from the rheological curves (Fig. 3).

In the miscible case, we study the displacement of an n-N fluid by an N one and a shift from a stable to an unstable transition. We compare experiments at different flow rates and with different glycerol (N) viscosity. We have observed that the mean velocity of the interface at the moment of the appearance of instabilities (v_c) is different for each flow rate. As the fluids are the same, we would expect the moment of the transitions to occur at the same mean velocity. At this moment, the mobility control parameter becomes higher than 1. It varies during the experiment because μ_2 increases when mean velocity decreases. The experimental results indicate that μ_2 are not only a function of vc but also of the initial condition of the front displacement (flow rate).

If we compare experiments with different viscosity (μ_1) at the same flow rate, v_c does not change significantly. This result would make us think that there are also additional mechanisms involved in the appearance of the instability.

In summary, with these experiments we expected to separate stable from unstable regions in the $\mu - \overline{v}$ plane. Even, if our results confirm the initial hypothesis, this separation is not accurate as, in the miscible case the influence of the mixing zone, which depends on the diffusion coefficient , is not considered whereas for immiscible displacements the role played by interfacial tension when controlling stability is not taken into account.

REFERENCES

[1]P. G. Saffman and G. I. Taylor, Proc. Roy Soc. A **245**, 312 (1958)

[2]L. Paterson, Phys. Fluids **28**, 26 (1985)

[3]G. M. Homsy, Annu. Rev. Fluid. Mech. **19**, 272 (1987)

[4]J. Chen, J.Fluid Mech. **201**, 223 (1989)

[5]S. D. Howison, J.Fluid. Mech. **167**, 271 (1987)

[6]S.Obernauer, A.D'Onofrio, M.Rosen, in " Fractals Reviews in the Natural and Applied Sciences" Ed. M.Novak. Chapman & Hall. 6, 56-62 (1995).

[7]J. Nittmann, G. Daccord and H. E. Stanley, Nature (London) **314**, 141 (1985)

[8]J. E. Sader, D. Y. C. Chan and B. Hughes, Phys. Rev. E. **49**, 420 (1994)

[9]K. Makino, M. Kawaguchi, K. Aoyama ant T. Kato, Phys. Fluids **7** (3), 455 (1995)

[10]H. Van Damme, F. Obrecht, P. Levitz, L. Gatineu and C. Laroche, Nature (London) **320**, 731 (1986)

[11]H. Zhao and J. V. Maher, Phys. Rev. E **47**, 4278 (1993)

[12]H. Zhao and J. V. Maher, Phys. Rev. A **45**, R8328 (1992)

[13]L. Kondic, M. J. Shelley and P. Palffy-Muhouay, Phis. Rev. Lett. **80**, 1436 (1998)

[14]W. L. Wilkinson, *Non-Newtonian Fluids, Fluid Mechanics, mixing and heat transfer*. (Pergamon Press, London, 1960) Vol. 1, p. 20

[15]W. Kozicki and C. Tiu, in *Encyclopedia of Fluid Mechanics, Rheology and non-Newtonian flows*. Edited by N. Cheremisinoff (Gulf Publishing Company, Houston, 1998) Vol. 7, p. 199

[16]Paterson L. J.Fluid Mech. 113,513 (1981)

[17]Obernauer S.,Drazer G., Rosen M.,Physica A 283, 187 (2000)

FIGURE CAPTIONS

FIGURE 1. The apparent viscosity of the polymer solution (1700ppm) as a function of the shear rate.

FIGURE 2

(a) The photography sequence corresponds to a n-N fluid displacing a N one (52cp) (immiscible case). The flow rate is 10 ml/min. The front is always unstable.

(b) The photography sequence corresponds to a non-N fluid displacing a N one (52cp) (immiscible case). The flow rate is 10 ml/min. The front was unstable and became stable at the end.

FIGURE 3. One curve corresponds to a typical pseudoplastic behavior and the other to a Newtonian one (η cte). Depending on the workint range in each experiment the viscosity ratio between the fluids could be larger or smaller than one. If the n-N fluid pushed the N fluid then in region A the displacement is always stable, in region C always unstable and in region B we could observe a transition from unstable to stable.

FIGURE 4. (a) Newtonian fluid (52cp) displacing non-Newtonian ones at Q=5ml/min. (b). The same fluids are inverted and the non-Newtonian fluid displaces the Newtonian (52cp) one at Q=5ml/min.

Unstable regimes with different final structure are observed.

FIGURE 5. The N fluid (52cp) pushes the n-N one at three different flow rates. (Q=0.5mil/min, Q=5mil/min, Q=50mil/min). At high flow rate the interface becomes unstable far from the injection point. In every case, the image was taken when the same volumes had been injected.

FIGURE 6. Temporal evolution of the position front. Comparison with the stable front. The N fluid (solution of glycerol of 52 cp) pushes a nN one.

FIGURE 7. (a) V_c as a function of the glycerol solution viscosity. (b) V_c as a function of the flow rate.

FIGURE 8. Comparison between experimental and predicted values.

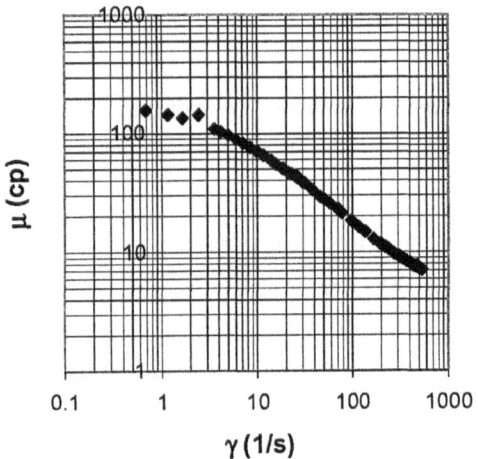

Figure 1

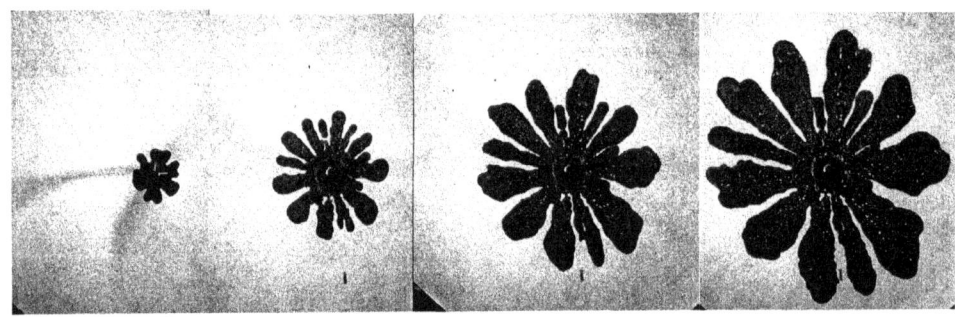

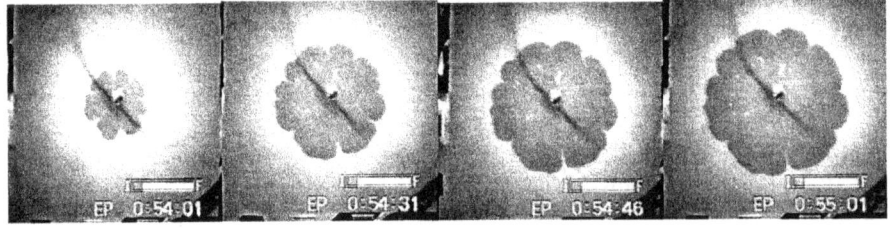

Figure 2

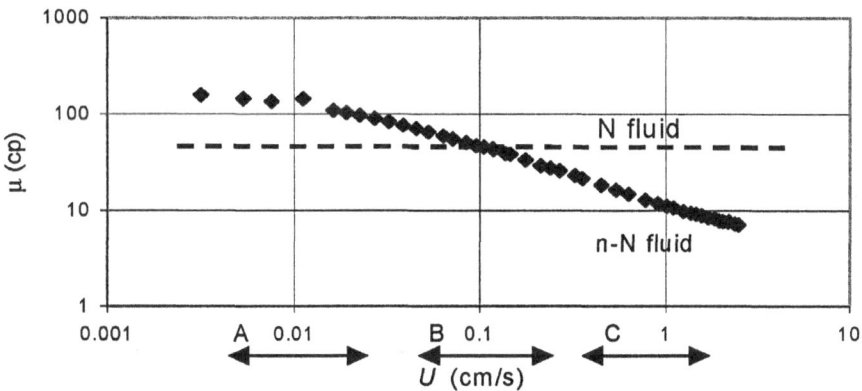

Figure 3

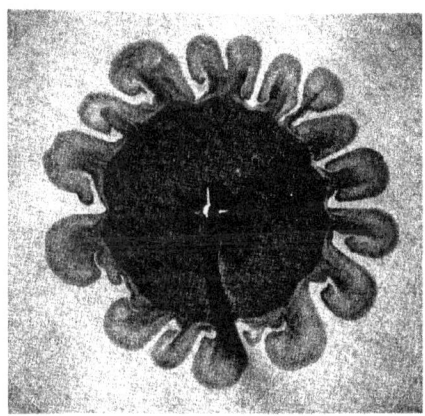
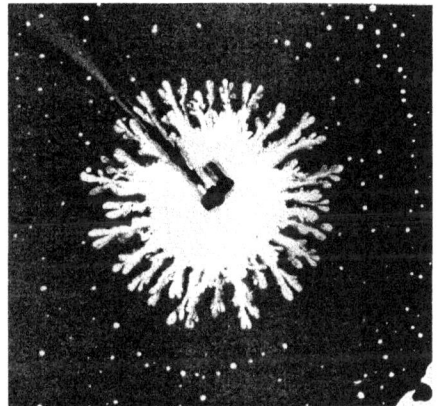

Figure 4

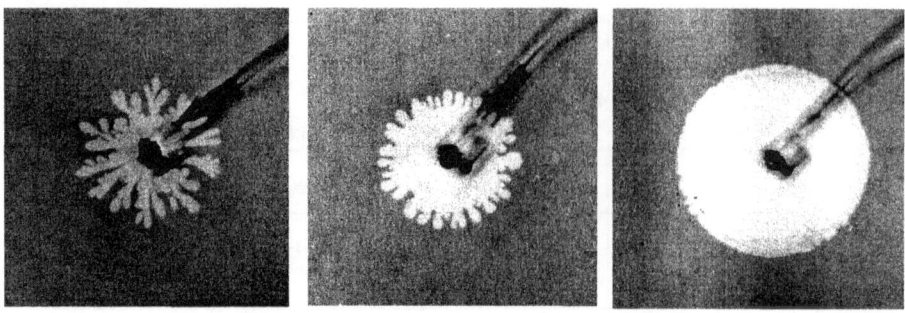

Figure 5

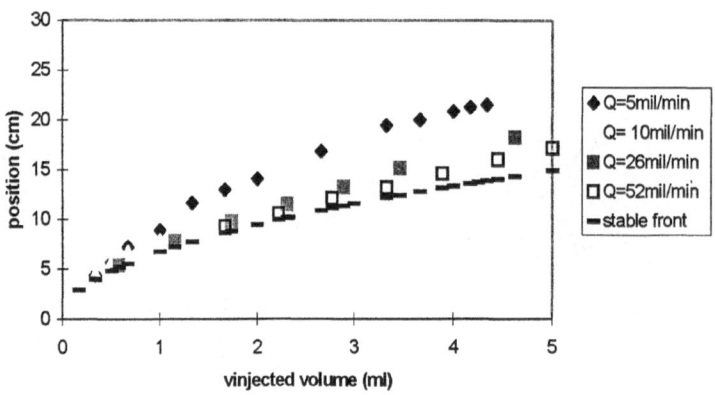

Figure 6

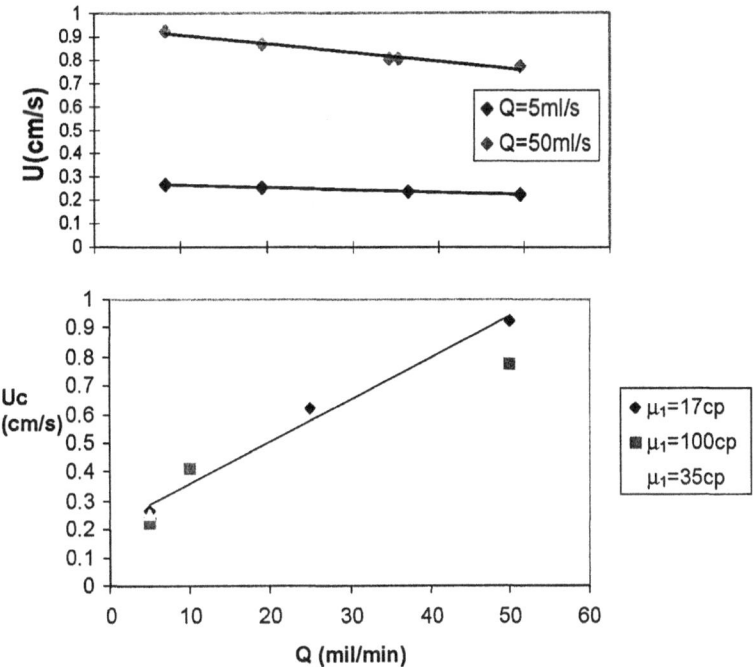

Figures 7(a) and 7(b)

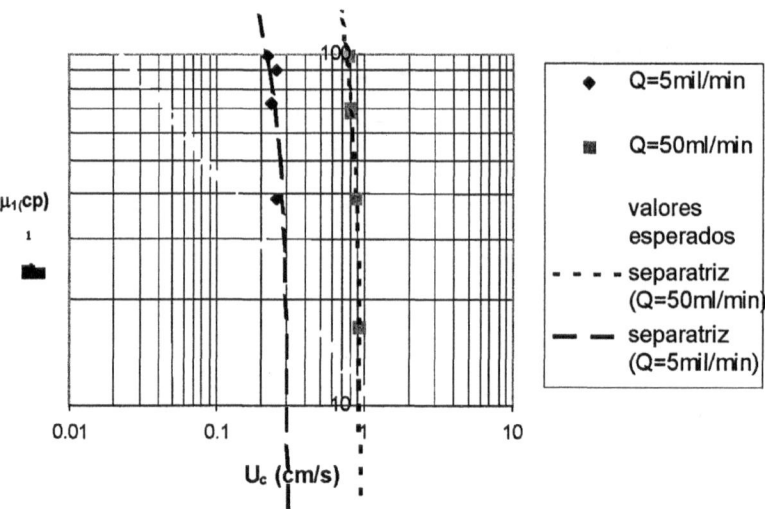

Figure 8

Force distribution in two dimensional sandpile

Manuel Olivares[a], Jean-Noel Roux[b] and Francisco Melo[a]

[a] *Departamento de Física de la Universidad de Santiago de Chile and Center for Advanced Interdisciplinary Research in Materials,*

Av. Ecuador 3493, Casilla 307 Correo 2 Santiago-Chile.

[b] *Laboratoire Central de Ponts et Chaussees, 58, boulevard Lefevre, 75732 Paris Cedex 15, France.*

We present an experimental study of the pressure on the base of a two dimensional sandpile. A force dip in the center of the pile can be artificially produced by changing the lateral conditions on the first row of the pile. This support the idea that arching, capable to carry the weight of the pile, can be induced externally. A numerical study is in good agreement with these finding, for similar experimental conditions

PACS numbers: 47.32.Cc, 47.35.+i,67.40.Vs

In recent years the physics of granular materials has received growing interest and the importance of these materials in a variety of industrial process such as agriculture, mining and pharmaceutical has been well established [1,2]. Despite this fact, a diversity of phenomena such as convection, surface waves, sound propagation as well as force distribution in these material are not well understood. Difficulties in the understanding of these phenomena arise as a consequence of the discrete character of grains that compose such material. Even for the apparently simplest static case, in which grains build up force network as a response to strain, many facts remains unsolved. One of the interesting feature of granular materials, which capture our attention here, is the apparently contradictory result that the force on the base of the sandpile has a minimum at its center; the so called "force dip". Our general understanding, however, tell us that it should be the opposite: since the pile center should support a maximum weight, the force should be a maximum. In fact, it is known [3] that in an perfect two dimensional hexagonal packing of identical particles the load acting on each grain at the bottom layer is constant. However, when a small packing disorder or polidispersity is introduced, strong fluctuations of the load acting on each particle appear.

O. Descalzi et al. (eds.), Instabilities and Nonequilibrium Structures VII & VIII, 297–303.

[4]. More important, if the boundary conditions are changed on a perfect pile, force contact network can be dramatically modified. In particular, S. Luding [4] has numerically shown that a small compression on the lowest row of the pile can favor lateral contact between grains. These contacts favor arching forming which in turn carries the pile weight at its center. As a result, a force dip can be induced in a controlled way.

Experimentally, the existence of the force dip in three dimensional sand pile has been controversial [5]. The effect was first observed by in an early work by Smid and Novosad [6]. Although in more recent works [7] the pressure dip is not clearly observed, it has become generally accepted that such dip depends strongly on the procedure used to built the sand pile up but little on the surface roughness of particles.

Here, we present an experimental and numerical study of a two dimensional hexagonal sandpile. We develop a simple experimental technique to automatically measure the load on each particle located at the bottom pile. Force distributions sensitivity to boundary conditions is studied by slightly compressing the lowest row of the pile. Results of our numerical simulations, using a different method than Luding, is in good agreement with our experiment.

The experimental setup in Fig. 1, consists of a two dimensional sandpile made of 800 stainless steel spherical particles $4mm$ in diameter. The pile which is $60degree$ angle is kept in place by two parallel glass walls, separated each other by steel bars about $4mm$ in thickness. The sand pile base is composed of 40 stainless steel cylinders $4mm$ in diameter and length. To observe the forces, we use a classical photoelastic method. In turn, the pile is located in contact with a large piece of a photoelastic material (Photoelastic Sheet $30cmx30cm$, PSM-4; Measurements Group, Inc.). Thus, the cylinders carrying the whole pile weight produce a deformation on this material that depend on the applied force. To measure the force, one of the glass walls was covered, in the region close to the pile base, with a polarizer and quarter-wave plate pair, arranged with an angle $\pi/4$ between them to produce circular polarized light. The opposite glass wall was covered with a similar pair arranged in reverse order and set crossed to the first one. With the photoelastic material used as a substrate for the pile and lighting conditions of our experiment the force sensitivity is a fraction of mg, where m is the mass of a single grain. Thus, we were able to observe the local stress distribution as the cylinders slightly deformed the photoelastic substrate. To modify boundary conditions at the pile base, we used two micrometric screws to control the displacement of the lateral bars, as shown on the lower panel of Fig. 2. Thus, bars displacement favor lateral contact between neighboring grains.

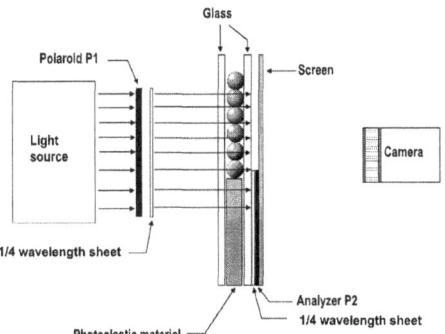

FIG. 1. Schematic view of our experimental setup. A conventional photoelastic method is applied to visualize the force on each individual particle on the base of a regular triangular pile.

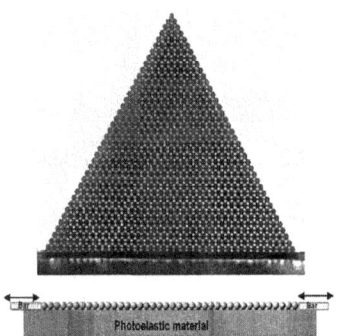

FIG. 2. Schematic view of two dimensional triangular sandpile. Lower panel indicate the pile base which is composed of cylinders instead spheres. In this way, a better linearity is obtained for the force as a function of the photoelastic light intensity.

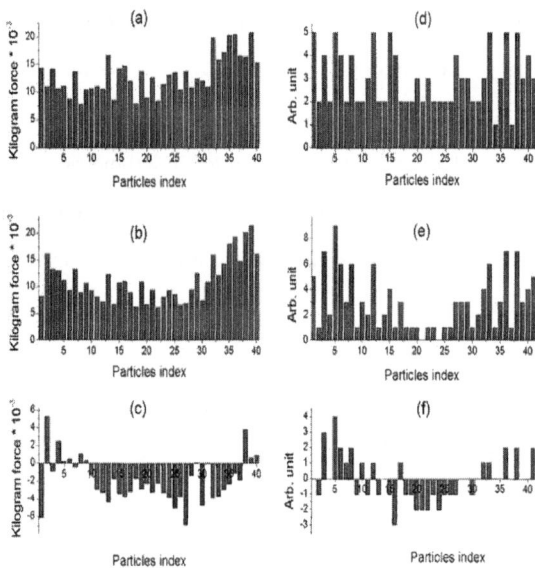

FIG. 3. a-c) Experimental force on the pile base; a) initial configuration; b) after a lateral displacement, $\delta = 0.3mm$ is applied; and c) force difference between previous configurations. d-f) Numerical results for similar configuration than a-c).

The initial force distribution of the pile, in Fig. 3a, is in average rather constant independent on position. It exhibits however strong fluctuations. We notice that in a perfect hexagonal lattice, as reported in reference [3], the load acting on each grain at the bottom layer is identical, $(n + 1)W/2$, W is the particle weight and $n(n + 1)/2$ is the total number of grains. However, this result is strongly singular in the sense that hard spheres can not be deformed to screen any deviation from a perfect hexagonal packing. Therefore, any small deviation of the lattice has strong effect, on the entire pile, producing large force fluctuations. In the experiment, the particles positions might vary in amount of $0.03mm$. Thus, the fact that average force is roughly constant, as seen in Fig. 3a., tell us that deviations from the perfect hexagonal packing can be considered as random: large force fluctuations are naturally present. In Fig. 3b, the force distribution after compressing the pile base $0.03mm$ and Fig. 3c is the force difference after applying such disturbance. Clearly a force dip has been induced. We believe that such dip is created when the lateral contacts between two neighboring grains become more active to carry forces. This lateral contact favor arching which in turn strongly modify the force distribution. This idea is well supported by molecular dynamics numerical simulations [4]. In order to obtain more precise information, in Fig. 3, we contrast experiments a)-c) to further numerical simulations d)-f).

Here our numerical results are obtained in similar conditions than experiments with a simulation method [8] that involves viscous lubricating forces. An application of this method is well described in reference [9]. The initial state of the pile, in Fig. 4a, is prepared by locating the particles one by one on the pile, until completing a triangular array. The random character of the triangular array is introduced here by considering a small polodispersity of the particles. Thus, a very small difference in particles diameter, chosen here equal to the typical experimental value of the compression of the pile, introduce the similar random behavior in the force distribution. The force distribution on the pile, after compression is represented in Fig. 4b. Thicker lines representing larger forces, mainly present at the border of the pile, are indicating that a force minimum appear at center pile as a result of the compression. When observing the force network of Fig 4b, it can be seen clearly that horizontal contact between the grain have been favored after compression. This kind of contact have replaced, alternatively at both sides of the pile, the typical 60 degrees contact of the perfect triangular pile.

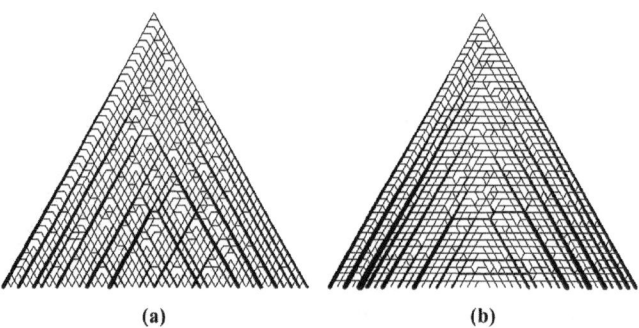

(a)　　　　　　**(b)**

FIG. 4. Schematic view of the force network obtained by a numerical simulation of a two dimensional sandpile. In figures, contact thickness is proportional to the force intensity. a) Force network for a initial condition for which the grain, whose size follow a given polidispersity, have been located in a triangular array. b) Force network after numerically compressing the lowest row of the pile.

In conclusion, our simple experimental set up allow us to accurately control lateral boundary conditions at the first row of the pile. Changes in these conditions naturally produce not only force fluctuations but also coherent formation of arches. We identify such arches as responsible of the force minimum observed at the pile base. Thus, the existence of such minimum is conditioned for procedure used to built sand pile up. Furthermore, the force dip is only espected to appear in experimental condition that favor horizontal contact among grains. This is in agreement with past experiments, since the force dip was not always observed.

This work was supported by Cátedra Presidencial en Ciencias (Chile) and the program ECOS France-Chili, N° C96E06, J-N. Roux acknowledges the program Fondap of Applied Mathematics of University of Chile for financial support.

[1] R.P. Behringer, Nonlinear Science Today, **3**, 3 (1993).

[2] R.M. Nedderman, Statics and Kinematics of Granular Materials, Cambridge Univ Press (1992).

[3] D. C. Hong, Phys. Rev. E. **470**, 760, (1993).

[4] S. Luding, Phys. Rev. E. **55**, 4720, (1997).

[5] See for instance, Hans-Georg Matutis, Granular Matter 1, 83-91 c Springer-Verlag (1998), and references within.

[6] J. Smid and J. Novosad, I. Chem. E. Symposium Series **63**, D3/v/1-D3/v/12, (1981).

[7] R. Brockbank, J. M. Huntley and R. C. Ball, J. Phys. II France, **7**, 1521, (1997).

[8] Details of this method will be discussed elsewhere. For general informations, see for instance, J. Moreau and M. Jean, Proc. 3rd. Biennial European Joint Conference on Engineering Systems Design and Analysis. July 1-4, (1996), Montpellier, France. J. Moreau. Eur. J. Mech. A/Solids **13**, n° 4-suppl., 93-114, (1994). M. Jean and J. Moreau, in Proceedings of Contact Mechanics International Symposium edited by A. Curnier (Presses Univ. Romandes, (1992) p. 31.

[9] S. Ouaguenoui and J. N. Roux, Europhys. Lett., **39** (2), 117, (1997).

Time-periodic forcing of Turing patterns in the Brusselator model

B. Peña and C. Pérez–García

Instituto de Física. Universidad de Navarra,
Irunlarrea, 1. 31008-Pamplona, Spain

Abstract

Experiments on periodic illumination of the CDIMA (chlorine dioxide-iodine-malonic acid) reaction have revealed that Turing patterns can be supressed for sufficiently high intensities [1]. The illumination modifies linearly the chemical kinetics. Here we present a preliminar work on Turing patterns under a time-periodic forcing of the control parameter in the Brusselator model. We have performed numerical simulations of the model under conditions for which reentrant hexagons appear. Surprisingly, the oscillating pattern can change its symmetry for a high enough amplitude forcing: a hexagonal pattern is replaced by oscillating squares, a kind of pattern still unobserved in chemical experiments. The theoretical mechanism underlying this change of symmetry is still under discussion.

Key words: Chemical kinetics, Turing patterns, Nonlinear dynamics.

1 Introduction

In recent years forcing of spatially extended systems has attracted the interest of many researchers in the framework of chemical reactions. Swinney *et al.* have investigated the effect of periodic forcing on a photosensitive form of the Belousov-Zhabotinsky reaction [2]. The experiments have revealed that induced resonances transform an initial spiral pattern in labyrinths or domain (Ising and Bloch) walls. Although the dynamics of the Brusselator cannot describe all the experimental features , it reproduces some important aspects as the transition to labyrinths [3]. It was also shown that the Brusselator model with a parametric forcing exhibits similar pattern and reproduces the resonance regions (Arnold's tongues) predicted theoretically. Some of these facts can be described by a general complex Ginzburg-Landau equation under resonant forcing [4].

O. Descalzi et al. (eds.), Instabilities and Nonequilibrium Structures VII & VIII, 305–311.
© 2004 *Kluwer Academic Publishers. Printed in the Netherlands.*

Other authors investigated the temporal forcing of Turing patterns arising in CDIMA reaction (*chlorine dioxide-iodine-malonic acid*). For small amplitude of light intensity, they obtained oscillating Turing structures, while they dissapeared for higher amplitudes [1]. This is even quantitatively described by the Lengyel-Epstein model, in which shinning effects are taken into account through an additive term [5]. Threshold of different solutions are strongly modified, but no new solutions were obtained.

In the present paper we investigate the effect of time-periodic forcing on the symmetry of Turing patterns in the Brusselator model. In Section 2 we present the model and shortly explain the dynamics without forcing. In Section 3 the parametric time–periodic forcing is studied by numerical simulations of the model. The Floquet analysis is carried out in Section 4 and finally we discuss the main results and summarize our conclusions in Section 5.

2 Stability of Turing patterns in the Brusselator Model

The Brusselator model, is a very simple model which exhibits Hopf and Turing instability. For this reason it has been used very often to understand the behaviour of patterns in the CDIMA reaction [6, 7, 8]. This model corresponds to the reactive scheme of Fig. 1 which, after suitable rescaling, results into the dimensionless equations:

$$\partial_t X = A - (B+1)X + X^2 Y + \nabla^2 X,$$
$$\partial_t Y = BX - X^2 Y + D\nabla^2 Y. \tag{1}$$

A and B represent the input reactants, which can be externally controlled, C and D are products continuously evacuated, and the dynamical species are U (activator) and V (substrate in autocatalytic reaction). Guided by the experimental findings we focus on parameter values for which stationary bands

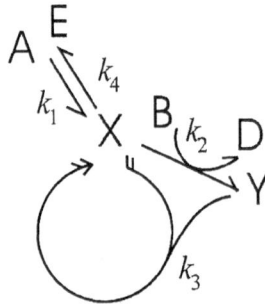

Fig. 1. Reaction scheme of Brusselator model. Arrows and tails indicate the stoichiometric coefficients of each reaction.

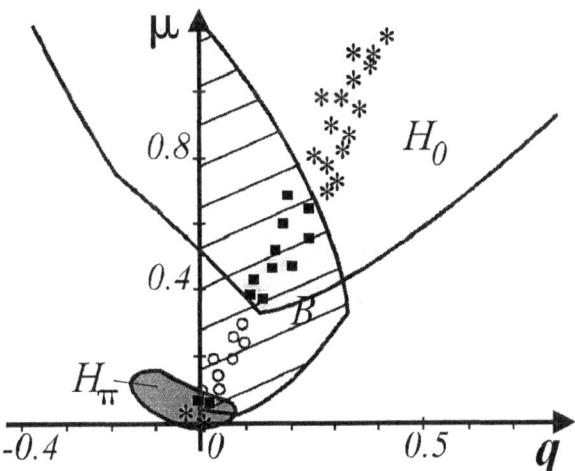

Fig. 2. Stability diagram without external forcing for $A = 4.5$ and $D = 8.0$ [7].

and two kind of hexagons (H_π and H_0) are obtained. In previous work we analyzed the stability of these solutions in the framework of amplitude equations [7, 8] and discussed the agreement between the analytical results and numerical simulations.

The diagram of Fig. 2 summarizes these results for $A = 4.5$ and $D = 8.0$, and will serve as a reference diagram for the present study. The horizontal axis represents the difference, $q = |k - k_c|$, between the wavenumber of the pattern and the critical value calculated at onset, $k_c = 1.26$. Grey and striped regions correspond to stable hexagons and bands, respectively. Near onset hexagons with total phase π (H_π) appear, while for higher values of the control parameter 0-hexagons are stabilized (H_0). Then, the maxima (white in the next figures) or the minima (black) of activator U are located in a hexagonal lattice. The wavenumbers of numerical solutions are represented by different symbols: asterisk for hexagons, circles for bands and squares for spatial mixtures of both patterns. The wavenumber selected in the pattern increases with the control parameter, in agreement with the higher linear growth of perturbations [7].

3   Time–periodic forcing

As mentioned the illumination effects on the CDIMA reaction seem quite well described through adding a linear term to the Lengyel–Epstein model [1, 5]. Instead, here we consider a temporal forcing through the control parameter B, by replacing $B \rightarrow B(1 + \gamma sin(\omega t))$ in Eqs. (1). In a continuously fed reactor this forcing could be achieved by a periodic change in one of the input reactants, a procedure likely difficult to implement in experiments.

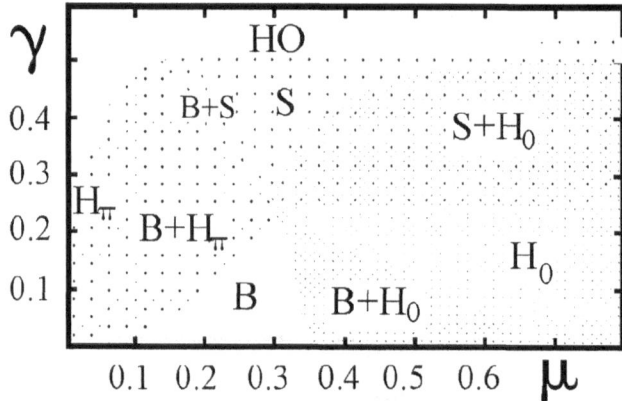

Fig. 3. Numerical solutions of the Brusselator model under time–forcing for $A = 4.5$, $D = 8$. and $\omega = 0.5$: HO for homogeneous oscillations, B stripes, H_π or H_0 for hexagons with total phase π or 0, respectively, and S for squares.

We integrated the model Eqs. (1) in $2D$ from different initial conditions (random, or gradually changing γ or μ). We used a pseudo–spectral method [9] with $\Delta x \sim 0.4$ and $\Delta t \sim 0.02$: the laplacian operators are discretized and the non-linear part is solved in the Fourier space at lower order.

In Fig. 3 the stability regions of each kind of pattern are drawn as function of supercriticality (defined as the normalized distance to onset, $\mu = \frac{B-B_c}{B_c}$) and forcing intensity γ. Near the Turing instability ($\mu \leq 0.05$) oscillating π-hexagons are stable for small γ, but they are replaced by a homogeneous state (HO) for higher values of γ. As μ is increased the situation becomes more involved. There is a wide region of bistability between oscillating stripes, B, and H_π, which spreads from $\gamma = 0$ to increasing values of μ. For a forcing strong enough, patterns of squares (S) stabilize for smaller γ as μ is increased. In agreement with the analysis without forcing, bands B are the sole solution in the interval $\mu \sim 0.1 - 0.35$ near the μ-axis, while they coexist with H_0-

Fig. 4. Front between squares and hexagons for $\mu = 0.3$, $\gamma = 0.3$, ($\omega = 4.5$).

hexagons for $\mu < 0.7$. For higher values of γ, squares appear also and they coexist with H_0 in an extended region. In these bistability regions, the final pattern depends on the initial condition and shows hysteresis phenomena. It is important to notice that these patterns are not standing waves, because the maxima and the minima do not interchange each other after a period T_0. In these oscillating patterns the wavenumber decreases when the forcing amplitude γ is increased.

A typical example of a front between hexagons and squares in a region of bistability is shown in Fig. 4. This kind of fronts seems metastable and after a sufficient time the mixed pattern is dominated by one of the symmetries (hexagons or squares). In Fig. 5 we show a temporal sequence of a pattern of squares for $\mu = 0.3$, $\gamma = 0.35$ and $\omega = 0.5$. It is constituted by domains with different orientation separated by defects which seem stationary in space.

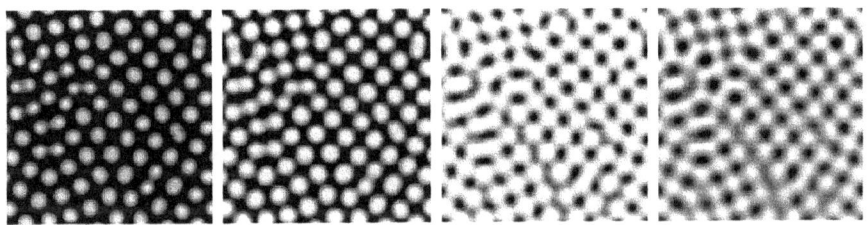

Fig. 5. Evolution of an irregular pattern of oscillating squares $\mu = 0.3$, $\gamma = 0.35$ and $\omega = 0.5$.

4 Floquet exponents

We guess that squares are favoured by a wavenumber selection process. But to deal with these oscillations a Floquet analysis is required. In the following paragraphs we sketch how this method can be adapted to our extended model. The Floquet method seeks for the stability of periodic solutions, considered as fixed points in the Poincaré map. If we denote by $\vec{c}_0(t) = (X_0(t), V_0(t), W_0(t))$ the reference solution , and $L(t)$ the linear matrix. Both are periodic and, therefore, invariant under a temporal translation $T : t \rightarrow t + T_0$. In absence of spatial effects, linear perturbations over a limit cycle $\vec{w}(t) = \vec{c}(t) + \vec{c}_0(t)$ obey the equation $\dot{\vec{w}} = L(t)\vec{w}$, which admits solutions in the form $\vec{w} = e^{\lambda t}\vec{v}$. In the simplest case of non–degenerated eigenvalues λ , vectors $\vec{v}$ are periodic with period T_0 [10]. When diffusion terms is included, the linear problem becomes $\dot{\vec{w}} = L(t)\vec{w} + D\nabla^2\vec{w}$, where D is a diagonal matrix which contains the diffusion coefficients and $L(t)$ is evaluated on the limit cycle. By expanding the perturbations in Fourier modes $\vec{w} = \sum \vec{q}_k e^{ikx}$ one obtains

$$\dot{\vec{q}}_k = \left[L(t) - Dk^2\right]\vec{q}_k. \qquad (2)$$

The Floquet theorem ensures that the solutions of Eq. (2) is of the form $\vec{q}_k = e^{\lambda_k t}\vec{u}_k(t)$, where $\vec{u}_k(t)$ is also periodic: $\vec{u}_k(t + T_0) = \vec{u}_k(t)$ [10]. Hence we obtain:

$$\vec{q}_k(t + T_0) = e^{\lambda_k(t+T_0)}\vec{u}_k(t + T_0) = e^{\lambda_k T_0}\vec{q}_k(t) \equiv \mathcal{M}_k\vec{q}_k(t) \qquad (3)$$

where the *Monodromy* matrix $\mathcal{M}_k$, gives the time evolution after a time period T_0. Consequently the linear stability analysis just requires the calculation of $\mathcal{M}_k$ because $e^{\lambda_k T_0}\vec{q}(t) = \mathcal{M}_k\vec{q}(t)$. Then, the eigenvalues of Eq. (2) are calculated from the eigenvalues of $\mathcal{M}_k$, denoted by σ_k (*Floquet multipliers*), because they are linked by the relationship $\Re[\lambda_k] = Ln\,|\sigma_k|\,/T_0$, in which λ_k stands for the *Floquet exponents*. If these are positive, i.e., if $\sigma_k > 1$ a mode with wavenumber k grows. The matrix $\mathcal{M}_k$ can be built in a simple manner. One takes the canonical initial conditions $\vec{q}_i(0) = (\delta_{i1}, \delta_{i2}, ..., \delta_{iN})^T$, con $i = 1, 2, ...N$, (as usual δ_{ij} stands for the *Kronecker delta*) and the vectors are calculated after a time period is elapsed, i.e., $\vec{q}_i(T_0) = \int_0^{T_0}[L(t) - Dk^2]\,\vec{q}_i(0)dt$. Hence $\mathcal{M}_k = [\vec{q}_1(T_0), \vec{q}_2(T_0), ..., \vec{q}_N(T_0)]$.

In our oscillating model this problem is solved by taking three initial values: one on the limit cycle, $\vec{c}_0(t)$, and two trivial vectors, given by $\vec{q}_i(0) = (\delta_{i1}, \delta_{i2})^T$ ($i = 1, 2$). At each computational time step the value $\vec{c}_0(t_i)$ is taken to calculate the linear problem:

$$L(t_i) - Dk^2 = \begin{pmatrix} -B(t_i) + 1 + 2X_0(t_i)Y_0(t_i) - k^2 & X_0(t_i)^2 \\ B(t_i) - 2X_0(t_i)Y_0(t_i) & -X_0(t_i)^2 - Dk^2 \end{pmatrix} \qquad (4)$$

After a period T_0, the matrix $\mathcal{M}_k$, and hence the Floquet exponents, is determined from $\vec{q}_i(T_0)$.

Preliminary results show that for small μ ($\mu \leq 0.05$) the higher Floquet exponent seem independent of γ. However, away from onset ($\mu \simeq 0.10$) the k with maximal growth slightly decreases as forcing is intensified, as obtained in numerical simulations.

5 Discussion

We have shown that the symmetry of Turing patterns can be modified by an external time-periodic forcing. Though the mechanism through which it occurs is still unclear, we think that this change of symmetry is related to the multistability in the Brusselator model and likely enhanced by a wavenumber selection mechanism. For $\gamma \neq 0$ the control parameter $\mu(t)$ becomes time-varying and induces a competition between hexagons (H_π and H_0) and bands which favours squares are as an intermediate state. This study should be completed from a stability analysis of the amplitude equations of the problem, a task that reveals to be rather involved.

Acknowledgements

The authors wish to thank Dr. A.P. Muñuzuri (Santiago de Compostela) for fruitful comments. This work has been supported by the MCyT (Spanish Government) under grant BFM2002-O1002 and by the PIUNA (Univ. Navarra). B.P. acknowledges a grant from the DAAD (German Government).

References

[1] A. Horváth, M. Dolnik, A. Muñuzuri, A. Zhabotinsky, and I. Epstein, Phys. Rev. Lett. **83**, 2950 (1999).

[2] Q. O. V. Petrov and H. L. Swinney, Nature **388**, 655 (1997).

[3] A. Lin, M. Bertram, K. Martinez, H. Swinney, A. Ardelea, and G. Carey, Phys. Rev. Lett. **84**, 4240 (2000).

[4] P. Coullet and K. Emilsson, Physica D **61**, 119 (1992).

[5] M. Dolnik, I. Berenstein, A. Zhabotinky, and I. Epstein, Phys. Rev. Lett. **87**, 238301 (2001).

[6] A. De Wit, Ph.D. thesis, Université Libre de Bruxelles (1993).

[7] B. Peña and C. Pérez-García, Europhys. Lett. **51**, 300 (2000).

[8] B. Peña and C. Pérez-García, Phys. Rev. E **64**, 56213 (2001).

[9] M. Bestehorn, *Strukturbildung durch Selbstorganisation in Flüssigkeiten und in chemischen Systemen* (Verlag Harri Deutsch, Stuttgart, 1995).

[10] H. Haken, *Advanced Synergetics: Instability Hierarchies of Self-Organizing Systems and Devices* (Springer Verlag, Berlin, 1987).

REALIZATIONS OF THE RANDOM REPRESENTATIONS OF THE NAVIER-STOKES EQUATIONS BY ORDINARY DIFFERENTIAL EQUATIONS

Diego L. Rapoport

Department of Applied Mechanics, Faculty of Engineering, Univ. of Buenos Aires,

Paseo Colón 850, Buenos Aires, Argentina; drapo@unq.edu.ar

Summary: We give the realizations by ordinary differential equations of the implicit random exact representations for the Navier-Stokes equations on smooth compact manifolds isometrically immersed in Euclidean spaces (viz. spheres, tori, euclidean spaces, etc.). We briefly discuss their relation with a hamiltonean system approach to the Navier-Stokes equations.

1   Introduction.

The reduction of the Navier-Stokes equations to ordinary differential equations has been the subject of intensive research; this is related to the construction of the inertial manifold for these equations [17], and is currently approached through dimensional analysis leading to modes suppresion when treating the Navier- Stokes equations as an evolution equation in appropiate functional spaces. This approach is unfortunately unwielding to actual computations, since the number of degrees of freedom continues to be rather big [10].

Stochastic approaches to the Navier-Stokes equations are not new [18]. In particular, diffusion processes on 2D have been applied to yield representations for the Navier- Stokes equations on the Euclidean plane [19]. Following a gauge-theoretical approach to characterize diffusion processes on smooth manifolds, the present author has given analytical implicit representations for Navier-Stokes equations in several instances: smooth compact manifolds isometrically immersed in Euclidean spaces (viz. spheres, tori, etc.) and as a particular case of this construction for Euclidean domains itself [3,4], for abstract compact manifolds without the need of isometrically embedding of them [4,5], for semispaces and smooth boundary manifolds [6], and finally given

313

O. Descalzi et al. (eds.), Instabilities and Nonequilibrium Structures VII & VIII, 313–331.

purely noise representations for which the nonlinearity can be subsumed into a geometric noise term [5]. From all this constructions it was relieved that the representations for Navier-Stokes are identical for any dimension (other than 1) with the exception of the non-generic Euclidean case.

In this article we shall present a realization by ordinary differential equations of the random representations for the Navier-Stokes equations, in the particular case of viscous incompressible fluids on smooth compact (and without boundary) manifolds isometrically immersed in Euclidean spaces, yet, for reasons of space we shall abstain of discussing the approximations to the pure-noise representations. Our constructions will follow the methodology of stochastic differential geometry that stems from the developing method of smooth curves due to Elie Cartan, extending it to the random development of Wiener processes, as the geometrical construction of the most general diffusion processes on general manifolds [7,16]. This is the method that allowed us to give random implicit invariant representations for the Navier-Stokes equations and for the equations of passive transport of fields along the fluids such as the kinematic dynamo of magnetohydrodynamics [4,5,6], and will now be used to approximate these random flows by ordinary differential equations at almost all times, following the same path that lead to extend the classical method to the random one.

2 Riemann-Cartan-Weyl Geometry of Diffusions

In this article M denotes a smooth compact orientable n-dimensional manifold (without boundary) provided with a linear connection (see [5,15]) described by a covariant derivative operator ∇ which we assume to be compatible with a given metric g on M, i.e. $\nabla g = 0$. Given a coordinate chart (x^α) $(\alpha = 1, \ldots, n)$ of M, a system of functions on M (the Christoffel symbols of ∇) are defined by $\nabla_{\frac{\partial}{\partial x^\beta}} \frac{\partial}{\partial x^\gamma} = \Gamma(x)^\alpha_{\beta\gamma} \frac{\partial}{\partial x^\alpha}$. The Christoffel coefficients of ∇ can be decomposed as:

$$\Gamma^\alpha_{\beta\gamma} = \left\{ \begin{matrix} \alpha \\ \beta\gamma \end{matrix} \right\} + \frac{1}{2} K^\alpha_{\beta\gamma}. \tag{1}$$

The first term in (1) stands for the metric Christoffel coefficients of the Levi-Civita connection ∇^g associated to g, i.e. $\{{}^\alpha_{\beta\gamma}\} = \frac{1}{2}(\frac{\partial}{\partial x^\beta}g_{\nu\gamma} + \frac{\partial}{\partial x^\gamma}g_{\beta\nu} - \frac{\partial}{\partial x^\nu}g_{\beta\gamma})g^{\alpha\nu}$, and

$$K^\alpha_{\beta\gamma} = T^\alpha_{\beta\gamma} + S^\alpha_{\beta\gamma} + S^\alpha_{\gamma\beta}, \tag{2}$$

is the cotorsion tensor, with $S^\alpha_{\beta\gamma} = g^{\alpha\nu}g_{\beta\kappa}T^\kappa_{\nu\gamma}$, and $T^\alpha_{\beta\gamma} = (\Gamma^\alpha_{\beta\gamma} - \Gamma^\alpha_{\gamma\beta})$ the skew-symmetric torsion tensor. We are interested in (one-half) the Laplacian operator associated to ∇, i.e. the operator acting on smooth functions on M defined as

$$H(\nabla) := 1/2\nabla^2 = 1/2 g^{\alpha\beta}\nabla_\alpha\nabla_\beta. \tag{3}$$

A straightforward computation shows that $H(\nabla)$ only depends in the trace of the torsion tensor and g, since it is

$$H(\nabla) = 1/2\triangle_g + \hat{Q}, \tag{4}$$

with $Q := Q_\beta dx^\beta = T^\nu_{\nu\beta}dx^\beta$ the trace-torsion one-form and where $\hat{Q}$ is the vector field associated to Q via g: $\hat{Q}(f) = g(Q, df)$, for any smooth function f defined on M. Finally, $\triangle_g$ is the Laplace-Beltrami operator of g: $\triangle_g f = \operatorname{div}_g \operatorname{grad} f$, $f \in C^\infty(M)$, with div_g the Riemannian divergence. Thus for any smooth function, we have $\triangle_g f = 1/[det(g)]^{\frac{1}{2}} g^{\alpha\beta}\frac{\partial}{\partial x^\beta}([det(g)]^{\frac{1}{2}}\frac{\partial}{\partial x^\alpha}f)$.

Consider the family of zero-th order differential operators acting on smooth k-forms, i.e. differential forms of degree k $(k = 0, \ldots, n)$ defined on M:

$$H_k(g, Q) := 1/2\triangle_k + L_{\hat{Q}}, \tag{5}$$

In the first summand of the r.h.s. of (5) we have the Hodge operator acting on k-forms:

$$\triangle_k = (d - \delta)^2 = -(d\delta + \delta d), \tag{6}$$

with d and δ the exterior differential and codifferential operators respectively, i.e. δ is the adjoint operator of d defined through the pairing of k-forms on M: $(\omega_1, \omega_2) := \int g(\omega_1, \omega_2) vol_g$, for arbitrary k-forms ω_1, ω_2, where $vol_g(x) = det(g(x))^{\frac{1}{2}}dx$ is the volume density. The last identity in (6) follows from the fact that $d^2 = 0$ so that $\delta^2 = 0$. Furthermore, the second term in (5) denotes the Lie-derivative with respect to the vectorfield $\hat{Q}$: $L_{\hat{Q}} = i_{\hat{Q}}d + di_{\hat{Q}}$, where $i_{\hat{Q}}$ is the interior product with respect to $\hat{Q}$: for arbitrary vectorfields $X_1, \ldots, X_{k-1}$ and ϕ a k-form defined on M, we have $(i_{\hat{Q}}\phi)(X_1, \ldots, X_{k-1}) = \phi(\hat{Q}, X_1, \ldots, X_{k-1})$. Then, for f a scalar field, $i_{\hat{Q}}f = 0$ and

$$L_{\hat{Q}}f = (i_{\hat{Q}}d + di_{\hat{Q}})f = i_{\hat{Q}}df = g(Q, df) = Q(f). \tag{7}$$

Since $\triangle_0 = (\nabla^g)^2 = \triangle_g$, we see that from the family defined in (5) we retrieve for scalar fields ($k = 0$) the operator $H(\nabla)$ defined in (3&4).

Proposition 1. Assume that g is non-degenerate. There is a one-to-one mapping

$$\nabla \leadsto H_k(g, Q) = 1/2\triangle_k + L_{\hat{Q}}$$

between the space of g-compatible affine connections ∇ with Christoffel coefficients of the form

$$\Gamma^{\alpha}_{\beta\gamma} = \left\{ \begin{array}{c} \alpha \\ \beta\gamma \end{array} \right\} + \frac{2}{(n-1)} \left\{ \delta^{\alpha}_{\beta} \, Q_{\gamma} \, - \, g_{\beta\gamma} \, Q^{\alpha} \right\}, n \neq 1 \tag{8}$$

and the space of elliptic second order differential operators on k-forms ($k = 0, \ldots, n$) with zero potential term.

The connections defined in (8) are called Riemann-Cartan-Weyl (RCW for short) connections [3,5,6,8,11,14,15]. The naming after Weyl of the trace-torsion is motivated by the fact that these geometries can be introduced through scale transformations which extend the Weyl transformations in the first ever conceived gauge theory. They are fundamental to the description of the motion of relativistic spinning particles in exterior gravitational fields [14] and to the existance of matter quantum fields enforcing the equivalence of the Maxwell equation and the massive non-linear Dirac-Hestenes equation for a spinor-field [8].

3 Riemann-Cartan-Weyl Diffusions of Differential forms

In this section we shall extend the correspondance of Proposition 1 to a correspondance between RCW connections and diffusion processes of k-forms ($k = 0, \ldots, n$) having $H_k(g, Q)$

as infinitesimal generators (i.g. for short, in the following).Thus, naturally we shall call these processes as *RCW diffusion processes*.

In the following we shall further assume that $Q = Q(\tau, x)$ is a time-dependant 1-form. The stochastic flow associated to the diffusion generated by $H_0(g, Q)$ has for sample paths the continuous curves $\tau \mapsto x_\tau \in M$ satisfying the Ito invariant non-degenerate s.d.e. (stochastic differential equation)

$$dx(\tau) = X(x(\tau))dW(\tau) + \hat{Q}(\tau, x(\tau))d\tau. \tag{9}$$

In this expression, $X : M \times R^m \to TM$ is such that $X(x) : R^m \to TM$ is linear for any $x \in M$, so that we write $X(x) = (X_i^\alpha(x))$ ($1 \leq \alpha \leq n$, $1 \leq i \leq m$) which satisfies $X_i^\alpha X_i^\beta = g^{\alpha\beta}$, and $\{W(\tau), \tau \geq 0\}$ is a standard Wiener process on R^m. Here τ denotes the time-evolution parameter of the diffusion (in a relativistic setting it should not be confused with the time variable), and for simplicity we shall assume always that $\tau \geq 0$. Consider the canonical Wiener space Ω of continuous maps $\omega : R \to R^m, \omega(0) = 0$, with the canonical realization of the Wiener process $W(\tau)(\omega) = \omega(\tau)$. The (stochastic) flow of the s.d.e. (9) is a mapping

$$F_\tau : M \times \Omega \to M, \quad \tau \geq 0, \tag{10}$$

such that for each $\omega \in \Omega$, the mapping $F_.(.\,, \omega) : [0, \infty) \times M \to M$, is continuous and such that $\{F_\tau(x) : \tau \geq 0\}$ is a solution of equation (9) with $F_0(x) = x$, for any $x \in M$.

Let us assume in the following that the components X_i^α, $\hat{Q}^\alpha$, $\alpha, \beta = 1, \ldots, n$ of the vector-fields X and $\hat{Q}$ on M in (9) are predictable functions which further belong to $C_b^{m,\epsilon}$ ($0 < \epsilon < 1$, m a non-negative integer), the space of Hoelder bounded continuous functions of degree $m \geq 1$ and exponent ϵ, and also that $\hat{Q}^\alpha(\tau) \in L^1(R)$, for any $\alpha = 1, \ldots, n$. With these regularity conditions, if we further assume that $x(\tau)$ is a semimartingale on a probability space $(\Omega, \mathcal{F}, P)$, then it follows that the flow of (9) has a modification (which with abuse of notation we denote as)

$$F_\tau(\omega) : M \to M, \quad F_\tau(\omega)(x) = F_\tau(x, \omega), \tag{11}$$

318

which is a diffeomorphism of class C^m, almost surely for $\tau \geq 0$ and $\omega \in \Omega$. We would like to point out that a similar result follows from working with Sobolev space regularity conditions instead of Hoelder continuity. Indeed, assume that the components of X and $\hat{Q}$, $X_i^\alpha \in H^{s+2}(M)$ and $\hat{Q}^\beta \in H^{s+1}(M)$, $1 \leq i \leq m$, $1 \leq \beta \leq n$, where the Sobolev space $H^s(M) = W^{2,s}(M)$ with $s > \frac{n}{2} + m$, $m \geq 1$. Then, the flow of (9) for fixed ω defines a diffeomorphism in $H^s(M, M)$, and hence by the Sobolev embedding theorem, a diffeomorphism in $C^m(M, M)$ [9].

Let us describe the (first) derivative (or *jacobian*) flow of (9), i.e. the stochastic process $\{v(\tau) := T_{x_0} F_\tau(v(0)) \in T_{F_\tau(x_0)} M, v(0) \in T_{x_0} M\}$; here $T_z M$ denotes the tangent space to M at z and $T_{x_0} F_\tau$ is the linear derivative of F_τ at x_0. The process $\{v_\tau, \tau \geq 0\}$ can be described [21] as the solution of the invariant Ito s.d.e. on TM:

$$dv(\tau) = \nabla^g \hat{Q}(\tau, v(\tau)) d\tau + \nabla^g X(v(\tau)) dW(\tau) \tag{12}$$

If we take U to be an open neighbourhood in M so that the tangent space on U is $TU = U \times R^n$, then $v(\tau) = (x(\tau), \tilde{v}(\tau))$ is described by the system given by integrating (9) and the covariant Ito s.d.e.

$$d\tilde{v}(\tau)(x(\tau)) = \nabla^g X(x(\tau))(\tilde{v}(\tau)) dW(\tau) + \nabla^g \hat{Q}(\tau, x(\tau))(\tilde{v}(\tau)) d\tau, \tag{13}$$

with initial condition $\tilde{v}(0) = v_0$. Thus, $\{v(\tau) = (x(\tau), \tilde{v}(\tau)), \tau \geq 0\}$ defines a random flow on TM.

4 Classical Flows and the Approximation of Riemann-Cartan-Weyl Diffusions

If we revise the history of the construction of invariant diffusion processes, we find that it stems from the extension of the classical method of development of classical flows due to Elie Cartan [7,15] [1], where the extension is carried out by replacing a classical (i.e. smooth) flow by

---

[1]This method has been used to ellaborate a theory of classical spinning particles on exterior gravitational fields without the need of introducing lagrangians or hamiltonean functions, just symplectic geometry and gauge

the random flow defined by the developing of a Wiener process, which is non-differentiable in every point. Furthermore, this gives the possibility of actually realizing these random flows by ordinary differential equations, i.e. by a classical flow. Yet, to actually carry out this program, it is preferable to pass to the Stratonovich representations, which are well known to have the same transformation rules in stochastic analysis that those of classical flows. Thus, instead of eqt. (9) we consider the Stratonovich representation (here denoted, as usual, by the symbol ∘) for it given by (see [7,15]):

$$
\begin{aligned}
dx(\tau) &= X(x(\tau)) \circ dW(\tau) + b^{Q,X}(\tau, x(\tau))d\tau, \\
&\text{where } b^{Q,X}(\tau, x(\tau)) = \hat{Q}(\tau, x(\tau)) + S(\nabla^g, X)(x(\tau)),
\end{aligned}
\tag{14}
$$

where the drift now contains an additional term, the Stratonovich correction term, given by

$$
S(\nabla^g, X) = \frac{1}{2}\mathrm{tr}(\nabla^g_X X),
\tag{15}
$$

where $\nabla^g_X X$, the Levi-Civita covariant derivative of X in the same direction and thus, it is a linear transformation from R^m valued on TM. Now we also represent the jacobian flow using the Stratonovich prescription

$$
d\tilde{v}(\tau) = \nabla^g X(x(\tau))(\tilde{v}(\tau)) \circ dW(\tau) + \nabla^g b^{Q,X}(\tau, x(\tau))(\tilde{v}(\tau))d\tau,
\tag{16}
$$

Now we shall construct the classical flow to approximate the random flow $\{x(\tau) : \tau \geq 0\}$. We start by constructing a polygonal approximation of the Wiener process. Thus, we set for each $n = 1, 2, \ldots$,

$$
W_n(\tau) = n[(\frac{j+1}{n} - \tau)W(\frac{j}{n}) + (\tau - \frac{j}{n})W(\frac{j+1}{n})], \text{if } \frac{j}{n} \leq \tau \leq \frac{j+1}{n}, j = 0, 1, \ldots
\tag{17}
$$

and we further consider the sequence $\{x_n(\tau)\}_{n \in N}$ satisfying

$$
\begin{aligned}
\frac{dx_n(\tau)}{d\tau} &= X(x_n(\tau))\frac{dW_n}{d\tau}(\tau) + b^{Q,X}(\tau, x_n(\tau)), \\
\frac{d\tilde{v}_n(\tau)}{d\tau}(x_n(\tau)) &= \nabla^g X(x_n(\tau))(\tilde{v}_n(\tau))\frac{dW_n}{d\tau}(\tau) + \nabla^g b^{Q,X}(\tau, x_n(\tau))(\tilde{v}_n(\tau)).
\end{aligned}
\tag{18}
$$

theoretical considerations [14]; remarkably, the Navier-Stokes equations admit a random hamiltonean which follows precisely from this historical construction which we shall present below [20].

Since $W_n(\tau)$ is differentiable a.e., thus $\{x_n(\tau) : x_n(0) = x(0)\}_{n \in N}$ is a sequence of flows obtained by integration of well defined ordinary differential equations on M, almost everywhere on τ, for all $W \in \Omega$. With the additional assumption that X and Q are smooth, then $\{x_n(\tau) : x_n(0) = x(0)\}$ defines for almost all τ, for all $W \in \Omega$, a flow of smooth diffeomorphisms of M, and thus, the flow $\{v_n(\tau) = (x_n(\tau), \tilde{v}_n(\tau)) : v_n(0) = (x(0), v(0))\}$ defines a flow of smooth diffeomorphisms of TM. In this case, this flow converges uniformly in probability, in the group of smooth diffeomorphisms of TM, to the the flow of random diffeomorphisms on TM defined by eqts. $(14-16)$ (see [7]).

5 Riemann-Cartan-Weyl Gradient Diffusions

Assume that there is an isometric immersion of an n-dimensional manifold M into a Euclidean space R^m given by the mapping $f : M \to R^m, f(x) = (f^1(x), \ldots, f^m(x))$. For example, $M = S^n, T^n$, the n-dimensional sphere or torus respectively, and f is an isometric embedding into R^{n+1}, or still $M = R^m$ with f given by the identity map. The existance of such a smooth immersion is proved by the Nash theorem in the compact manifold case, yet the result is known to be valid as well for non compact manifolds [1]. Assume further that $X(x) : R^m \to T_x M$, is the orthogonal projection of R^m onto $T_x M$ the tangent space at x to M, considered as a subset of R^m. Then, if $e_1, \ldots, e_m$ denotes the standard basis of R^m, we have

$$X(x) = X^i(x)e_i, \text{ with } X^i(x) = \text{grad } f^i(x), i = 1, \ldots, m. \qquad (19)$$

We are interested in the RCW *gradient* diffusion processes on compact manifolds isometrically immersed in Euclidean space, given by (9) with the diffusion tensor X given by (19). We shall now give the Ito formula for k-forms on compact manifolds which are isometrically immersed in Euclidean space. Recall that the k-th exterior product of a vector field v is written as $\Lambda^k v = v \wedge v \wedge \ldots \wedge v$ (k times). We further denote by $C_c^{1,2}(\Lambda^p(R \times M))$ the space of time-dependant p-forms on M continuously differentiable with respect to the time variable and of class C^2 with respect to the M variable and of compact support with its derivatives.

Theorem 1 (Ito Formula for k − forms [20]). Let M be isometrically immersed in R^m as in (20). Let $V_0 \in \Lambda^k T_{x_0} M$, $0 \leq k \leq n$. Set $V_\tau = \Lambda^k (TF_\tau)(V_0)$. Then $\partial_\tau + H_k(g, \hat{Q})$ is the i.g. (with domain of definition the differential forms of degree k in $C_c^{1,2}(\Lambda^p R \times M)$) of $\{V_\tau : \tau \geq 0\}$.

Remarks 2. Therefore, starting from the flow $\{F_\tau : \tau \geq 0\}$ of the s.d.e. (12) (or its Stratonovich version given by eqt. (14)) with i.g. given by $\partial_\tau + H_0(g, Q)$, we construct (fibered on it) the derived velocity process $\{v(\tau) : \tau \geq 0\}$ given by (12) (or (9&13), with the diffusion tensor given by (19), or still, its Stratonovich version given by eqts. $(14 - 16)$) which has $\partial_\tau + H_1(g, Q)$ for i.g.. Finally, if we consider the diffusion processes of differential forms of degree $k \geq 1$, we further get that $\partial_\tau + H_k(g, Q)$ is the i.g. of the process $\{\Lambda^k v(\tau) : \tau \geq 0\}$, on the Grassmannian bundle $\Lambda^k TM$, $(k = 0, \ldots, n)$. Note that consistent with our notation, and since $\Lambda^0(TM) = M$ we have that $\Lambda^0 v(\tau) \equiv x(\tau), \forall \tau \geq 0$. In particular, $\partial_\tau + H_2(g, Q)$ is the i.g. of the stochastic process $\{v(\tau) \wedge v(\tau) : \tau \geq 0\}$ on $TM \wedge TM$.

The Ito formula will turn out to be the key instrument for writing down random representations for the solutions of linear transport equations, and in particular, for the vorticity as well as for the velocity of an incompressible fluid.

Consider in a smooth manifold M isometrically immersed in Euclidean space, the following initial value problem: We want to solve

$$\frac{\partial}{\partial \tau} \beta = H_p(g, Q)\beta_\tau, \tag{20}$$

with given

$$\beta(0, x) = \beta_0(x), \tag{21}$$

for an arbitrary time-dependant p-form β defined on M which belongs to $C_c^{1,2}(\Lambda^p(R \times M))$. Then, the formal solution of this problem is as follows: Consider the stochastic differential equation given by running backwards in time eqt. (9)[2]:

$$dx^{\tau,s,x} = X(x^{\tau,s,x})dW(s) + \hat{Q}(\tau - s, x^{\tau,s,x})ds, x^{\tau,0,x} = x \in M. \tag{22}$$

---

[2]We can, of course, solve this problem by running the Stratonovich version.

and the derived velocity process $\{v^{\tau,s,v(x)}, v^{\tau,0,v(x)} = v(x) \in T_x M, 0 \leq s \leq \tau\}$ which in a coordinate system we write as $v^{\tau,s,v(x)} = (x^{\tau,s,x}, \tilde{v}^{\tau,s,v(x)})$ verifying (22) and the s.d.e.

$$d\tilde{v}^{\tau,s,v(x)} = \nabla^g X(x^{\tau,s,x})(\tilde{v}^{\tau,s,v(x)})dW(s) + \nabla^g \hat{Q}(\tau - s, x^{\tau,s,x})(\tilde{v}^{\tau,s,v(x)})ds, \tilde{v}^{\tau,0,v(x)} = v(x). \quad (23)$$

Notice that this system is nothing else that the jacobian process running backwards in time until the beginning.

Theorem 2 ([3,4]). The formal solution of the initial value problem (20&21) is

$$\beta(\tau, x)(\Lambda^p v(x)) = E_x[\beta_0(x^{\tau,\tau,x})(\Lambda^p \tilde{v}^{\tau,\tau,v(x)})]. \quad (24)$$

where the l.h.s. $\Lambda^p v(x)$ denotes the exterior product of p linearly independant tangent vectors at x, and in the r.h.s. $\Lambda^p v^{\tau,\tau,v(x)}$ denotes the exterior product of the flows having initial condition given by them.

Proof. It follows from the Ito formula.

6 The Navier-Stokes Equation and Riemann-Cartan-Weyl Gradient Diffusions

In the sequel, M is a compact orientable n-manifold (without boundary) provided with a Riemannian metric g. Further, M has a "velocity" 1-form $u_\tau(x) = u(\tau, x)$ satisfying the incompressibility condition

$$\delta u_\tau = -div(\hat{u}_\tau) = 0, \forall \tau \quad (25)$$

and the covariant Navier-Stokes equations which from $[1, 2, 3, 4, 17]$ we know to be

$$\frac{\partial u}{\partial \tau} = P[2\nu\triangle_1 - L_{\hat{u}_\tau}]u_\tau$$

where P denotes the projection operator of the space of 1-forms into the δ-closed component of its Hodge-Helmholtz decomposition; taking in account the definition (5) we rewrite in the concise form

$$\frac{\partial u}{\partial \tau} = PH_1(2\nu g, \frac{-1}{2\nu}u_\tau)u_\tau. \quad (26)$$

Let us introduce a new variable: the vorticity time-dependant 2-form

$$\Omega_\tau = du_\tau. \tag{27}$$

Now, if we know Ω_τ for any $\tau \geq 0$, we can obtain u_τ by inverting the definition (27). Namely, applying δ to (27) and in further account of (6) we obtain the Poisson-de Rham equation

$$\triangle_1 u_\tau \equiv H_1(g, 0)u_\tau = -\delta\Omega_\tau. \tag{28}$$

Thus, the vorticity Ω_τ is a source for the velocity one-form u_τ, for all τ; in the case that M is an euclidean domain, (14) is integrated to give the Biot-Savart law of Fluid Mechanics as described in [3,19]. Now, apply d to eqt. (26) to obtain (for the details see [4,5]) the evolution equation for the vorticity

$$\frac{\partial \Omega_\tau}{\partial \tau} = H_2(2\nu g, \frac{-1}{2\nu}u_\tau)\Omega_\tau. \tag{29}$$

Theorem 3([**4, 5**]). Given a compact orientable Riemannian manifold with metric g, the Navier-Stokes equation (26) for an incompressible fluid with velocity one-form $u = u(\tau, x)$ such that $\delta u_\tau = 0$, assuming sufficiently regular conditions, is equivalent to a diffusion process for the vorticity given by (29) with u_τ satisfying the Poisson-de Rham equation (28). The RCW connection on M generating this process is determined by the metric $2\nu g$ and a trace-torsion 1-form given by $-u/2\nu$.

7   Integration of the Navier-Stokes equation for the vorticity:

In the following we assume additional conditions on M, namely that it is isometrically immersed in an Euclidean space, so that the diffusion tensor is given in terms of the immersion f by $X = \nabla f$.

Let u denote a solution of (26) and consider the flow $\{F_\tau : \tau \geq 0\}$ of the s.d.e. whose i.g. is $\frac{\partial}{\partial \tau} + H_0(2\nu g, \frac{-1}{2\nu}u)$; from (9) and Theorem 1 we know that this is the flow defined by integrating the non-autonomous Ito s.d.e.

$$dx(\tau) = [2\nu]^{\frac{1}{2}} X(x(\tau))dW(\tau) - \hat{u}(\tau, x(\tau))d\tau, x(0) = x, 0 \leq \tau. \tag{30}$$

324

We shall assume in the following that X and $\hat{u}$ have the regularity conditions stated in Section 3 so that the random flow of (16) is a diffeomorphism of M of class C^m.

Theorem 4. Equation (30) is a random Lagrangian representation for the fluid particles positions, i.e. $x(\tau)$ is the random position of the particles of the incompressible fluid whose velocity obeys (26).

Remarks 3. Assuming thus a continuum of particles, we obtain an *exact* representation of the flow of a *exact* solution of NS. Notice that when we set $\nu = 0$, the classical flow of the velocity yields the solution of the *Euler* equations for *inviscid* fluids.

Now if we express the random Lagrangian flow in Stratonovich form

$$dx(\tau) = [2\nu]^{\frac{1}{2}}X(x(\tau)) \circ dW(\tau) + b^{-u,X}(\tau, x(\tau))d\tau,$$

$$\text{with } b^{-u,X}(\tau, x(\tau)) = \nu\nabla^g_X X(x(\tau)) - \hat{u}(\tau, x(\tau))), \tag{31}$$

we can approximate in the group of diffeomorphisms of M this flow by considering the sequence of ordinary differential equations

$$\frac{dx_n}{d\tau}(\tau) = [2\nu]^{\frac{1}{2}}X(x_n(\tau))\frac{dW_n}{d\tau}(\tau) + b^{-u,X}(\tau, x_n(\tau)), \tag{32}$$

and the jacobian flow on TM, $\{v(\tau) = (x(\tau), \tilde{v}(\tau))$ with $\tilde{v}(\tau)$ satisfying the Stratonovich eqts.

$$d\tilde{v}(\tau)(x(\tau)) = [2\nu]^{\frac{1}{2}}\nabla^g X(x(\tau))(\tilde{v}(\tau)) \circ dW(\tau) + \nabla^g b^{-u,X}(\tau, x(\tau))(\tilde{v}(\tau))d\tau, \tag{33}$$

can be approximated by $\{x_n(\tau), \tilde{v}_n(\tau))\}$ given by integrating

$$\frac{d\tilde{v}_n(\tau)}{d\tau}(x_n(\tau)) = [2\nu]^{\frac{1}{2}}\nabla^g X(x_n(\tau))(\tilde{v}_n(\tau))\frac{dW_n}{d\tau}(\tau) + \nabla^g b^{-u,X}(\tau, x_n(\tau))(\tilde{v}_n(\tau)). \tag{34}$$

Thus, we have a system of ordinary differential equations a.e. for τ, and if u is of class C^m ($m \geq 1$), this flow converges uniformly in probability, in the group of diffeomorphisms of class C^m of M to the random diffeomorphism flow, of the same class, that integrates Navier-Stokes. In the next section we shall see how we can approximate using this classical flow, the random implicit representations for Navier-Stokes equations.

7.1 Initial Value Problem for the Vorticity

Let us find the form of the strong solution (whenever it exists) of the initial value problem for $\Omega(\tau, x)$ satisfying (29) with initial condition $\Omega(0, x) = \Omega_0(x) = du_0(x)$. For this, we run backwards in time the random lagrangian flow (31): For each $\tau \geq 0$ consider the s.d.e. (with $s \in [0, \tau]$):

$$dx^{\tau,s,x} = [2\nu]^{\frac{1}{2}} X(x^{\tau,s,x}) \circ dW(s) + b^{-u,X}(\tau - s, x^{\tau,s,x}) ds, x^{\tau,0,x} = x. \tag{35}$$

and the derived velocity process $\{v^{\tau,s,v(x)} : v^{\tau,0,v(x)} = v(x) \in T_x M, 0 \leq s \leq \tau\}$ which in a coordinate system we write as $v^{\tau,s,v(x)} = (x^{\tau,s,x}, \tilde{v}^{\tau,s,v(x)})$ verifying (35) and the s.d.e.

$$
\begin{aligned}
d\tilde{v}^{\tau,s,v(x)} &= [2\nu]^{\frac{1}{2}} \nabla^g X(x^{\tau,s,x})(\tilde{v}^{\tau,s,v(x)}) \circ dW(s) \\
&+ \nabla^g b^{-u,X}(\tau - s, x^{\tau,s,x})(\tilde{v}^{\tau,s,v(x)}) ds, \tilde{v}_0^{\tau,0,v(x)} = v(x) \in T_x M. \tag{36}
\end{aligned}
$$

Let $v^1(x)$ and $v^2(x)$ non-zero linearly independant vectors in $T_x M$, be initial conditions for the flow $\tilde{v}^{\tau,x,v(x)}$.

Theorem 5. If there is a $C^{1,2}$ (i.e. continuously differentiable in the time variable $\tau \in [0, T)$, and of class C^2 in the space variable) solution $\tilde{\Omega}_\tau(x)$ of the initial value problem, it is

$$\tilde{\Omega}_\tau(v^1(x) \wedge v^2(x)) = E_x[\Omega_0(x^{\tau,\tau,x})(\tilde{v}^{\tau,\tau,v^1(x)} \wedge \tilde{v}^{\tau,\tau,v^2(x)})], \tag{37}$$

where E_x denotes the expectation value with respect to the measure on $\{x^{\tau,\tau,x} : \tau \geq 0\}$.

Proof. It is evident from Theorems 1 and 2.

Now we can approximate this equation by taking the jacobian flow $(x_n^{\tau,s,x}, \tilde{v}_n^{\tau,s,v(x)})$ in the r.h.s. of expression (37), where

$$
\begin{aligned}
\frac{dx_n^{\tau,s,x}}{ds} &= [2\nu]^{\frac{1}{2}} X(x_n^{\tau,s,x}) \frac{dW(s)}{ds} + b^{-u,X}(\tau - s, x_n^{\tau,s,x}), x_n^{\tau,0,x} = x, \\
\frac{d\tilde{v}_n^{\tau,s,v(x)}}{ds} &= [2\nu]^{\frac{1}{2}} \nabla^g X(x_n^{\tau,s,x})(\tilde{v}_n^{\tau,s,v(x)}) \frac{dW(s)}{ds} \\
&+ \nabla^g b^{-u,X}(\tau - s, x_n^{\tau,s,x})(\tilde{v}_n^{\tau,s,v(x)}) ds, \tilde{v}_n^{\tau,0,v(x)} = v(x) \in T_x M \\
\text{where } \frac{dW(s)}{ds} &= 2^n \{W(\frac{[2^n s/\tau] + 1}{2^n}) - W(\frac{[2^n s/\tau]}{2^n})\}, s \in [0, \tau], (\tau > 0), \tag{38}
\end{aligned}
$$

with $[z]$ the integer part of $z \in (0,1]$, is the Stroock & Varadhan polygonal approximation [7]. Since this is an o.d.e. a.e. on τ, for each *fixed* realization W in Wiener space, the expectation value is redundant in (37) for a fixed realization (otherwise, we have to average with the standard Gaussian on Wiener space), and we can write the formal expression for each *fixed* realization W:

$$\tilde{\Omega}_\tau(v^1(x) \wedge v^2(x)) = lim_{n\to\infty}\Omega_0((x_n^{\tau,\tau,x})(\tilde{v}_n^{\tau,\tau,v^1(x)} \wedge \tilde{v}_n^{\tau,\tau,v^2(x)})]. \tag{39}$$

We can improve the convergence by taking different realizations and running the corresponding flows to further take an arithmetic mean on the realizations, as a kind of Monte-Carlo expression on a.e. differentiable curves [21]. It appears to be an interest problem to estimate the quality of this convergence which in principle is not worse than the Monte-Carlo method.

8 Integration of the Poisson-de Rham equation

We are finally left with the Dirichlet problem posed by eqt. (28), for which we need to write in Stratonovich form the flow generated by $H_1(g,0)$, which is nothing else that the jacobian flow associated to the restriction of placing $u \equiv 0$ in eqt. (30) further reparametrized by multiplication by $(2\nu)^{\frac{-1}{2}}$:

$$\begin{aligned} d\tilde{x}^{s,x} &= X(\tilde{x}^{s,x})dW(s) \equiv X(\tilde{x}^{s,x}) \circ dW(s) + S(\nabla^g, X)(x^{s,x})ds, x^{0,x} = x \\ d\tilde{v}^{s,v(x)}(\tilde{x}^{s,x}) &= \nabla^g X(x^{s,x})(\tilde{v}^{s,v(x)}) \circ dW(s) + \nabla^g S(\nabla^g, X)(\tilde{x}^{s,x})(\tilde{v}^{s,v(x)})ds, \\ \tilde{v}^{0,v(x)} &= v(x) \in T_x M. \end{aligned} \tag{40}$$

The formal solution on a coordinate system where for each open set we have the boundary condition described by a 1-form ϕ such that $\delta\phi = 0$, then the expression for the velocity is

$$\tilde{u}_\tau(x)(v(x)) = E_x^B[\phi(\tilde{x}^{\tau_e,x})(\tilde{v}^{\tau_e,v(x)}) + \int_0^{\tau_e} \frac{1}{2}\delta\Omega_\tau(\tilde{x}^{s,x})(\tilde{v}^{s,v(x)}(s))ds] \tag{41}$$

where τ_e is the first-exit time of the process of this coordinate neighborhoood, and where the expectation value is with respect to $p^g(s,x,y)$ the transition density of the s.d.e. (40), i.e.the fundamental solution of the heat equation on M:

$$\partial_\tau p(y) = 1/2\triangle_g p(y) \tag{42}$$

with

$$p(s, x, -) = \delta_x, s \to 0^+ \tag{43}$$

where δ_x is the Dirac delta on x and E_x denotes the expectation value with respect to the measure (if it exists) on $\{x^{\tau,x} : \tau \geq 0\}$. Now we can approximate the velocity by running the approximating sequence of ordinary flows (32, 34) with $u \equiv 0$ and reparametrized by $(2\nu)^{\frac{-1}{2}}$ as explained above, and still we approximate the expression (41) as before.

This completes the realization of the Navier-Stokes equations, through ordinary differential equations.

Remark 4. The case of smooth boundary manifolds and Euclidean semispace [6], requires a different technique which will be presented elsewhere, while the pure-noise representations in [5] are approximated by the same method as above.

9 Conclusions

We have presented an implicit scheme by which we have reduced the Navier-Stokes equations to running smooth flows associated to ordinary differential equations on the same manifold M; this is a very different approach to the usual scheme in the construction of the inertial manifold [10], and in principle should be amenable to numerical implementations. The present approach, with some modifications to work in the bundle of orthonormal frames, can be used to carry the same approximations for the smooth boundary case with no-slip conditions for the velocity on the boundary, which we have integrated implicitly in [6]. Our approximations scheme by ordinary differential equations is invariant, i.e. independant of coordinate systems, just as the whole construction here presented. As for its applications, it applies to any fluctuational process described in invariant manner, say, an invariant theory of thermodynamical systems in equilibrium [15] or still far from equilibrium [11], just as the one constructed by the author in the same geometrical approach presented here, or still in a Riemannian geometry approach as originally conceived by Graham [13], and further persued by Descalzi, Tirapegui and collaborators [12]. Methodologically, it is nothing else than the extension of the E.Cartan developing method for classical motions, to its random realm [7,15] . Thus it applies as well to quantum mechanics and gravitation, as described in this geometrical setting by the author [8,15,22]. Remarkably enough, these constructions lead to an invariant hamiltonean description on T^*M, the cotangent manifold, as a "random phase space" extension of classical mechanics. In the particular case of fluid-dynamics, which we have dealt with in this article, it is well known that the Euler equation for perfect fluids has a hamiltonean structure derived from a symplectic 2-form on the (infinite-dimensional) Lie-algebra of the group of volume preserving diffeomorphisms. The present random flows do not preserve the Riemannian volume, with the exception of the case of Euclidean space, so in general the extension of a hamiltonean theory to the random flows is not trivial. Surprisingly, the random flows do preserve the *canonical* symplectic structure on T^*M which can be built associated to these random flows, in going to the approximation by the ordinary differential equations which we have presented above. Thus, it is *finite*-dimensional in essence, in distinction to the Euler equation. This random hamiltonean formalism applies to all random flows, quantum, thermodynamical or due to viscosity, or still determined by whatever characteristic "diffusion" constant the particular theory may be affected, and thus it is not specific to the Navier-Stokes equations; we shall present these results elsewhere.

Acknowledgements

Our sincere gratitude to the Organizers of the Workshop and to Profs. O. Descalzi, J. Martinez & E. Tirapegui, for the possibility of contributing to this series, and an always pleasant meeting and stay in Valparaíso.

References

[1] M. Taylor, *Partial Differential Equations*, vols. I and III, Springer-Verlag, Berlin, 1995.

[2] V.I. Arnold and B.A. Khesin, *Topological Methods in Hydrodynamics*, Springer, New York, 1999.V.I Arnold, Sur la géometrie différentielle des groupes de Lie de dimension infinie et ses applications

a l'hydrodynamique des fluides parfaites, Ann. Inst. Fourier **16**, (1966), 319-361. D. Ebin & J. Marsden, Annals of Maths. **92** (1971), 102-163.

[3] D.L. Rapoport, Random diffeomorphisms and integration of the classical Navier-Stokes equations, *Rep.Math.Phys.***49** ,no.1, 1-27 (2002).

[4] D.L. Rapoport, Random representations for viscous fluids and the passive magnetic fields transported by them, in *Proceedings, Third International Conference on Dynamical Systems and Differential Equations, Kennesaw, May 2000*, Discrete & Cont.Dyn.Sys. special issue, S.Hu (ed.), 2000.

[5] D.L. Rapoport, On the geometry of the random representations for viscous fluids and a remarkable pure noise representation, *Rep.Math.Phys.***50, no.2, 211-247 (2002).

[6] D.L. Rapoport, Martingale problem approach to the representations of the Navier-Stokes equations on smooth boundary manifolds and semispace, *Rand.Oper.Stoch.Eqts.* **11**, no.2 (2003).

[7] P.Malliavin, *Stochastic Analysis*, Springer Verlag, 1998; J.M. Bismut, *Mécanique Analytique*, Springer LNM **866**, 1982.

[8] D.L.Rapoport, Torsion, Brownian motion, spinor fields, quantum mechanics and fluid-dynamics, I & II, In *IX th. Marcel Grossmann Meeting in Relativity, Gravitation and Field Theories, Rome, 2000*, V.Gurzadyan et al (eds), World Sc., Singapore, 2003 & www.icra.it/MG/mg9/Proceedings/Proceedings.html.

[9] P.Baxendale and K.D.Elworthy, Flows of Stochastic Dynamical Systems, Z.Wahrschein.verw.Gebiete 65, 245-267 (1983). K. Kunita, *Stochastic Flows and Stochastic Differential Equations*, Cambridge Univ. Press, Cambridge (U.K.), 1994.

[10] C. Foias, O.Manley, R.Rosa & R.Temam, *The Navier-Stokes equations and Turbulence*, Encyclopedia of Mathematics, Cambridge Univ. Press, 2001.

[11] D. Rapoport, Covariant thermodynamics and the ergodic theory of stochastic and quantum flows, in *Instabilities and Nonequilibrium Structures, V*, E. Tirapegui (ed.), Kluwer, & *Instabilities and Nonequilibrium Structures IX* O. Descalzi et al (edts.), Kluwer, Dordrecht.

[12] O. Descalzi, S. Martinez & E. Tirapegui, Thermodynamical potentials for non-equilibrium systems, *Chaos, Solitons and Fractals* **12** (2001), 2619-2630 and references therein.

[13] R. Graham, In: F. Moss et al (edts.), *Noise in nonlinear dynamical systems, I*, Cambridge Univ. Press, 1989.

[14] D. Rapoport & S. Sternberg, On the interactions of spin with torsion, Annals of Physics **158** (1984), 447-475.

[15] D. Rapoport, Non-riemannian geometry and random dynamics of infinite-particle systems, in *Instabilities and Nonequilibrium Structures IX*, Proceedings, Vina del Mar, December 2001, O. Descalzi, J. Martinez and S. Rica (edts.), Kluwer 2003.

[16] D. Rapoport, Riemann-Cartan-Weyl Quantum Geometry, II: Cartan Stochastic Copying method, Fokker-Planck Operator and Maxwell-de Rham equations, Intern. J. Theor. Phys. **36**, no. 10, 2115-2152 (1997).

[17] R. Temam, *Infinite Dimensional Dynamical Systems in Mechanics and Physics*, Springer Verlag, New York, 1988.

[18] O. Reynolds, On the dynamical theory of turbulent incompressible fluids and the determination of the criterion, Phil. Trans. Royal Soc. London A, **186** (1894), 123-161. S.B. Pope, *Turbulence*, Cambridge Univ. Press, Cambridge (U.K.), 2000.

[19] A. Chorin, *Turbulence and Vorticity*, Springer Verlag Series in Applied Mathematics, 1995, New York. C. Marchioro & M Pulvirenti, *Mathematical Theory of Incompressible Nonviscous Fluids*, Springer Verlag, 1994. K. Gustafson & J. Sethian (edts.): *Vortex Methods and Vortex Motions*, SIAM, Philadelphia, 1991.

[20] K.D. Elworthy, Stochastic Flows on Riemannian Manifolds, in *Diffusion Processes and Related Problems in Analysis*, M.Pinsky et al (edts.), vol. II, Birkhauser, 1992.

[21] G.N. Milstein, *Numerical Integration of Stochastic Differential Equations*, Kluwer Series in Mathematics and its applications, 1995. P. Kloeden & E. Platen, *Numerical Solutions of Stochastic Differential Equations*, Applications of Mathematics (Stochastic Modelling and Applied Probability) vol. **23**, Springer Verlag, 1995.

[22] D. Rapoport, Cartan-Weyl Geometries, Brownian motion of multiforms, and quantum gravity; submitted to *Clifford Algebras and its Applications to Mathematical Physics, Proceedings of the Fifth*

International Conference, Tennessee Polytechnic, May 2002, R.Ablamowicz & W. Baylis (edts.), 2002.

Characterization of the Dynamical Evolution of Electroencephalogram Time Series.

O. A. Rosso, S. Blanco and A. Figliola

Instituto de Cálculo,

Facultad de Ciencias Exactas y Naturales, Universidad de Buenos Aires.

Pabellón II, Ciudad Universitaria.

(1428) Buenos Aires, Argentina.

Abstract

The entropy defined from the Wavelet transform is a measure of the order/disorder degree presents in a time series, and in consequence this entropy evaluates over EEG time series gives information about changes in the underlying dynamical process. The Wavelet Entropy (WS) results independent of the signal energy and it is parameter free and independent of the stationarity of the time series. In this work we analyzed the results of the time evolution of WS performed over epileptic EEG records provide by depth electrodes.

1 Introduction

The clinical interpretation of EEG attempts to link pathological features (clinical symptomatology) with the visual inspection and pattern recognition of the EEG. Although this traditional analysis is quite useful, the visual inspection of the EEG is subjective and hardly allows any systematization. In order to overcome this, quantitative EEG (qEEG) analysis introduces objective measures reflecting the characteristics of the brain activity as well as the associated dynamics. In the characterization of the time evolution of the EEG associated dynamics, some results have been recently reported by some research groups using quantifiers based on measurements of nonlinear dynamics applied to EEG epileptic records. They analyze the EEG during the seizure onset and the background EEG (pre- and post-ictal), following the temporal evolution of the signal complexity (associated with the measurement of the correlation dimension D_2) [1,2] and the chaoticness degree (throughout the largest Lyapunov exponent, Λ_{max}) [3,4]. The main findings from these reports are a reduction of these two quantifiers during the ictal stage, suggesting that a transition takes place at the seizure onset in the dynamical behavior of the neural network from a complex behavior to a simpler one. Furthermore, some researchers report a significant decrease of these quantifiers a few minutes before the seizure-onset raising the hypothesis that epileptic seizure could be predicted.

The basic requirement for the nonlinear dynamics metric tools can be applied to experimental data is the stationarity of the time series, that means the time series is representative of a unique and stable attractor. Also for the evaluation of D_2 and Λ_{max} long time recordings are required, because they are defined as asymptotic properties of the attractor. By applying a static measurement of D_2 and Λ_{max} to selected portions of brain electrical activity (EEG time series), satisfying all the mathematical hypothesis requirement, different states could be characterized [5,6]. These metric invariant require the computation of some previous parameters and

O. Descalzi et al. (eds.), Instabilities and Nonequilibrium Structures VII & VIII, 333–338.

© 2004 *Kluwer Academic Publishers. Printed in the Netherlands.*

also degrades rapidly with additive noise, many times giving vague results. Unfortunately, for long recordings of neural mass activity we can not assume a stable attractor and, therefore we can not come straightforward conclusions concerning dynamics characterization. Moreover when one is interested in different state transitions, specially if the moments of the transitions are not strictly defined. In these cases, one can relax the stationary requirement of the time series and introduce two quantifiers: the *dimensionality* and the *chaoticity*. In this way, computing their values in slide time windows (with or without overlapping) the EEG associated dynamics could be accessed [1-4]. They formally are defined as D_2 and Λ_{max}, but they are not equivalents, because they violate the basic mathematical hypothesis of stationarity.

Another group of qEEG methods are the based on time-frequency analysis. Recently we introduce quantifiers based on the Orthogonal Discrete Wavelet Transform (ODWT) [7]. This method, does not make assumptions about record stationarity (they do not need reconstruct any attractor of the dynamical system) and they work only with their measurable response, that is the time series. Taking as a basic element the Wavelet Transform of the EEG signal, we define two quantifiers: the *Relative Wavelet Energy* (RWE) and the *Wavelet entropy* (WS). The RWE give information about the relative energy associated with the different frequency bands present in the EEG and their corresponding degree of importance. WS bring us information about the degree of order/disorder associated with a multi-frequency signal response. In consequence the time evolution of the WS could give information about the dynamics associated with the EEG records. In this work we present the analysis of the dynamical evolution of an epileptic EEG time series taken the WS as basic tool.

2 Clinical Data

Our method was applied to the EEG recording from inter-ictal and ictal brain activity of refractory epileptic patient prone to surgical treatment. This information was simultaneously correlated with clinical symptomatology. According to the visual assessment of the EEG seizure recording, this patient presented an epileptogenic focus in the left hemisphere corresponding to Hypocampus with immediate propagation to Gyrus Cyngular and Motor Supplementary Area and to the right contralateral homologous area. In Fig.1 the EEG signal ($\omega_{sample} = 250\,Hz$) for 64 *sec* corresponding to one depth electrode in the left hypocampus region is shown. From a visual inspection, it is clear that around the second 10 the epileptic seizure starts, and finishes around the second 54. The use of depth electrodes provides records where the noise and artifact contamination effects are minimized.

3 Wavelet Entropy

The wavelet analysis is a method which relies on the introduction of an appropriate basis and a characterization of the signal by the distribution of amplitude in the basis. If the wavelet are required to form a proper orthogonal basis, it has the advantage that an arbitrary function can be uniquely decomposed and the decomposition can be inverted [8,9].

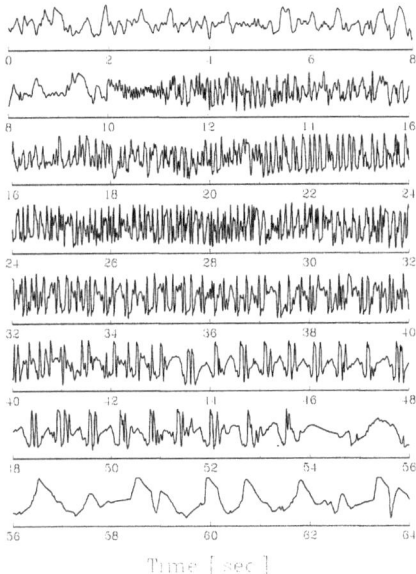

Figure 1: Recording of the EEG signal corresponding to a contact in the epileptogenic region, left Hypocampus. Epileptic seizure start around 10 *sec* and finished at 55 *sec*.

The correlated *decimated discrete wavelet transform* (DWT) provides a nonredundant representation of the signal and its values constitute the coefficients in a wavelet series. These wavelet coefficients give us full information in a simple way and a direct estimation for local energies at the different scales. Moreover, the information can be organized according a hierarchical scheme of nested subspaces called multiresolution analysis in $L^2(\mathcal{R})$ [8,9]. In the present work we employ orthogonal cubic spline functions as mother wavelets. Among several alternatives, cubic spline functions are symmetric and combine in a suitable proportion smoothness with numerical advantages and they have become a recommendable tool to represent natural signals.

In the following we assume that the signal is given by the sampled values $S = \{s_0(n), \; n = 1, \cdots, M\}$, which correspond to an uniform time grid with sampling time t_s. For simplicity we suppose that the sampling rate is $t_s = 1$. If the decomposition is carried out over all resolutions levels, $N = \log_2(M)$, the wavelet expansion will be:

$$S(t) \;=\; \sum_{j=-N}^{-1} \sum_k C_j(k) \, \psi_{j,k}(t) \;=\; \sum_{j=-N}^{-1} r_j(t) \,, \tag{1}$$

where wavelet coefficients $C_j(k)$ can be interpreted as the local residual errors between successive signal approximations at scales j and $j+1$, and $r_j(t)$ is the *detail signal* at scale j. It contains the information of the signal $S(t)$ corresponding with the frequencies $2^{j-1}\omega_s \leq |\omega| \leq 2^j \omega_s$.

Since the wavelet family $\{\psi_{j,k}(t)\}$ is an *orthonormal* basis for $L^2(\mathcal{R})$, the concept of energy is linked with the usual notions derived from the Fourier theory. Then, the wavelet coefficients are given by $C_j(k) = \langle S, \psi_{j,k} \rangle$, the energy at each resolution level $j = -1, \cdots, -N$, will be the energy of the detail signal

$$E_j = \|r_j\|^2 = \sum_k |C_j(k)|^2 \tag{2}$$

and the energy at each sampled time k will be

$$E(k) = \sum_{j=-N}^{-1} |C_j(k)|^2 . \tag{3}$$

In consequence the total energy can be obtained by

$$E_{tot} = \|S\|^2 = \sum_{j<0} \sum_k |C_j(k)|^2 = \sum_{j<0} E_j . \tag{4}$$

Then, the normalized values, which represent the *Relative Wavelet Energy*

$$\rho_j = E_j / E_{tot} \tag{5}$$

for the resolution level $j = -1, -2, \cdots, -N$, define by scales the probability distribution of the energy. Clearly, $\sum_{j<0} \rho_j = 1$ and the distribution $\{\rho_j\}$ can be considered as a time-scale density.

The Shannon entropy gives an useful criteria for analyzing and comparing this probabilistic distribution, since it well known, it provides a measure of the information of any distribution. We define the total *Wavelet-Entropy* as

$$WS = - \sum_{j<0} p_j \cdot \log_2[p_j] . \tag{6}$$

In order to study their temporal evolution, we divide the signal under analysis in slide temporal windows of L *data* length and for each interval we evaluate the WS using eq. (6) and assign the obtained value to the end of time window. The minimum length of the temporal window will be one including at least one wavelet coefficient in every scale.

The WS can be thinking as a measurement of the degree of order/disorder of the signal, and in consequence could give some information about the underlying dynamical process associated with the signal. In fact, a very ordered process could be thought as a periodic mono-frequency signal (signal with a narrow band spectrum). A wavelet representation of such a signal, will be greatly resolved in one unique wavelet resolution level. That means all RWE will be almost zero except for the wavelet resolution level which include the representative signal frequency. For this special level the RWE will be almost one and in consequence the total WS will be near zero or present a very low value. By other hand, a signal generated by a totally random process could be taken as representative of very disordered behavior. In this case one can expect that a very large number of frequencies will be necessary to represent such a signal (signals with a broad band spectrum). This kind of signal will have a wavelet representation with significant contributions in all frequency band considered in their analysis. Moreover, one could expect that all the contributions will be of the same order. Consequently the RWE will be almost equal for all the resolutions levels and the WS will take their maximum value.

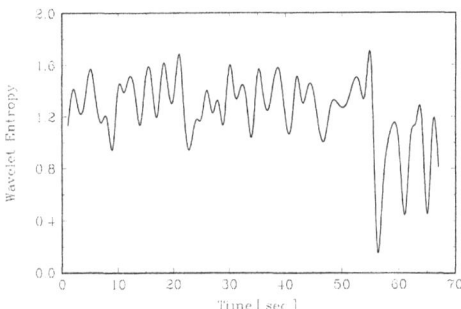

Figure 2: Time evolution of WS corresponding to the EEG signal display in Fig. 1.

4   Results and Conclusions

The time evolution of WS corresponding to the EEG signal (Fig.1) are display in Fig.2. The wavelet entropy was evaluated in sliding non overlapping time windows of length $L = 256$ $data \equiv$ 1 sec. From figures 1 and 2 we can observe that the relevant changes in the EEG are put in evidence by the WS. That is the beginning of the seizure is represented by a decrement in the in the WS around second 10, and the end of the seizure is detected by a strong decrement around second 52. This last decrement can be associated with the very slow activity characteristic of the "neuronal fatigue" of the post-ictal stage. The relative minima are related with evident morphological changes in the EEG signal. Such as the minimum around second 24 and 48 seconds. This minima correspond to the appearance spike trends with different natural frequencies.

A significative decay in the WS was observed at the beginning and during the ictal stage indicating a more rhythmical and ordered behavior compatible with a dynamic process of synchronization in the brain activity. This behavior may be thought as induced by one hypothetical epileptic focus. The use of the quantifiers based on time-frequency together with the clinical patient history and the visual assessment of the EEG, can contribute to the identification of the source of epileptic seizures and to a better understanding of it dynamics. Certainly, the use of these quantifiers is not intent to replace conventional EEG analysis, but rather to complement it and also to provide further insights into the underlying mechanisms of ictal patterns.

The WS gives a measure of the order/disorder of the system in time. Therefore the WS is a good parameter in order to detect dynamical changes in the system behavior as well as a form of quantify then. In addition the WS has the following advantages over the following parameters: *i)* the spectral entropy, because it is capable of detecting changes in a non stationary signal due to the localization characteristics of the wavelet transform; *ii)* dimensions and Lyapunov exponents (which are only defined for stationary behaviors), or dimensionality and chaoticity (stationary constrains removed) due to the computational time is significantly lower since the algorithm involves the use of fast wavelet transforms in a multiresolution framework; *iii)* contaminating noises (if they are basically concentrated in some frequency bands) contributions can be easily eliminated; and finally and very important, *iv)* the WS is parameter free.

Acknowledgments

This work was partially supported by the Consejo Nacional de Investigaciones Científicas y Técnicas (CONICET, Argentina), PIP 0029/98 and Fundación Alberto J. Roemmers (Argentina). One of the authors (O.A.R.) is very grateful to the organizer of the workshop for their very kind hospitality during their stay in Chile.
E-mail: rosso@ic.fcen.uba.ar; blanco@ic.fcen.uba.ar; figliola@ic.fcen.uba.ar

References

[1] M. C. Casdagli, L. D. Iasemedis, R. S. Savit, R. L. Gilmore, S. N. Roper, and J. C. Sackellares, Electroenceph. Clin. Neurophysiol. **102**, 98 (1997).

[2] K. Lehnertz and C. E. Elger, Phys. Rev. Lett. **80**, 5019 (1998).

[3] L. D. Iasemedis, J. C. Sackellares, H. P. Zaveri, and W. J. Williams, Brain Topography **2**, 187 (1990).

[4] L. D. Iasemedis and J. C. Sackellares, in *Measuring Chaos in Human Brain*, edited by D. Duke and W. Pritchards (World Scientific, Singapure, 1991), pp. 49 – 89.

[5] S. Blanco, A. Figliola, S. Kochen, and O. A. Rosso, IEEE Eng. Med. and Biol. Mag. **16**, 83 (1997).

[6] E. Başar, *Brain Function and Oscillations.I: Brain Oscillations, Principles and Approaches* (Springer, Berlin, 1998).

[7] S. Blanco, A. Figliola, R. Quian Quiroga, O. A. Rosso and E. Serrano, Phys. Rev. E **57**, 932 (1998).

[8] I. Daubechies, *Ten Lectures on Wavelets* (SIAM, Philadelphia, 1992).

[9] S. Mallat, *A Wavelet Tour of Signal Processing*, 2nd. Edition, (Academic Press, San Diego, 1999).

EXACT RESULTS IN TRAPPING REACTIONS FOR MOBILE PARTICLES AND A SINGLE TRAP

Alejandro Sánchez[*], Miguel A. Rodriguez[†] and Horacio S. Wio[‡]

Instituto de Física de Cantabria, CSIC and Universidad de Cantabria, Santander, Spain
†*Grupo de Física Estadística, Centro Atómico Bariloche (CNEA) and Instituto Balseiro (UNC), 8400 San Carlos de Bariloche, Argentina*

Abstract. We have exploited a stochastic model for the description of diffusion controlled reactions for trapping processes with a single trap in a one–dimensional lattice, in order to describe in an exact way the particle distribution as seen from the trap, being it fix or mobile, as well as for perfect or imperfect reactions. From this we obtain the exact result for the total number of absorbed particles. Moreover we get a formal expression for the nearest-neighbor-particle distribution that gives the known results in the limit cases of fixed trap and fixed particles.

One quantitative measure of the tendency of low-dimensional reacting systems to segregate is the distance d (in a trapping system $A + B \to B$, with several A and one B) of the B particle (or trap) to the nearest unreacted A particle.

Several analyses of this quantity in the limited cases when B is stationary and the As diffuse [1, 2, 3] and the opposite case of As stationary and B mobile [4] have been published. In the first limit case the analysis yields that the average nearest-neighbor-distance grows asymptotically as $\langle d \rangle \sim t^{1/4}$, while in the second the result is $\langle d \rangle \sim t^{1/2}$. There are simulations in the intermediate region [5] but no theory. These simulations indicate that $\langle d \rangle \sim t^{\alpha}$, where α is some value that interpolates between $1/2$ and $1/4$ and is a function of the relation between D_A and D_B. Moreover the flux at the trap has a $t^{-1/2}$ dependence.

Here we present a study of this problem within the framework of a recently introduced model [6]. The model equation for the evolution for $N(x,t)$, the density of the diffusing A particles, with a given trap realization $\varepsilon(t)$, is the following

$$\frac{\partial}{\partial t}N(x,t) = D_A \frac{\partial^2}{\partial x^2}N(x,t) - \gamma \delta\left[x - \varepsilon(t)\right]N(x,t), \tag{1}$$

where γ is a constant which measures the reaction probability.

---

[1] E-mail address: asanchez@ifca.unican.es
[2] E-mail address: rodrigma@ifca.unican.es
[3] E-mail address: wio@cab.cnea.gov.ar

O. Descalzi et al. (eds.), Instabilities and Nonequilibrium Structures VII & VIII, 339–343.
© 2004 *Kluwer Academic Publishers. Printed in the Netherlands.*

Firstly, we take the integral form of Eq. (1):

$$N(x,t) = n_0 - \gamma \int_0^t dt' \int_{-\infty}^{+\infty} dx' G(x,t|x',t') \delta[x' - \varepsilon(t')] N(x',t'), \qquad (2)$$

where $G(x,t|x',t')$ is the propagator of the diffusive particles.

In order to change our reference system from the laboratory to another one fixed to the trap, we make the following change of variable $x = \varepsilon(t) + z$. The form of the equation, using the new variable is

$$\tilde{N}(z,t) = n_0 - \gamma \int_0^t dt' G[\varepsilon(t) + z, t|\varepsilon(t'), t'] \tilde{N}(0,t'), \qquad (3)$$

where $\tilde{N}(z,t) = N[\varepsilon(t) + z, t]$.

We perform the average over the process $\varepsilon(t)$ and finally obtain

$$\tilde{n}(z,t) = n_0 - \gamma \int_0^t dt' \int_{-\infty}^{+\infty} dz' \tilde{G}(z,t|z',t') \delta(z') \tilde{n}(z',t'), \qquad (4)$$

where we call $\tilde{n}(x,t) = \langle \tilde{N}(x,t) \rangle$ and $\tilde{G}(z,t|z',t')$ is a diffusion propagator with diffusivity $D = D_A + D_B$.

Comparison of Eq. (4) with Eq. (2) clearly indicates that the form of the distribution $\tilde{n}(z,t)$ is analogous to the distribution $n(x,t)$ from a fixed trap at the origin [6] but replacing D_A by D, i. e.,

$$\tilde{n}(x,t) = n_0 \left[\operatorname{erf}\left(\frac{|x|}{\sqrt{4Dt}} \right) + e^{\frac{\gamma|x|}{2D} + \frac{\gamma^2 t}{4D}} \operatorname{erfc}\left(\frac{|x|}{\sqrt{4Dt}} + \gamma\sqrt{\frac{t}{4D}} \right) \right]. \qquad (5)$$

This is an exact result valid for all cases. We have tested previous equation compared it with computer simulations for several paramters. In Fig. 1 we present a representative situation.

We can also calculate the number of traps absorbed N_{ABS} integrating $n_0 - \tilde{n}(x,t)$. The result is

$$N_{ABS} = 4n_0 \left[\sqrt{\frac{Dt}{\pi}} + \frac{D}{\gamma} e^{\frac{\gamma^2 t}{4D}} \operatorname{erfc}\left(\gamma\sqrt{\frac{t}{4D}} \right) - \frac{D}{\gamma} \right], \qquad (6)$$

where two last terms vanish for perfect trapping rendering $t^{1/2}$ dependence in the complete time regime. This result agrees with previous simulations of the flux at the trap with perfect absorption [5] which is given by $d_t N_{ABS} = 2n_0 \sqrt{D/\pi} t^{-1/2}$. In Fig. 2 we show the comparison with simulations for different values of γ. The ageement is again excellent up to times where finite time effects are noctable.

Now, we generalize the procedure used by Redner [2] in the case of a fixed trap. We consider a particular realization of the process $\varepsilon(t)$ and an average over the particle random walks. The density, as viewed from the trap, is $\tilde{N}(x,t)$. Thus, following [2], the probability of not finding any particle in $(-d,d)$ is

$$P(d,t) = \left(1 - \frac{\int_{-d}^{d} \tilde{N}(x,t)dx}{\int_{-L}^{L} \tilde{N}(x,t)dx} \right)^{\int_{-L}^{L} \tilde{N}(x,t)dx}, \qquad (7)$$

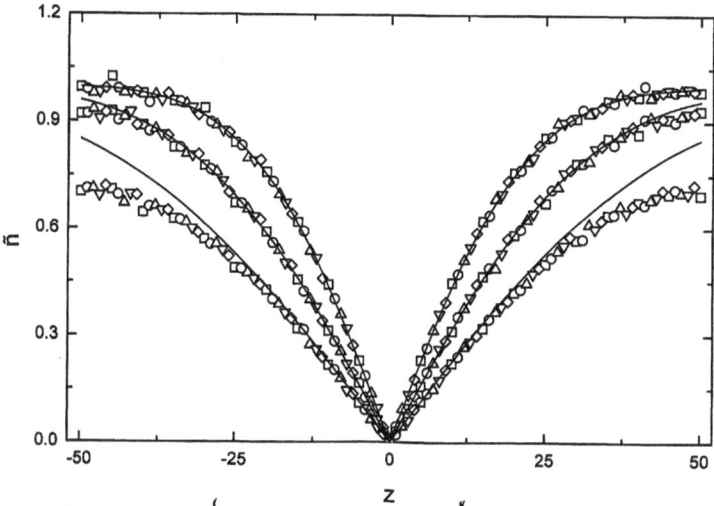

FIGURE 1. Density of A particles in the reference frame of the trap for perfect absorption. The solid line corresponds to our model results and the points are the results of simulations for different D_A values: $D_A = 1$ (squares), $= 0.75$ (circles), $= 0.5$ (up triangles), $= 0.25$ (down triangles) and $= 0$ (diamonds); in such way that $D = 1$ in all cases. The different curves corresponds to times (from top to bottom): $t = 150$, 300 and 600.

where $2L$ is the lattice length. The averages over $\varepsilon(t)$ must be taken at the end, hence, after doing $L \to \infty$, we have the pdf differentiating $P(d,t)$ respect to d and averaging. The final result is

$$
\begin{aligned}
p(d,t) &= -\partial_d \left\langle \exp\left(-\int_{-d}^{d} \tilde{N}(x,t)dx \right) \right\rangle \\
&= \left\langle \left[\tilde{N}(d,t) + \tilde{N}(-d,t) \right] \exp\left(-\int_{-d}^{d} \tilde{N}(x,t)dx \right) \right\rangle,
\end{aligned}
\tag{8}
$$

where the brackets indicate averages over $\varepsilon(t)$. The last expression reduces to the known results for both $D_A = 0$ and $D_B = 0$.

As a simple approximation we expand average in Eq. (8) using the method of cumulants and closing the expansion to a determined order. Closing up first order the pdf can be written as

$$
p(d,t) = 2\tilde{n}(d,t) \exp\left(-2\int_{0}^{d} \tilde{n}(x,t)dx \right)
\tag{9}
$$

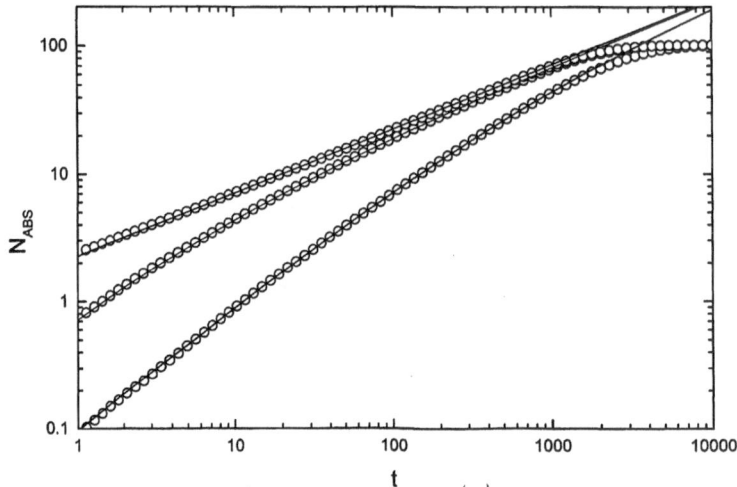

FIGURE 2. Log–log plot of absorbed particles for different values of absorption. From top to bottom: $\gamma \to \infty$, $\gamma = 1$ and $\gamma = 0.1$. The points are the result of simulations while the solid line corresponds to our model. The diffusivities used are $D_A = D_B = 0.5$.

and an explicit expression can be given since $\tilde{n}$ (Eq. (5)) and the integral involved can be exactly calculated. Within this approach the absorption of diffusing particles with a coefficient D_A due to a diffusing trap of coefficient D_B is equivalent to the case of diffusing particles with coefficient $D = D_A + D_B$ absorbed by a fixed trap. This approximation is valid for low diffusivity of the trap, i. e. D_B small. Fig. 3 shows comparison with simulations. We note that $p(d,t)$ is exact for $D_B = 0$.

A thorough discussion of this study can be found in [7].

Acknowledgements Financial support from Ministerio de Educación y Ciencia, Spain (Project DGICyT No, PB93-0054-C02-02), and from CONICET, Argentina (Project PIP-4953/96) are greatly acknowledged.

REFERENCES

1. G. H. Weiss, R. Kopelman and S. Havlin, Phys. Rev. A **39**, 466 (1989)
2. S. Redner and D. Ben-Avraham, J. Phys. A **23**, L1169 (1990).
3. H. Taitelbaum, R. Kopelman, G. H. Weiss and S. Havlin; Phys. Rev A **41**, 3116 (1990).
4. D. Ben-Avraham and G. H. Weiss, Phys. Rev. A **39**, 6436 (1989).

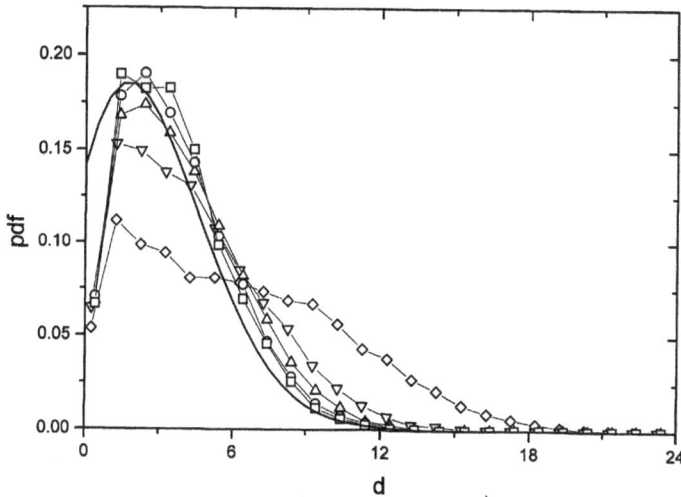

FIGURE 3. Log–log plot of absorbed particles for different values of absorption. From top to bottom: $\gamma \to \infty$, $\gamma = 1$ and $\gamma = 0.1$. The points are the result of simulations while the solid line corresponds to our model. The diffusivities used are $D_A = D_B = 0.5$.

5. R. Schoonover, D. Ben-Avraham, S. Havlin, R. Kopelman and G. H. Weiss, Physica A **171**, 232 (1991).
6. M. A. Rodriguez, G. Abramson, H. S. Wio and A. Bru, Phys. Rev. E **48**, 829 (1993).
7. A. Sanchez, M.A. Rodriguez and H.S. Wio, Phys. Rev. E **57**, 6390-6397 (1998).

Vortex-dipole surface wave interactions in deep water

Francisco Vivanco and Francisco Melo

Departamento de Física de la Universidad de Santiago de Chile and Center for Advances

Interdisciplinary Research in Materials

Av. Ecuador 3493, Casilla 307 Correo 2 Santiago-Chile

We describe a simple experiment to study the interaction between surface waves and vertical vorticity in the deep water regime. The vortex circulation introduces dislocations on the wavefront which can be explained by the advection of the propagating wavefront, due to the fluid motion. The analogy between wave vortex interactions and the Aharonov-Bohm effect is explored further by considering the case of surface waves interacting with a vorticity dipole.

I. INTRODUCTION

In 1959, Aharanov and Bohm [1] discovered that the magnetic vector potential influences the dynamics of a charged particle in a region where the magnetic field vanishes. Since then a large amount of experimental and theoretical work has been devoted to the study of this effect. However, the main issue about the physical reality of the magnetic vector potential in Quantum Mechanics is still subject of discussion [2]. Here we present a brief study of a surface wave interacting with filamentary vortex emerging at the surface of deep water. As pointed out in a remarkable work of Berry and coworkers [3], this problem is the Classical Mechanics analog of the Aharonov-Bohm effect. One difference however is that in the fluid case the analogous of the magnetic vector potential, the velocity field, can be directly measured. Thus, this simple mechanical configuration offer an interesting tool to illustrate several Quantum Mechanics phenomena. Another interesting aspect is that surface waves are useful to characterize static and dynamics of surface vorticity of importance in rivers, lakes and seas [4,5]. In addition, it is accepted that coherent structures such as vortex filaments are prominent features of turbulent flows [6]. In this sense, some progress has been done to obtain valuable information about vorticity fields, in two dimensional

O. Descalzi et al. (eds.), Instabilities and Nonequilibrium Structures VII & VIII, 345–354.
© 2004 *Kluwer Academic Publishers. Printed in the Netherlands.*

flows, by looking at their interaction with an acoustical wave. For instance, Baudet and collaborators [7] have characterized a von Kármán array of vortices and studied the effect of vorticity on a plane fronted ultrasonic wave. Also Fink and coworkers [8], with a single vortex have shown, by using time reversal techniques, that this method is useful to measure vortex core size and vortex circulation as well. However, a full visualization of the scattered wave in the acoustical case is still difficult to achieve. In the case of surface waves, the advantage relies in the fact that these waves are much easier to detect that ultrasonic waves. Our paper is organized as follows, in section II we remind the main experimental results of a single vortex interacting with a surface wave. The main idea behind the dislocations appearance on the wavefront is introduced in section III to complement section II and illustrate connections with the Quantum Mechanics case. Experiments on two vortex wave interaction are presented in section IV which also includes qualitative explanations based on Quantum Mechanics concepts. Finally, a brief discussion is given in the concluding remarks section.

II. SURFACE WAVE SINGLE-VORTEX INTERACTION

In the experiment, decribed in a previous work [9], a plane fronted wave is excited by moving horizontally a rigid dipper at the surface of a water tank. For the case of a single vortex, by a using a particle tracking method, we have found that the fluid at the vortex core rotates approximately as a rigid body, whereas the tangential velocity far from the core decays as $1/r$. Thus Vortex circulation, $\Gamma = \int \vec{V}(\vec{r}) \cdot d\vec{r}$ is approximately constant except within vortex core.

Fig. 1 shows interference patterns produced by surface waves interacting with a vortex, for constant frequencies and increasing vortex circulation. To visualize the pattern the water surface was illuminated from above with a parallel beam. The light reflected on the wavy surface of water forms caustic lines on a horizontal screen located just above water tank. In all figures, the incident wave propagates from left to right and the vortex circulation is counterclockwise. It is observed that following a line parallel to the incident wavevector and containing the vortex core, the wavefront has a discontinuity which corresponds to a jump in the phase of the wave. Due to the fluid motion in the upper side of figures the wave is compressed with respect to the incident one whereas, the wave in the lower part is dilated. The global effect of the advection is to produce phase discontinuities in the wave front. As pointed out by Berry et. al. [3], such a discontinuity, similar to a dislocation of atomic planes in a crystal, are currently present in the wavefunction of electrons interacting with a magnetic potential decreasing as $1/r$.

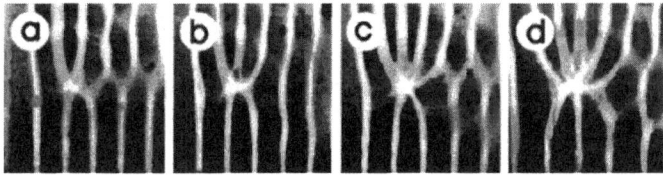

FIG. 1. Snapshots of dislocations resulting on a surface wave propagating from left to right ($\lambda = 1.4cm$), interacting with a counter rotating vortex. a) $b/\lambda = 0.5$, b) $b/\lambda = 1.0$, c) $b/\lambda = 1.5$ and d) $b/\lambda = 2.0$.

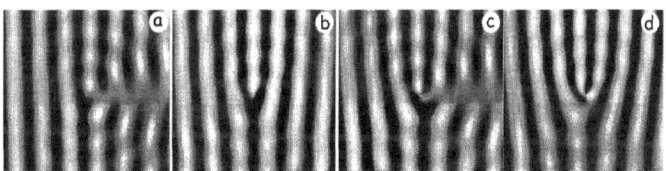

FIG. 2. Snapshots of dislocations resulting on electron wavefunction propagating from left to right, interacting with a counter rotating vector potential. a) $b/\lambda = 1/2$, b) $b/\lambda = 1$, c) $b/\lambda = 3/2$ and d) $b/\lambda = 2$.

In previous works, we have quantified the Burgers vector by measuring the relative phase of the dislocated wave with respect to the incident one. It was observed that phase shift has a jump when crossing the dislocation. Measurements of such jump phase, $\Delta\phi$, give directly Burgers vector b which is associated to wavefront dislocations as $\Delta\phi/2\pi = b/\lambda$, where λ is the wavelength. Experimental results show that b/λ is independent on frequency but increases as Γ increases. Therefore, Burgers vector is, in agreement with previous results [3], a function not only of f or λ, but also a function of Γ, satisfying the relation, $b/\lambda \sim \Gamma/V_g\lambda$

III. WAVE ADVECTION

In the following we will show, in a simple manner, that the existence of dislocations in the wavefront is a result of the compression and a dilation of the wave induced by the vortex flow when the wave crosses the vorticity region. Without loosing generality, let us first consider a surface wave moving in a region of gradient of speed in the same direction that the wave propagation, x.

The frequency in the laboratory frame then reads,

$$\omega(x) = \omega_V(x) + kV(x) \tag{1}$$

where $V(x)$ is the fluid velocity k wavevector which is linked to ω_V trough the relation dispersion of surface waves in deep water. Since the phase of the wave is conserved $\omega(x)$ must be a constant independent of the coordinate x. This constrain relies on the fact that there is neither sources nor sink of front waves within the interest region. Differentiating the former relation, we obtain the wavevector variation in space,

$$\frac{dk}{dx} = -k\frac{dV/dx}{V_g + V}, \tag{2}$$

where V_g is the surface wave group velocity. Thus the wave front is compressed if the gradient of speed is positive whereas it is dilated in the opposite case. In the case of $V_g << V$, which is not always satisfied in our system, the total variation of the local k, with respect to the incident wavevector k_0, at a given position $x > 0$, reads

$$\Delta k = \int_0^x \frac{dk}{dx'}dx' \sim -k_0\frac{V(x)}{V_g}. \tag{3}$$

Where, without loosing generality, we have assumed that $V = 0$ and $dV/dx = 0$ for $x < 0$. The total variation of the wave phase $\Delta\phi_x$, with respect to the incident wavefront, at the position x, is therefore,

$$\Delta\phi_x = -\int_0^x k_0 \frac{V(x')}{V_g} dx' \tag{4}$$

This formula can be easily written for the case the wavevector, $\vec{k}$ makes an arbitrary angle with the flow velocity $\vec{V}(\vec{x})$,

$$\Delta\phi_x/2\pi = -\int \frac{1}{V_g\lambda}\vec{V}(\vec{x}) \cdot d\vec{x} \tag{5}$$

For the case of a vortex, this relation is useful to compute the total phase shift of the wave in a circuit including the vortex core as

$$\Delta\phi_x/2\pi = \frac{b}{\lambda} = \frac{1}{V_g\lambda} \oint \vec{V}(\vec{x}) \cdot d\vec{x} = \frac{\Gamma}{V_g\lambda}, \tag{6}$$

which is the same formula as the one deduced rigorously by Coste et al. [10], in the deep water regime for low Mach Numbers, $V/V_g << 1$.

As pointed out early, these kind of phenomena are not specific to surface wave system. Indeed, our problem possesses a close analogy with the Aharonov-Bohm effect. The simplest form of this effect, analogous to surface wave vortex interaction problem, occurs when a beam of particles of charge q and mass m is incident normally on a long thin cylinder containing a magnetic field $\vec{B}(x)$ (the analogous of the vorticity field) parallel to its axis. Outside the cylinder of a finite radius, (analogous of the vortex core) the vector potential $\vec{A}(x)$ (the velocity field) is an azimuthal vector whose absolute value decreases as $1/r$. Of course $\vec{B} = 0$ outside the cylinder. Fig. 2 shows patterns produced by the interaction of the electron wavefunction with filamentary magnetic field, for increasing the magnetic flux, Φ (the circulation Γ). These patterns, obtained numerically by using the Berry approach for the Aharonov-Bohm effect discussed in reference [3], should be contrasted to the ones presented in Fig. 1. In this case, the electron wavefunction has also a discontinuity corresponding to a dislocation whose Burgers vector obeys to $b/\lambda = \Phi/hV_e\lambda$ [3]. Here V_e and λ are the group velocity and the wavelength of the electrons wavefunction respectively and h is the Plank constant. This relation is obviously the analog of equation (6) in which Γ is replaced by Φ/h.

Thus, surface waves interacting with a filamentary vortex is analogous to the simplest form of the Aharonov-Bohm effect. Such analogy can be useful to illustrate some particular features of the physics of the vortex-electron interaction in Quantum Mechanics. For instance, this system can be an useful tool to study similar phenomena that usually appear in superconductor or superfluid [11]

IV. TWO VORTEX WAVE INTERACTION

To illustrate the variety of experimental configurations that can be explored with this system, we investigate qualitatively the two vortex-wave interaction problem. In order to produce two stable vortex, in our water tank apparatus, the two counter rotating vortex are generated by extracting the water throughout two holes, $4mm$ in diameter and located a distance $d = 2cm$ apart (see Fig. 3). The outcoming fluid is reinjected homogeneously in the radial direction trough a gap ring shaped, $6cm$ in diameter. Thus, the incoming radial flow does not inject any vertical vorticity to the system. One way of control the vortex rotation sense is by slightly breaking the symmetry of the radial flow along an axis parallel to the one that contains the vortex dipole.

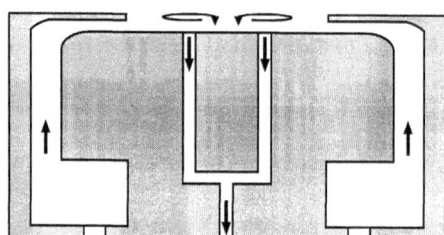

FIG. 3. Schema of the system to generate the two vortex configuration.

Two vortex-wave interaction is illustrated in Fig. 4. In Fig. 4a and 4b, the vortex dipole is perpendicular to the incident wavevector, we notice that similarly to what occur in the case of one vortex, the wavefront exhibit a phase shift that is proportional to Γ. However, here the dipole does not break the symmetry of the wavefront. The whole wavefront is advanced, Fig. 4a, or retarded Fig 4b, over a distance defined by the vortex separation d. When the dipole is parallel to $\vec{k}$, as in Fig. 4c, the situation is rather different, the phase shift induced by the first vortex is compensated by the second vortex in such a way that no net effect is observed in the wavefront far from the vortex. The general case illustrated in Fig. 4d, shows that the effect of the dipole on the wave front is over a distance that corresponds to the absolute value of $dsin(\vec{k}, \vec{d})$, where $\vec{d}$ goes from the negative to the positive vortex. This relation suggests that $\vec{k} \times \vec{d}$ is the relevant quantity for the dipole effect. To our knowledge, up to now, no theory, to contrast to our results, has been derived.

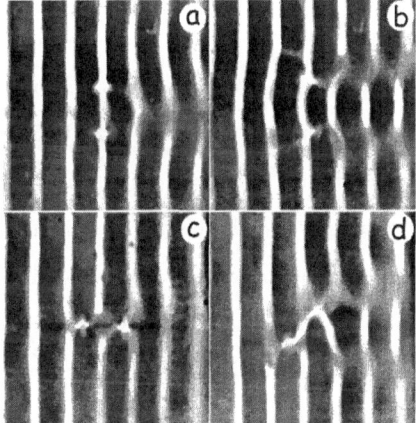

FIG. 4. Snapshots for several configurations. In all pannels, the wave propagates from left to right. The wavevector $\vec{k}$, contained on the figure plane, is then positive. In a) and b) the dipole axis, also contained on the figure plane, is perpendicular to the wavevector. In a) the dipole is positive and negative in b). In c) the dipole is antiparallel to $\vec{k}$. In d) it makes an angle of about $-3\pi/4$ with $\vec{k}$.

V. CONCLUDING REMARKS

In conclusion, we have found that simple visualization of a plane wave interacting with a single vortex reveals dislocations in the wave front and the presence of a scattered wave. The effect of the vortex is twofold: the advection of the wave due to vortex circulation produces a phase shift responsible for dislocations, the vortex core in turn is responsible for the scattered wave. We have checked that the Burgers vector in the regime accessible to our experiment, is consistent with our simple analysis and the Quantum Mechanics analog introduced by Berry.

On the other hand, the cross section resulted an asymmetric function with respect to an axis containing the vortex core and parallel to the incident wavevector. This results contrast with calculations [3], for the Quantum Mechanics case. The difference arises as a consequence of the rigid character of the solenoid. In fact, as opposed to the case studied by Berry, the surface wave penetrates the vortex core, the wave is then deflected by the core similarly to what occurs for a light ray when it crosses a region of varying index refraction.

Our investigations exploring configurations with more than one vortex support the interpretation of dislocations as dilation and compression induced by the flow gradient in the wavefront. Two counter rotating vortex have only a local effect on the wave front when the dipole coincides with propagation direction. When the dipole is perpendicular to the wavevector a symmetric phase shift is induced in the wavefront and the effective effect is the superposition of each vortex separately. The general case is more complicated, however our results suggests a relation of $\vec{k} \times \vec{d}$ type for the range of the wavefront distortion. Finally, let us emphasize that more complex vorticity configurations can be explored and much learned to make contact with Quantum Mechanics problems usually not accessible experimentally.

We would like to thank P. Umbanhowar, E. Tirapegui and J-C. Géminard for helpful discussions. This work was supported by a Cátedra Presidencial en Ciencias and Fondecyt under Project: 1990913.

[1] Aharonov Y. & Bohm D., "Significance of Electromagnetic Potentials in the Quantum Theory", Phys. Rev. Lett. **115**, 485-489 (1959).

[2] For a review, see for instance M. Peshkin and A. Tonomura, "The Aharonov-Bohm effect", Lecture Notes in Physics 340, Springer (1989).

[3] M. V. Berry, R. G. Chambers, M. D. Large, C. Upstill and J. C. Walmsley, "Wavefront Dislocations in the Aharonov-Bohm Effect and its Water Wave Analogue", Eur. J. Phys. **Vol. 1**, 154 (1980).

[4] J-H. Shyu and O. M. Phillips, "The Blockage of Gravity and Capillary Waves by Longer Waves and Currents", J. Fluid. Mech. **217**, 115 (1990).

[5] M. S. Longuet-Higgins, "Surface Manifestations of Turbulent Flow", J. Fluid. Mech. **308**, 15 (1996).

[6] U. Frisch, *Turbulence*, Cambridge University Press (1995), Ch. 8.

[7] C. Baudet, S. Ciliberto and J. F. Pinton, "Spectral analysis of the von Kármán flow using ultrasound scattering", Phys. Rev. Lett. **67**, 193-195 (1991)

[8] P. Roux, J. de Rosny, M. Tanter and M. Fink, "The Aharonov-Bohm Effect Revisited by an Acoustic Time-Reversal Mirror", Phys. Rev. Lett., **D 70**, 3170.(1997). M. Fink, Physics Today, March (1997).

[9] F. Vivanco, F. Melo, C. Coste and F. Lund, "Surface Wave Scattering by a Vertical Vortex and the Symmetry of the Aharonov-Bohm effect", Phys. Rev. Lett. **83**, 1966 (1999).

[10] C. Coste, M. Umeki and F.Lund, "Scattering of dislocated wavefronts by vertical vorticity and the Aharonov Bohm effect I: Shallow water waves", (preprint); C. Coste and F. Lund, "Scattering of dislocated wavefronts by vertical vorticity and the Aharonov Bohm effect II: Dispersive case", (preprint).

[11] See for instance: C. Nore et al, "Scattering of First Sound by Superfluid Vortices", Phys. Rev.

354

Lett. **72**, 2593, (1994) and P. Ao and D. Thouless, "Berry's Phase and the Magnus Force for a Vortex Line in a Superconductor", Phys. Rev. Lett. **70**, 2158-2161 (1993).

NON LINEAR DYNAMICS, PATTERN FORMATION
AND MATERIALS SCIENCE.

by

Daniel Walgraef [†]

Center for Nonlinear Phenomena and Complex Systems,
Université Libre de Bruxelles, CP 231,
B - 1050 Brussels, Belgium

Abstract

Spatio-temporal pattern formation in physico-chemical systems far from thermal equilibrium has long been a puzzling phenomenon. Until the last decade, understanding pattern selection and stability mechanisms was considered as a challenge. Fortunately, thanks to intensive theoretical and experimental research, a unified framework is now available to study pattern formation phenomena. It has been successfully applied to several systems, in different fields, such as hydrodynamics, chemistry, and nonlinear optics. They are now being applied to various types of materials instabilities, and will hopefully lead to a better understanding of phenomena such as the formation and evolution of defect microstructures in plastically deformed or irradiated materials, the formation and symmetries of regular deformation patterns in surfaces and thin films under laser irradiation, the role and the control of instabilities in surface modification technologies, etc. In this context, defect microstructures appear as the result of defect motion and nonlinear interactions, which naturally destabilize uniform distributions. The applicability of the methods of nonlinear dynamics to materials instabilities is analyzed, and an appropriate methodology is proposed. The importance of nonlinear analysis beyond instability thresholds in the determination of pattern selection and stability is emphasized. Several examples are discussed, with references to relevant reviews and technical publications.

[†] Director of Research at the Belgian National Fund for Scientific Research.

355

O. Descalzi et al. (eds.), Instabilities and Nonequilibrium Structures VII & VIII, 355–370.
© 2004 *Kluwer Academic Publishers. Printed in the Netherlands.*

1 Introduction

Although the search for simplicity has always been an important trend of scientific activities, today, in physics and chemistry related to research in materials a new trend is clear : the degree of complexity is increasing. By the way, this is not only true for materials research, but for every field where systems are operating in far from thermal equilibrium conditions [1, 2]. These systems are able to develop spontaneous self-organization processes and produce all kinds of microstructures. Hence, understanding the origin of spatio-temporal order in systems far from thermal equilibrium and the selection mechanisms of spatial structures and their symmetries is a major theme of present day research into the structures of continuous matter. The development of methods for producing spatially ordered microstructures in solids by non-equilibrium methods opens the door to many technological applications.

Let me also emphasize that complexity and self-organization are seductive and that these concepts are at the origin of a true scientific revolution. Indeed, the appearance of order which usually has been associated with equilibrium phase transitions appears to be characteristic of systems far from thermal equilibrium. This phenomenon which was considered exceptional at first now appears to be the rule in driven systems [1, 2, 3]. The chemical oscillations obtained in the Belousov-Zhabotinskii reaction were initially rejected by a large number of chemists, on the basis on erroneous thermodynamic considerations. Now, and mainly thanks to the work of I.Prigogine and the Brussels school [1, 2, 3], these oscillations and related phenomena (waves, chaos, etc.) are the subject of intensive research and new classes of chemical oscillators have been recently discovered. Even living organisms have long been considered as the result of chance rather than necessity. Such points of view are now abandoned under the overwhelming influence of spatio-temporal organization phenomena in various domains ranging from physics to biology via chemistry, nonlinear optics, and materials science [10, 11].

Today, materials science is undergoing a complete revolution. Indeed, by the use of new technologies (laser and particle irradiation, ion implantation, ultrafast quenches, etc.) it is possible to escape from the tyranny of the phase diagram and to process new materials with unusual properties. One can unfortunately not present here all the examples that come to mind to illustrate this evolution of materials science. Let me however briefly describe a few among the most fascinating ones :

- new materials such as quasi-crystals, high Tc superconductors, semiconductor heterostructures and superlattices are typical examples of complex structures. The quasi-crystals present symmetries totally unexpected and even forbidden a decade ago; high temperature superconductors are made of simple layers but their properties result from complex atomic bondings and charge transfers.

- it is now possible to produce complex structures or composites that simultaneously satisfy very diverse requirements. To do so one has to control the material on length scales that vary from the atomic to the micrometer level. Self-organization should be a precious ally for the design of such materials.

- many materials are used today in critical conditions : they are deformed, corroded, irradiated, etc., and, in these conditions, complex phenomena arise from the presence of excesses of defects which are able to organize themselves in regular structures and modify the physico-chemical properties of the materials.

Materials science is essentially related to experimental physics and technological applications, but, there is today an increasing demand for a better understanding of new materials from a more fundamental point of view. In order to describe and understand such materials, dynamical concepts related to nonequilibrium phenomena, irreversible thermodynamics, nonlinear dynamics, and bifurcation theory, are required [6, 9].

Before discussing the importance of nonlinear and self-organization phenomena in materials science, let me point out that the development of a theoretical framework to describe and interpret self-organization phenomena was made easier by the progress of thermodynamics of irreversible processes and by the introduction of the concept of dissipative structure. In this context it is clear that the nonlinearities of the dynamics and the distance from thermal equilibrium are at the origin of to spatio-temporal organization [1].

Similar phenomena appear in very different systems : spiral waves in chemical systems (but also in the cortex or cardiac activity), the aggregation of micro-organisms, and convective rolls associated with hydrodynamical instabilities in normal fluids and liquid crystals [5]. These varied appearances show that these phenomena are not induced by the microscopic properties of the systems but are triggered by collective effects including a large number of individuals (atoms, molecules, cells, etc.).

The role of fluctuations is also very important in such circumstances. Effectively, near instability points, the space and time scales are so large that the structures are particularly sensitive to even small fluctuations. When different states are simultaneously stable beyond an instability, such fluctuations or small external fields may affect the pattern selection mechanisms. Furthermore, in the case of spatial patterns, the position and orientation of the structure which are described by phase variables are usually fixed by the boundary conditions in small systems. This is of course not the case in large systems where phase fluctuations may trigger the nucleation of defects analogous to dislocations and disclinations. These effects show the importance of a stochastic description of self-organization phenomena far from equilibrium [8].

In pioneering fields, such as hydrodynamics or nonlinear chemistry, the comparison between theoretical predictions and experimental observations has long been qualitative but has reached the quantitative level recently [4]. This is because of new experimental methods using laser and computer technology and of theoretical progress based on the theory of dynamical systems, on bifurcation calculus, and on the development of supercomputers which make numerical simulations feasible.

While quantitative and systematic experimental analysis followed theoretical analysis in the case of nonlinear chemistry, the evolution has been quite different in the field of hydrodynamics. Despite the fact that convective instabilities and turbulence have been studied for more than a century , definite progress in understanding pattern formation, selection and stability, and the origin of chaotic behavior were achieved only recently It is worth noting that these problems present severe difficulties. From the experimental point of view, the absence of any operational definition of turbulence, the lack of sensitivity of traditional measurement techniques to the temporal behavior of hydrodynamical flows, and a poor resolution of boundary effects limited the progress until the last decade. From the theoretical viewpoint, a major difficulty has been finding analytic solutions because of the complexity of the Navier-Stokes equations.

Significant progress have been achieved in the experimental analysis of instabilities and hydrodynamical flows because of new techniques (laser velocimetry, cryogenic techniques, image processing, etc.) , the systematic use of computer science in data processing and experiment control, and the linkage with new theoretical approaches based on instability and bifurcation theory. On the other hand, the study of the succession of instabilities obtained by increasing the bifurcation parameter requires nonlinear analysis which extends far beyond the classical studies in the field. Hence a few relatively simple systems (Rayleigh-Benard, Taylor-Couette, Benard-Marangoni, etc.) became very popular as prototypes of complex behavior where nonlinear theories of pattern formation may easily be tested.

Although the Rayleigh-Benard type of instabilities have been discussed at length in the literature and are still providing new challenges for theorists and experimentalists [7], some of their basic aspects bear reviewing. When a thin horizontal layer of fluid is heated from below or cooled from above, a temperature gradient is generated across the sample. For small gradients, the fluid remains in a conductive state but, on increasing the temperature difference between the horizontal fluid boundaries, the gradient may reach a threshold where this conductive state becomes unstable. Beyond this threshold (instability or bifurcation point), convection sets as cellular structures associated with periodic spatial variations of the hydrodynamic fluid velocity field and of the temperature field.

Several types of structures may be obtained according to the working conditions: rolls, hexagons, squares, traveling or standing waves. On increasing further the bifurcation parameter, these patterns may in turn become unstable causing successive bifurcations to occur driving the system to chaos.

From the theoretical point of view, while the first bifurcation may easily be determined from the Navier-Stokes equation, it is a formidable task to determine the behavior of the system beyond the hydrodynamic instabilities with these equations. Fortunately, the derivation of amplitude equations for the patterns led to definite progress in the study of their formation, selection and stability properties. These equations which are usually of the Ginzburg-Landau type, correspond to reduced versions of the complete dynamics which contain all the symmetries of the problem. They may be solved more easily and describe correctly the dynamics of the system on long space-time scales close to the bifurcation point.

Thanks to the permanent interactions between theory, experiment and numerical analysis, significant progress have been made during the past 20 years on the mathematical methods of nonlinear dynamics and in the understanding of simple fluids instabilities. It has become quite clear that such instabilities manifest themselves in the form of various patterns which vary from the simple to the complex. More recently, the growth in the body of knowledge of liquid-crystal hydrodynamics furnished an exciting ground for further experimental observations on the nature of transitions from one pattern structure to another. Nonlinear interactions at the micro level can explain, to a large degree, the onset and propagation of instabilities at the macro level. Instabilities are saturated through the formation of what is now commonly known as "dissipative" structures. The geometry and properties of these structures can be well explained by a competition between local and non local transport reactions. This framework appears to be quite general, at least conceptually, and can be seen in many physical phenomena (e.g. laser-material interactions, energetic particle-material interactions, magnetic fluids, plastic instabilities, plasma and electric systems). General observations of these vastly diverse physical systems show striking similarities in the nature and occurrence of patterns as manifestations of instabilities.

It appears now that instabilities and patterns occur all the time in materials science. They affect the properties of the materials and hence, need to be understood and controlled. This is valid from pattern formation in metallurgy during the solidification processes to the nucleation of defect microstructures in deformed, corroded or irradiated metals and alloys. Let me only touch upon three examples :

- Dendrites have been observed at length as a typical microstructure in the growth of solids from vapor (snow flakes) or solution (metal casting). After decades of observation we still do not fully understand how crystalline structures and dendrites grow. With the necessity of developing more efficient materials processing methods, it becomes essential to understand the parameters that control the microstructure. In this framework, self-organization may play an important role.

- During the last decade, the field of superlattices, modulated or incommensurate structures has widened from superconductors to metals, insulators and polymer mixtures. Layered materials present unexpectedly rich behaviors and totally new phenomena. Here also the control of the structure is essential and it would be particularly interesting if the system could control himself its wavelength or modulation period, or its layer thickness. In layered superconductors as in other materials, defects are always present and control such important quantities as transition temperatures or critical current densities. One can thus not avoid the search for a better understanding of defect behavior in solids.

- It is effectively well known that defects play an important role in determining materials properties. Point defects play a major role in all macroscopic material properties which are related to atomic diffusion mechanisms, and to electronic properties in semiconductors. Line defects, or dislocations, are unquestionably recognized as the basic elements which lead to plasticity and fracture. While the individual properties of solid state defects are at an advanced level, studies of their collective behavior, which is essential in determining their influence at the macroscopic level, is still elementary. Nonetheless, significant progress has been made in the area of dislocation dynamics and plastic instabilities over the past several years, and the importance of nonlinear phenomena has also been assessed in this field.

2 Non Linear Phenomena in Modern Materials Science

The processing and utilization of materials devoted to advanced applications occurs in far from thermal equilibrium conditions. Hence the understanding of the related materials behaviors requires a combination of atomic physics, nonequilibrium statistical mechanics, nonlinear dynamics and process engineering. Up to now nonlinear science has proven useful in various aspects of materials research from pattern formation during solidification and growth phenomena to defect dynamics, fracture and crack propagation or corrosion.

Let me consider for example the solidification process of metallic alloys. In this process, crystallites or grains are formed which look like a collection of snowflakes since each grain is formed by a dendritic process. The speed at which the dendrites grow and the regularity and spacing of their sidebranches determines a pattern that is called the microstructure of the solid material. This microstructure governs many physico-chemical properties of the material such as its mechanical strength and its response to heating and deformation.

The possibility of predicting and controlling microstructure formation is thus an important issue for metallurgists. The development of automated methods for the processing of materials with made to order properties ultimately depends on our understanding of this problem in nonequilibrium pattern formation. Major advances in the microstructure problem occured only recently and one understands better the contradiction between the fact that, at one side, it is well known from experiments that the dendritic growth rate, the sharpness of the dendrite tip, the spacing of its sidebranches and almost all properties of the free dendrite are determined uniquely by the undercooling of the melt, and that, at the other side, simple considerations of steady-state heat diffusion essentially predict the opposite, i.e. that a wide range of growth modes varying from thick to sharp dendrites are equally possible. The question of mode selection is thus of particular importance in this problem. After almost two decades of intensive research on this issue, it appears now that dendritic crystal growth should be controlled by weak effects that had long been considered as negligeable (e.g. weak anisotropies, small changes in the surface tension of the crystal, thermal fluctuations in the melt). The understanding of how tiny perturbations are amplified in these systems is an important topic of nonlinear science which required a subtle combination of analytical and numerical studies.

The solution of the free dendrite problem is only a first step towards a full understanding of practical solidification processes. The next one is probably the understanding of the directional solidification instabilities. In this industrially useful process the initially flat interface between solid and melt becomes unstable and, after a few secondary instabilities towards cellular patterns, finally forms an array of dendrites. This front instability occurs when the interface is forced to move relative to the gradient temperature in which it is sitting. The challenge to the metallurgist is, for example, to control and reproduce the spacing of the final array, the challenge to the theorist being to predict it, this process being highly non linear.Furthermore, in real industrial processes, a great deal of energy is added to the system to melt the alloy and consequently, the melt undergoes turbulent convection which should be controlled since it strongly affect the dendritic array. Hence, the microstructure problem is at the forefront of physics and engineering science but also of modern mathematics.

Other nonlinear phenomena are of importance either in the design of new materials, mainly of the layered type, or in the damage occuring to materials used in hostile environments. In the first case, the aim is, for example, to produce heterostructures superlattices with special semiconducting, magnetic or super-conducting properties. This implies the mastering of sophisticated techniques such as ion implantation, chemical vapor deposition, sputtering, laser annealing, evaporation, molecular beam epitaxy, etc. where much remains to be understood, both from experimental and theoretical points of view. In the second case, the behavior of the material is mainly dictated by the behavior of its defects. For example, it is well known that plasticity is related to dislocation motion, that the swelling of irradiated metals and alloys is induced by the formation of vacancy loops and void or bubble lattices. In fact, in driven materials, defects are produced in excess, and, due to the coupling between their mobility and their interactions, uniform distributions become usually unstable and defect microstructures are formed. Recently, the understanding of defect patterning made great progress by the combined use of numerical simulations and of dynamical models of the reaction-transport type. This made possible, for example, the description of strain localization and plastic instabilities, of microstructure evolution during particle irradiation, of void lattice formation, but much remains to be done to solve very practical problems such as the influence of dislocation microstructures on crack propagation or the effect of the crystal symmetries on the void lattices.

I will now list and review some domains of materials science where the methods of nonlinear dynamics and pattern formation theory have been applied, sometimes preliminarily. I will not discuss the technical aspects of these problems, which are now currently available in the literature. I will rather give the reader who is willing to enter in this field the relevant references

3 Application of Nonlinear Dynamics in Research and Development Projects

3.1 Surface Modification Technologies

Today, surface modification technologies play an essential role in materials processing and protection. They allow for example the processing of thin films with original physico-chemical properties or of coatings which enhance the materials resistance to chemo-mechanical agressions and damage. Due to the complexity and non linearity of the processes involved in such methods as chemical or physical vapor deposition (CVD, PVD), plasma processing, thermal laser deposition and etching, morphological instabilities easily arise which may affect positively or negatively the growth or the surface structure of the films or coatings. It would thus be essential to understand and master these instabilities in order to optimize the processing methods and the materials performances. This preoccupation is also related to the fact that, although materials research is essentially dealing with applications and experiments, there is an increasing demand for a better understanding of novel materials from a more fundamental point of view. This trend led to the development of a rapidly increasing activity dedicated to modelling materials processing and behavior which, in turn, requires concepts related to nonequilibrium phenomena, irreversible thermodynamics, nonlinear dynamics and mechanics.

Up to now, this modelling activity has been mostly restricted to what may be called computational materials science which involve numerical simulations at the microscopic atomistic level. They include electronic structure calculations, Monte Carlo and molecular dynamics simulations, or cellular automata models, and are essential in explaining and predicting the properties of materials such as hetero- or nanostructures, of surface morphology and of various processing methods based on atomic deposition.

However, many of these new aspects of materials development involve instabilities induced by collective behaviors in atomic or molecular populations. The description of such instabilities and of their effect on materials properties, microstructures, defect behavior etc. , with microscopic methods present severe difficulties related, on one side, to the complexity of the computations and their memory requirements, and, on the other side, to the divergence of the length and space scales close to instabilities. Hence, mesoscopic descriptions leading to kinetic equations of the reaction-diffusion type for continuous variables such as atom or defect concentrations, deformation fields, surface bending or curvature, etc. could be of particular interest in the analysis of surface instabilities since they may benefit from the important progress obtained during the last decade in the study of complex systems in the framework of nonlinear dynamics and instability theory.

In surface technologies based on vapor deposition or plasma processing, nonlinear approaches are particularly well adapted to the modelling of the systems behavior. Effectively, the basic elements of any dynamical model should be :

- the forced flow of reactants over the substrate
- the adsorption of the active species on the substrate
- autocatalytic chemical reactions between the adsorbed species and their surface diffusion
- desorption of species from the substrate
- forced flow evacuation of the system

Such models are known to contain the necessary ingredients for pattern formation or growth instabilities. In fact, modelisation of heterogeneous catalysis phenomena have been undertaken with success along these lines in order to describe the formation of spatio-temporal structures or chemically induced phase transitions on the surface of the catalyst [16]. One should be able to achieve similar results in the present case and to describe and predict the morphological or physico-chemical properties of the surface as functions of experimental variables such as temperature, flow characterisitcs, substrate or deposit compositions, etc.

It is possible to describe along these lines real surface processing experiments showing for example transitions from rough to smooth surfaces, or transitions between microstructures of different symmetries.

3.2   Laser Irradiation and Thin Film Deformation

Laser surface interaction is a field where patterning phenomena are overwhelming and where the methods of nonlinear dynamics will hopefully lead to a better understanding of the mechanisms of pattern formation, selection and stability. This is particularly significant when one considers the technological importance of this field. Effectively, strong laser radiation induces structural and morphological changes in matter which are responsible for the degradation of light emitting devices, the cumulative laser damage of optical components, the nonuniform melting of semiconductor surfaces, to cite only a few of these aspects. Furthermore, laser annealing and fast recrystallization may lead to special types of structures including molten and crystalline phases, and laser assisted thin film deposition processes should also be in the mainstream of this activity [12, 14, 13].

The main instability mechanism in laser irradiated materials is due to the coupling between defect dynamics and the deformation of the surface. Effectively, on one side, strong laser irradiation generates an excess of defects in the surface layer. Example are : electron-hole pairs in strongly absorbing semi-conductors, interstitials and vacancies in thin film deposition, coating and laser annealing, voids or dislocation loops in prolonged irradiation. On the other side, the defect field give rise to strong deformations of the subsurface layer of the material. Finally, it is the coupling between defect generation, diffusion and the deformation field which leads to pattern forming instabilities [15]. As a result, the dynamical description of such phenomena should be based on the dynamics of the defect field in the layer, the elastic continuum of the host material described by the displacement vector with the appropriate boundary conditions (bulk, surface or thin film), both dynamics being coupled through the defect- strain interaction. Elementary dynamical models have already been derived in this framework, and were able to reproduce the qualitative aspects of the generation of deformation patterns in thin metallic films under extended and focused laser irradiation [17, 18]. Examples of the results obtained for focused laser irradiation [18], and that agree with experimental observations [13], are displayed in figure 1. This effort should be pursued, in order to reach a quantitative determination of the instability thresholds, but, above all of the patterns nucleation rates, selected symmetries and stability ranges, in function of materials parameters and experimental set-ups.

3.3 Defect microstructures and damage in irradiated metals and alloys.

It has been extensively shown in various works devoted to materials instabilities, that the coupling between reaction and transport may induce pattern forming instabilities in defect distributions in driven crystalline materials. For example, uniform distributions of point defects such as interstitials and vacancies in irradiated materials, e.g. in fusion reactors or in space devices such as rocket parts, may become unstable and lead to the precipitation of solid solutions, to the nucleation of voids and of void lattices [19, 20, 21, 22]. In this framework, it is shown, within the framework of a dynamical model of the resction-diffusion type [23], that vacancy loop ordering (cf. fig.2) occurs under very general conditions in irradiated metals and alloys. It mainly results from a Turing-like instability induced by the different mobilities and bias in the migration of point defects to line defects such as vacancy loops or network dislocations. According to the generic properties of such pattern forming instability, structures with different symmetries, such as bcc lattices or wall structures, may be simultaneously stable in well defined parameter ranges. However, at sufficiently high values of the bifurcation parameter, the only stable stucture corresponds to planar arrays [23].

Here, one has to take into account that the microstructure is continuously evolving. In the case of fusion reactors, the analysis of the dynamical model [23] predicts that, in the case of stainless steel under normal irradiation conditions, the effective bifurcation parameter, the irradiation dose, is continuously increasing with time or irradiation dose. Hence, after an initial phase where loop clustering should occur in the form of bcc lattices, the system should

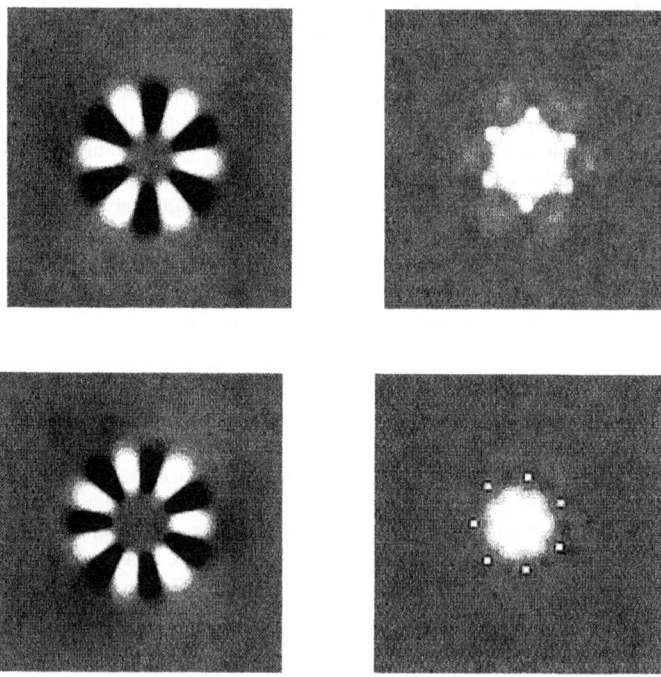

Figure 1: Deformation field (left) and upper surface vacancy density, obtained by numerical integration of a dynamical model describing the coupled evolution of these two variables in thin films under focused laser irradiation [17,18]. The laser intensity increases from top to bottom, and so do the number of petals of these rose patterns, as predicted by the theory, and in agreement with experimental observations.

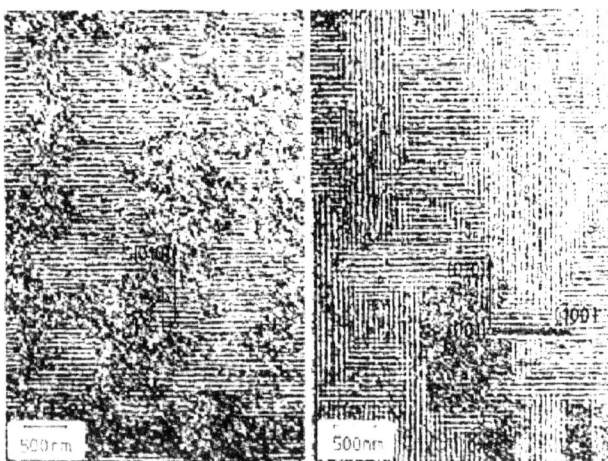

Figure 2: Dislocation microstructures (periodic arrays of vacancy loops) in Copper irradiated with protons, with irradiation dose of 2*dpa* (courtesy of W.Jaeger).

evolve towards a uniform state followed by a bcc structure and should finally reach a planar wall structure, even for isotropic defect mobilities. These structures could then be in non parallel orientations, i.e. with a structure different from the structure of the host lattice. This behavior, which seems consistent with experimental observations, should also occur in accelerator conditions either for annealed and cold worked steels. In the case of anisotropic interstitial diffusion, planar structures should also be the rule, but in this case, the orientation of the walls should be determined by the anisotropy (cf. fig.3). For cold worked steels in reactor conditions, the bifurcation parameter cannot reach the instability threshold. However, since bcc structures may appear subcritically, 3D loop clustering could occur transitorily, as observed experimentally [22], but the final defect distributions should be uniform [25].

Hence, since the symmetry of the defect structures is a crucial issue in irradiated materials, the present discussion shows that a careful study of the post-bifurcation regime is needed to test the relevance of particular kinetic models to the interpretation of experimental observations [24]. Furthermore, we think that a coherent description of the materials instabilities that induce the spatio-temporal organization of defect populations will hopefully lead to a deeper understanding of the behavior of irradiated materials, either in fusion reactors or in space research. Indeed, due to the strong nonequilibrium conditions under which irradiation induced defect patterning occur, classical mechanical or thermodynamical considerations are not sufficient to interpret these phenomena and we need to study them in the framework of nonlinear dynamics and instability theory. In particular, we showed that, despite the huge complexity of defect dynamics, even in the case of phenomenological models, valuable information can be obtained via the reduced dynamics near instability points leading to a realistic description of the pattern selection and stability properties in the post-bifurcation regime.

3.4 Dislocation Dynamics, Plastic Instabilities and Fracture.

Strain localization and dislocation microstructure formation are typical features of the plastic deformation of metals and alloys . Plastic deformation occurs by the glide of dislocations, and, although the dislocation distributions are rather uniform at its onset, they usually become unstable when deformation proceeds and undergo successive transitions towards various types of microstructures such as cells, deformation bands, persistent slip bands, labyrinth structures, etc [26, 27, 28]. This phenomenon is experimentally well documented , but, despite a huge number of theoretical investigations and modelling attempts, it is still poorly undertstood .

Several attempts have been made to understand and describe the formation and the properties of dislocation patterns. They roughly belong to three types of approaches. The first one is mesoscopic and consists in the derivation of rate equations describing the dynamical evolution of dislocation densities [29, 30]. The other ones are numerical and based on lattice gas [31, 9] or molecular dynamics techniques [32, 33]. Of course, the outcome of first approach strongly depends on the physical processes included in the dynamics, and it is sometimes hard to know the ones that play the key role in dislocation patterning, although our experience of other pattern forming phenomena tells us that many qualitative aspects of pattern formation, selection and stability do not depend on the details of the microscopic dynamics, but rather on more general properties such as symmetries, bifurcation classes, etc. .

On the other hand, numerical approaches have greatly helped to clarify the mechanisms of dislocation patterning. However they require heavy and time consuming computations and, although they may provide checks on the role and importance of the different physical processes involved in dislocation dynamics, they usually do not lead, by essence, to the qualitative descriptions that are required for a global understanding of the phenomenon.

Hence, it would be extremely useful to study with these methods the various problems related, not only to the deformation, but also to the fracture of metals and alloys during fatigue (persistent slip band formation, crack nucleation and propagation, effect of dislocation microstructures on crack velocity, etc.), the cracking of composites materials used in high temperature applications (e.g. derivation of fracture thresholds), the self-organization of voids and bubbles in irradiated ones, etc. [33, 34, 35, 36].

4 Conclusion

Many aspects of the behavior of materials under extreme non-equilibrium conditions, and particularly under the influence of radiation and deformation, are now modeled through several types of advanced analytical and numerical methods, such as:

- rate theory and model equations of the reaction-transport type for defect dynamics and microstructure evolution,

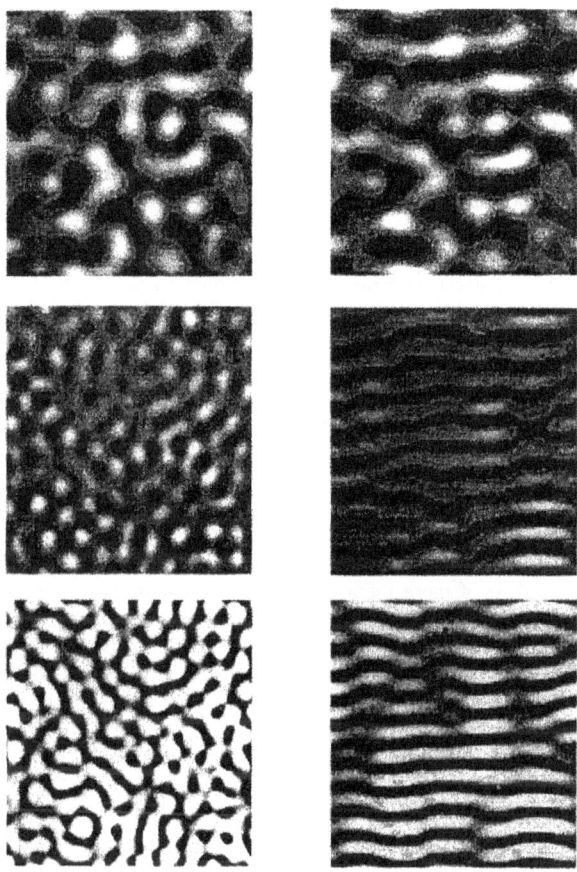

Figure 3: Dislocation microstructures (periodic arrays of vacancy loops) obtained numerically from the dynamical model analyzed in [25], and describing proton irradiated annealed stainless steel in accelerator conditions, for increasing dose (from top to bottom) and isotropic (left column) and uniaxial (right column) interstitial diffusion (courtesy of J.Lauzeral).

- Fokker-Planck theory applied to irradiated materials and the formation of thin films,
- derivation of dislocation models of the micromechanics of plasticity and fracture.
- study of pattern formation, selection and stability with analytical and numerical methods including bifurcation and multiple scale analysis, amplitude equations, nonequilibrium statistical mechanics,
- computer simulations of the dynamics of collision cascades and ballistic effects in solids under irradiation or sputtering, etc.

Considering the results already obtained, and briefly discussed in the preceeding sections, one may definitely hope that, by the combination of bifurcation analysis, amplitude equation formalism and numerical simulations, significant progress may be expected in the understanding and prediction of the effects of materials instabilities on the macroscopic behavior of driven or degrading solids. In particular, the formation of defect microstructures may be qualitatively described by simple dynamical models based on defect interaction and motion, eventually coupled with the deformation field of the material. It appears clearly that pattern selection and stability strongly depends either on linear and nonlinear mechanisms.

As in other complex pattern forming systems studied so far, such as hydrodynamic or chemical systems, it is essential, in each case, to proceed in close interaction with experimental studies. Appropriate dynamical models may be derived for each material instability. One may then determine the conditions for instability, and instability thresholds. Special care has to be taken on performing non-linear analysis beyond instability thresholds, especially in the presence of particular boundary effects, induced, for example, by residual surface stresses, grain boundaries, etc. . Detailed analytical and numerical analysis of pattern formation, selection and stability beyond the instabilities, should assess the effects of pattern formation on the macroscopic behavior of the materials. Furthemore, the influence of experimental procedures, external fields and geometrical effects on the instabilities and patterns may easily be incorporated in such a description. It is the comparison with experimental analysis that determines if modelisation needs to be be improved, or if more refined analytical and numerical studies are needed. In this framework, experimental programmes systematically designed to test proposed dynamical models are particularly important. Finally, as a follow-up of these analysis, new experimental programs may then be devised for specific technological problems

References

[1] G. Nicolis and I. Prigogine, Self-Organization in Non Equilibrium Systems, Wiley, New York, 1977.

[2] G. Nicolis and I. Prigogine, Exploring Complexity, W. H. Freeman, New York, 1989.

[3] H. Haken, Advanced Synergetics, Springer, Berlin, 1983.

[4] H. Swinney and J.P. Gollub, Hydrodynamic Instabilities, Springer, Berlin, 1984.

[5] F. Busse and L. Kramer, eds., Nonlinear Evolution of Spatio-Temporal Structures in Dissipative Continuous Systems, Plenum, New York, 1990.

[6] D. Walgraef and N.M. Ghoniem, eds., Patterns. Defects and Materials Instabilities, Kluwer Academic Publishers, Dordrecht, 1990.

[7] P. Manneville, Dissipative Structures and Weak Turbulence, Academic Press, Boston, 1990.

[8] F. Baras and D. Walgraef, eds., Nonequilibrium Chemical Dynamics : from experiment to microscopic simulation, Physica **A 188** , 1992.

[9] G. Martin and L.P. Kubin, eds., Nonlinear Phenomena in Materials Science I & II, Transtech, Aedermannsdorf (Switzerland) , (1988, 1992).

[10] M.C. Cross and P.C. Hohenberg, Rev.Mod.Phys. **65** , 854, 1993.

[11] D.Walgraef, Spatio-temporal Pattern Formation with examples from Physics, Chemistry and Materials Science, Springer-Verlag, New York, 1996.

[12] R.Kossowsky and S.C.Singhal eds., Surface Engineering, Surface Modification of Materials, Martinus Nijhoff, Dordrecht, 1984.

[13] D.Bauerle, Chemical Processing with Lasers, Springer Ser.Mat.Sc. **1**, Springer-Verlag, Berlin, 1986.

[14] L.Laude, D.Bauerle and M.Wautelet eds., Interfaces under Laser Irradiation, Martinus Nijhoff, Dordrecht, 1987.

[15] V.I.Emel'yanov, Laser physics **2**, 389 (1992).

[16] R.Imbihl, in Nonequilibrium Chemical Dynamics : from experiment to microscopic simulation, F. Baras and D. Walgraef, eds.,Physica **A 188**, 34-46 (1992).

[17] N.M.Ghoniem, J.Lauzeral and D.Walgraef, Phys.Rev **B 56**, 15361-15376 (1997).

[18] J.Lauzeral, D.Walgraef and N.M.Ghoniem, Phys.Rev.Lett. **79**, 2706-9 (1997).

[19] J.H.Evans, Nature, **229**, 403, (1971); Rad. Effects, **10**, 55, (1971).

[20] G.L.Kulcinski, J.L.Brimhall and H.E.Kissinger, Production of Voids in Pure Metals by High-Energy Heavy-Ion Bombardment, Proc. 1971 International Conference on Radiation-Induced Voids in Metals, Albany, New York, June 1971, USAEC, CONF-710601, NTIS, (April 1972), 465.

[21] F.W.Wiffen, The Effect of Alloying and Purity on the Formation and Ordering of Voids in BCC Metals, Ref. 2, p. 386.

370

[22] W.Jaeger, P.Ehrhart and W.Schilling, in Nonlinear Phenomena in Materials Science, G.Martin and l.P.Kubin eds., Transtech, Aedermannsdorf (Switzerland), 279, (1988).

[23] N.M.Ghoniem and D.Walgraef, Modelling Simul. Mater. Sci. Eng., 1(5), 569, (1993).

[24] D.Walgraef, in Reactive Phase Formation at Interfaces and Diffusion Processes, Y.Limoge and J.L.Bocquet eds., Materials Science Forum, 155-156, Trans Tech, Switzerland, pp. 401-408, (1994).

[25] D. Walgraef and N.M.Ghoniem, Phys.Rev. B 52, 3951, (1995).

[26] H. Mughrabi, F. Ackermann and K. Herz, In: E.T. Fong (Editor), Fatigue Mechanisms, Proceedings of an ASTM-NBS-NSF Symposium. ASTM-STP675, Kansas City, p. 69-105, 1979.

[27] C. Laird, in Fatigue and Microstructures, ASM Materials Science Seminars, St. Louis, Missouri, pp. 149 - 203, 1978.

[28] F. Ackermann, L.P. Kubin, J. Lepinoux and H. Mughrabi, Acta Metall. , 32, 715 (1984).

[29] D. Walgraef and E.C. Aifantis, Int. J. Eng. Sci., 24, 1798, (1986).

[30] D. Walgraef and E. C. Aifantis, Int. J. Eng. Sci., 23, 1351, 1359, 1364 (1986).

[31] J.Lepinoux and L.P.Kubin, Scripta Met.21, 833, (1987).

[32] N.M. Ghoniem, J. R. Matthews, and R. J. Amodeo, Res Mechanica, 29, 197 (1990).

[33] N.M. Ghoniem, in Non-Linear Phenomena in Materials Science II, L. Kubin and G. Martin, Eds., Kluwer Academic Publishers, 1992, pp. 429 - 444.

[34] J.C.Charmet, S.Roux and E.Guyon eds., Disorder and Fracture, NATO ASI Series 235, Plenum Press, New York, 1989.

[35] H.Herrmann and S.Roux, Statistical Models for the Fracture of Disordered Solids, North Holland, 1990.

[36] J.M.Salazar and D. Walgraef, in Plasticity and Fracture, Instabilities in Materials, N.M.Ghoniem ed., AMD-vol 200, MD-vol 57, p.179, The American Society of Mechanical Engineers, NY (1995).

NEW RESULTS FOR DIFFUSION-LIMITED REACTIONS WITHIN A STOCHASTIC MODEL

Horacio S. Wio[*], Miguel A. Rodriguez[‡] and Alejandro Sánchez[‡]

*Grupo de Física Estadística, Centro Atómico Bariloche (CNEA) and Instituto Balseiro (UNC),
8400 San Carlos de Bariloche, Argentina
†Instituto de Física de Cantabria, CSIC and Universidad de Cantabria, Santander, Spain

Abstract. In this contribution we study the time behaviour of a couple of diffusion-limited reaction processes in a one-dimensional system by means of a recently intoduced stochastic approach.

The first problem corresponds to the behaviour of the reaction front between initially separated reactants (reaction $A + B \to C$). The asymptotic results of the stochastic scheme give a rather good agreement with simulations, regarding the time power indexes of scaling for the height and width of the distribution of C particles.

The second problem corresponds to the study of a trapping process where the traps (particles B), besides being mobile, have a variable number. We analyze two cases related with the coupled reactions: $A + B \to B, B + C \to C$, and $A + B \to B, B + C \to 0$. It is shown that the time evolution of the traps strongly influences the kinetics of the trapping process, yielding qualitatively different behaviour in both cases. The results of our model, adapted from the stochastic one for simple trapping, have been compared with simulations yielding good qualitative agreement.

INTRODUCTION

The study of irreversible bimolecular diffusion-reaction systems within the diffusion-limited regime has attracted enormous interest for well over a decade. This was motivated by the "anomalous" kinetic laws that govern the evolution of these chemical reactions as, in low dimensional systems, they depart from the standard mean field rate equations [1,2]. Although exact solutions are hard to obtain, several asymptotic results for the particle concentration decay have been obtained by means of simulations and/or analytical procedures, through heuristic arguments or by scaling analysis. So far, the different aspects studied cover the influence of dimensionality, conservation laws, segregation properties [3-6], statistics of nearest neighbor distances [7-9], effects of noise and/or disorder [10,11], etc. As has been indicated in almost all references, this process could be used to model several different physical and chemical systems.

In this contribution we review some results obtained within a recently introduced stochastic model [12-19]. This scheme, that covers the whole regime from the diffusion-

---

[1] E-mail address: `wio@cab.cnea.gov.ar`.

[2] E-mail address: `rodrigma@ifca.unican.es`.

[3] E-mail address: `asanchez@ifca.unican.es`.

O. Descalzi et al. (eds.), Instabilities and Nonequilibrium Structures VII & VIII, 371–384.
© 2004 Kluwer Academic Publishers. Printed in the Netherlands.

up to the reaction-controlled limit, was initially introduced for the study of neutron diffusion problems with absorption by "small" (and mobile) absorbers [20,21].

The main characteristics of the referred model are:

1- It naturally includes the possibility of imperfect reactions or imperfect trapping.

2- It is the continuous limit of the corresponding master equation usually employed in simulations.

3- It shows to be an appropriate framework in order to obtain analytic, exact or approximate, results.

4- It offers the possibility to analyze not only the asymptotic long time regime, but also the short and intermediate time regimes.

In the next section we briefly review the characteristics of our model. After that, we present some results for the problem of the behaviour of the reaction front in annihilation reactions ($A + B \rightarrow 0$) as well as the application of this model to the problem of trapping with a variable (time-dependent) number of traps ($A + B \rightarrow B$, with $n_B = n_B(t)$). In the last section me make a general discussion of these results and on future work.

REVIEW OF THE STOCHASTIC MODEL

Here we make a brief review of our scheme. The complete details of the method can be found in Refs.[13-19].

We consider two kinds of particles, A and B, with independent motion and with the possibility of annihilation when they are in contact. We assume that the motion of an individual particle is a Markovian process with conditional probability $G_A(r,t|r',t')$ for the A particles and $G_B(r,t|r',t')$ for the B ones. These probabilities fulfill the evolution equations:

$$\dot{G}_A = L_A G_A, \tag{1}$$

$$\dot{G}_B = L_B G_B, \tag{2}$$

with $G_A(r,t_0|r',t_0) = G_B(r,t_0|r',t_0) = \delta(r-r')$ as initial conditions, and L_A, L_B being linear operators. The Markovian character, which is essential in this formulation, makes it possible to factorize the one particle joint probability in the usual way.

Each particle moves independently, but its survival depends on the evolution of the other particles. We identify each particle with a numerical index. Since the number of particles is not conserved, the set of indices corresponding to the existing particles varies with time, $\{i(t)\}$. The density of particles of one kind at time t is given by:

$$n_{A,B}(r,t) = \left\langle \sum_{\{i_{A,B}(t)\}} \delta(r - r_{i_{A,B}}(t)) \right\rangle, \tag{3}$$

where the brackets $\langle \cdots \rangle$ indicate an average over the motion process and over all possible annihilation processes. Also, the density functions defined by:

$$n_{A,B}(r,t'|t) = \langle \sum_{\{i_{A,B}(t)\}} \delta(r - r_{i_{A,B}}(t')) \rangle. \tag{4}$$

correspond to the density of particles at time t' that survive until time t.

To formulate the evolution equations, we define $n_{A,B}(r,t|\{B,A\})$ as the density of the A or B particles conditioned to the occurrence of a process $\{r_{i_B}(t)\}$ or $\{r_{i_A}(t)\}$ in the evolution of the A and B particles. Hence, the evolution of these densities is governed by:

$$\dot{n}_{A,B}(x,t\{B,A\}) = L_{A,B}n_{A,B}(x,t|\{B,A\}) - \gamma \sum_{\{i_{B,A}(t)\}} \delta(x - x_{i_{B,A}}(t))n_{A,B}(x,t|\{B,A\}). \quad (5)$$

Here γ is the probability per time unit for the reaction to take place in each (collision) event. This equation is strictly correct in dimension one, but in higher dimensions we need to take into account the size of the particle (in dimension one we assume point absorbers) [16,35]. The equations are coupled in the sense that the target particles act like an external absorber. The rate of absorption for each particle is then the same only in average. In other words, the annihilation of one particle does not necessarily imply the annihilation of the other. This is the main approximation of the method that, as the simulations indicate [13-19], is excellent even when the reaction is imperfect.

The most naive approximation for the average of Eq.(5) consists of the simple factorization of the reaction term: $\left\langle \sum_{\{i(t)\}} \delta\left(x - x_{i_{B,A}}(t)\right) n_B(z|\{A\}) \right\rangle \approx n_A(z)n_B(z)$. This corresponds to a mean field approximation which is only valid at short time intervals. An exact method for the averaging of this type of equations is given in References [16,17]. With such a method it is possible to find a systematic expansion of products and convolutions of conditional probabilities and mean densities. Here we use this method with the same notation used therein and in reference [16,35].

For the sake of simplicity we only consider one of Eqs.(5). The other equation requires an identical treatment. First we take the integral form of Eq.(5), iterate it and multiply by $\sum_{\{i(t)\}} \delta(x - x_{i_B}(t))$. Averaging this equation, taking into account the factorization properties of the joint probability, we obtain:

$$\frac{\partial n_A}{\partial t} = L_A n_A - \gamma \quad_A, \quad (6)$$

where the *absorption function* $_A(z) \equiv \left\langle \sum_{\{i\}} \delta(z)n_A(z \mid \{B\}) \right\rangle$ is given by the expansion as was shown in references [16,35]. We will not repeat those results here but only indicate that, since the density is a monotonically decreasing function of time, the longtime behaviour will be dominated by the lowest orders in $n_B(t)$, that can be summed giving:

$$_A(z) \approx T(z,\underline{z_1})n_B(z_1|t)G_{Az,\underline{z'}}\delta(t')n_A(z'), \quad (7)$$

with

$$T(z,z') = \delta(z-z') - \gamma G_{Az,z'}G_{Bz,z'} + \gamma^2 G_{Az,\underline{z_1}}G_{Bz,z_1}G_{Az_1,z'}G_{Az_1,z'} + \cdots, \quad (8)$$

where we have used the following shorthand notation: we denote by z_i the pair (x_i,t_i), $\underline{z_i}$ indicates integration over the variables x_i,t_i for all functions containing them, $G_{z,z'}$

is $G(x,t|x',t')$. The remaining terms can also be expressed in terms of these functions. Also, in order to get a good approximation for short times we must keep the contribution of the highest orders in n_B. Hence, neglecting the crossed terms of mixed products of G_B and n_B, $\;_A(z)$ will be the solution of the following integral equation:

$$_A(z) = \quad T(z,\underline{z_1})n_B(z_1|t)G_{Az_1,\underline{z'}}\delta(t')n_A(z') + \\ + T(\overline{z},\underline{z'_1})n_B(z'_1|z)G_{Az'_1,\underline{z_1}}\;_A(z_1). \tag{9}$$

After some more replacements and rearrangements we finally obtain an equation for $n_A(z)$:

$$\frac{\partial n_A}{\partial t} = L_A n_A - \gamma T(z,\underline{z_1})n_B(z_1\mid t)n_A(z_1). \tag{10}$$

valid for short and long times. A completely similar equation is obtained for the averaged density of B particles:

$$\frac{\partial n_B}{\partial t} = L_B n_B - \gamma T(z,\underline{z_1})n_A(z_1\mid t)n_B(z_1). \tag{11}$$

Equations (10) and (11) are the main result obtained in reference [16,17] and will be our starting point in the study of the problem of initially separated reactants. As pointed out earlier, the derivation of $n(z_1\mid t)$ as a function of $n(z)$ is not possible in general making it necessary to resort to some kind of approximation. This will depend on each particular problem, but, as a first approach, two alternatives are possible, either $n(z_1\mid t) \sim n(z_1)$ or $n(z_1\mid t) \sim n(z)$. The first (a kind of sudden approximation) seems to be better adapted to problems in which diffusion is the main phenomenon. The second is an adiabatic approximation valid when the shape of the density varies slowly.

BEHAVIOUR OF THE REACTION FRONT

The recent interest raised by the kinetics of the recombination process $A + B \to C$ (C inert and immobile) is due to the fact that segregation of like particles occurs, leading to anomalous reaction rate laws [1,2]. A particular aspect that has been studied through experiments, theory and simulations, is the situation in which the reactants are initially separated in space [22-33]. In such a case the system develops a reaction front at the interface separating the reactants, which is marked by the concentration of the C particles. As was discussed, the behaviour of such a reaction front should be of relevance for a variety of biological, chemical and physical systems [34].

Usually, the theoretical analysis of the behaviour of this front is done in terms of diffusion–reaction equations where the reaction terms are modeled according to chemical kinetics [22-30]. Recently, renormalization group techniques have also been used, starting from a master equation describing the process and transforming it into a second quantized version [33].

Here we use a different, more accesible, framework, whose starting point is a stochastic equation for the density of a species of particles in which the other species acts as an absorber with a given absorption shape function. To this end, we will apply the scheme reviewed in the previous section [35] to the reaction front problem described above and we will show how the exponents indicating the time behaviour of the front width and height that result from the present analysis are in rather good agreement with those known from simulations [24,29,31].

In order to apply our scheme to the problem of the reaction front we will consider the case of a set of particles A and B with initially separated uniform densities: $n_A(x, t = 0) = n_B(-x, t = 0) = n_0\theta(x)$, and with the same diffusion coefficient D. L_A and L_B are diffusion operators $L_A = L_B = \Delta$ and the corresponding Green function has the usual form $G_A = G_B = (4\pi Dt)^{-\frac{d}{2}}\exp(-r^2/4Dt)$. Since we restrict our analysis to a one dimensional situation, we can use the point absorber approach (that is, the limit $\varepsilon \to 0$). In such a case, the convoluted functions $n_A^H(z_1 \mid t)$ and G_b^H become ordinary functions $n(x,t)$ and $G_B(x,t \mid x',t')$. In order to proceed with our calculation, we will partially follow the approach of Ref. [29a]. We define the function $F(x,t) = n_A(x,t) - n_B(x,t)$, that fulfills the equation

$$\frac{\partial F}{\partial t} = D\frac{\partial^2 F}{\partial x^2}, \tag{12}$$

which, due to the indicated initial conditions, has the solution

$$F(x,t) = n_0\text{erf}\left(\frac{x}{\sqrt{4Dt}}\right). \tag{13}$$

We now rewrite the concentrations of A and B particles as

$$n_A(x,t) = G_1(x,t) + \delta n(x,t), \tag{14}$$

$$n_B(x,t) = G_2(x,t) + \delta n(x,t), \tag{15}$$

where

$$G_1(x,t) = \begin{cases} n_0\text{erf}[\frac{x}{\sqrt{4Dt}}] & x > 0 \\ 0 & x < 0, \end{cases} \tag{16}$$

and also $G_2(x,t) = G_1(-x,t)$. In terms of these functions we can write $F(x,t)$ and $F(\mid x \mid,t)$ as:

$$F(x,t) = G_1(x,t) - G_2(x,t),$$

$$F(\mid x \mid,t) = G_1(x,t) + G_2(x,t).$$

The function defined by $S(x,t) = n_A(x,t) + n_B(x,t)$ must be a continous one given by:

$$S(x,t) = F(\mid x \mid,t) + 2\delta n(x,t)$$

Substitution of the forms of $n_A(x,t)$ and $n_B(-x,t)$ into equations (10) and (11), reduces this set of equations to the following equation for $\delta n(x,t)$ alone,

$$\frac{\partial \delta n}{\partial t} = D\frac{\partial^2 \delta n}{\partial x^2} \quad - \quad \gamma_e T(z,\underline{z_1})F(\mid x_1 \mid,t_1)\delta n(z_1 \mid t)$$
$$- \quad \gamma_e T(z,\underline{z_1})\delta n(z_1)\delta n(z_1 \mid t), \tag{17}$$

with the boundary condition:

$$\delta n_x(0^\pm, t) = \mp \frac{n_0}{4\sqrt{Dt}},$$

which comes from the continuity of the function $S(x,t)$ at the origin.

We have indicated in references [32,35] that it is the *absorption function*, i.e. the reaction term of (17), that is the relevant function for determining the evolution of the concentration of the product C. Here, as we have done there, we also need to resort to Fourier-Laplace transforming this term. Hence, instead of dealing with T through the series in (8) we consider its Fourier-Laplace transform $T(k,s)$. In the present case it reduces to:

$$T(k,s) = \frac{\frac{1}{\sqrt{2\pi}}}{1 + \frac{\gamma_\varepsilon \pi}{\sqrt{(Dk)^2 + 2sD}}}. \tag{18}$$

At this point it is possible to show that the *sudden* approximation described above (i.e.: $\delta n(z_1 \mid t) = \delta n(z_1)$) is equivalent to, and yields the same results as, a *mean field* approach. However, guessing that the shape of the δn varies subdiffusively (that is, the typical time scale is slower than $t^{1/2}$), we must resort to the *adiabatic* approximation (i.e.: $\delta n(z_1 \mid t) = \delta n(x,t)$). Hence, we can follow the line of work indicated in reference [32], resorting to the Fourier-Laplace transform of $T(z,z_1)F(\mid x_1 \mid, t_1)$, yielding the result

$$FL\left\{T(z,\underline{z_1})F(\mid x_1 \mid, t_1)\right\} = n_0 T(k,s)\frac{1}{s}[\sqrt{2\pi}\delta(k) - \frac{\sqrt{(\frac{2s}{\pi D})}}{\frac{s}{D} + k^2}], \tag{19}$$

where $FL\{.\}$ means the Fourier-Laplace transformation. We then find that the inverse Fourier transform gives

$$L\left\{T(z,\underline{z_1})F(\mid x_1 \mid, t_1)\right\} = \frac{n_0}{s}(T(k=0,s) - T(k=i\sqrt{\frac{s}{D}}, s)\exp(-\sqrt{\frac{s}{D}}\mid x \mid)), \tag{20}$$

which asymptotically ($s \to 0$), and in the approximation $x << \sqrt{4Dt}$, reads:

$$L\left\{T(z,\underline{z_1})F(\mid x_1 \mid, t_1)\right\} \sim \pi^{-\frac{3}{2}}(1 - \frac{1}{\sqrt{2}})\frac{n_0}{\gamma}(D/s)^{\frac{1}{2}}. \tag{21}$$

Then, Eq.(17) can be assymptotically written as :

$$\frac{\partial \delta n}{\partial t} = D\frac{\partial^2 \delta n}{\partial x^2} - \frac{a}{\sqrt{t}}\delta n - \gamma_\varepsilon T(z,\underline{z_1})\delta n(z_1)\delta n(x,t), \tag{22}$$

with $a = (1 - \frac{1}{\sqrt{2}})n_0\sqrt{D}$.

An approximate solution of this equation can be obtained using a scaling method. Let us assume the scaling function $\delta n(x,t) \sim t^{-\gamma}g(\frac{x}{t^\alpha})$. The asymptotic behavior of $T(z,z_1)\delta n(z_1)$ can be obtained as before using the Fourier-Laplace transformation:

$$FL\{T(z,\underline{z_1})\delta n(z_1)\} \simeq T(k,s)s^{\gamma-1}k^{-1}G(k/s^\alpha). \tag{23}$$

Substituting here Eq.(18), taking the asymptotic limit ($s \to 0$) and taking the inverse transformation, we obtain

$$T(z, \underline{z_1})\delta n(z_1) \simeq t^{-\gamma - \alpha} g\left(\frac{x}{t^\alpha}\right). \tag{24}$$

After substituting the scaling function into Eq.(22) and taking into account the last equation we find that a solution exists of Eq.(22) with the following values for the scaling exponents: $\alpha = \gamma = \frac{1}{4}$. Moreover, the scaling function fulfills the equation ($u = x/t^\alpha$):

$$D\frac{d^2}{du^2}g - ag - vg^2 = 0, \tag{25}$$

with the boundary condition $g_u(0^\pm) = \mp\frac{n_0}{4\sqrt{D}}$. The constant v could be determined after solving Eq.(25).

The knowledge of the asymptotic behavior of δn allows us to determine all the relevant quantities. Particularly we can determine the so called reaction rate $R(t)$ [24,29,31] that is the most common quantity determined in simulations. In the present case

$$R \simeq \frac{a}{t^{1/2}}\delta n \simeq t^{-3/4}g(xt^{-1/4}). \tag{26}$$

Concerning the scaling exponents, this result is in complete agreement with the most recent and refined simulations [31]. However, the asymptotic form of the scaling function ($g \sim \exp(-\sqrt{a/D}u)$) resulting from Eq.(25) does not reproduce the gaussian form found in the indicated simulations. This fact is not unexpected, on one hand because the neglected (crossed) terms in the series defining $_A(z)$ [35], that are clearly not relevant in the homogeneous case, could play a non negligible role in the present case; while in the other hand we have the non-uniform convergence of the scaling limit to the true solution [36].

TRAPPING WITH A TIME-DEPENDENT NUMBER OF TRAPS

In this section we address the problem of a trapping reaction (symbolically written $A + B \to B$) in a one dimensional system of diffusing A particles and B traps, but in the case where the number of traps is time dependent. This dependence can arise because the traps participate in another reaction or because they are externally controlled. Such a situation, which has not been treated previously in the scientific literature, besides its interest in relation with several problems related to heterogeneous reactions and catalysis, shows some peculiarities that makes relevant its study on its own right [37]. Here we will consider the following two related situations: (a) $A + B \to B$, $B + C \to C$ (double trapping). (b) $A + B \to B$, $B + C \to 0$ (trapping with annihilated traps). In both cases it is clear that the second reaction will not be affected at all by the first one. This allows us to exploit known results for trapping and annihilation. In case (b), as usual, we restrict ourselves to equal initial densities of B and C particles in the annihilation reaction.

Here, we show the results of Monte Carlo simulations made for both cases, and comparisons with the result of a mean field evaluation and with our stochastic model, adapted to the present situation.

Firstly we present the mean field results in both situations. In case (a) the solution is given by

$$n_B(t) = n_{B0} \exp\left(-\gamma_{BC} \, n_C \, t\right) \tag{27}$$

$$n_A(t) = n_{A0} \exp\left[-\frac{\gamma_{AB} \, n_{B0}}{\gamma_{BC} \, n_C}\left(1 - e^{-\gamma_{BC} \, n_C \, t}\right)\right]. \tag{28}$$

For case (b), the solution is

$$n_B(t) = n_{B0}\left(1 + \gamma_{BC} \, n_{B0} \, t\right)^{-1} \tag{29}$$

$$n_A(t) = n_{A0}\left(1 + \gamma_{BC} \, n_{B0} \, t\right)^{-\frac{\gamma_{AB}}{\gamma_{BC}}}. \tag{30}$$

Here $\gamma_{AB,BC}$ are the reaction rates of each reaction, $n_{A0,B0}$ the initial densities of A, B particles and n_C the (fixed) C density for the case (a).

It is worth remarking that the qualitative behaviour of the indicated solutions is clearly different. The most remarkable aspect is that asymptotically they reach completely different limits. In the first case we have

$$n_A(t \to \infty) = n_{A0} \exp\left(-\frac{\gamma_{AB} \, n_{B0}}{\gamma_{BC} \, n_C}\right). \tag{31}$$

This finite value is in contrast with the second case where we have total extinction for A particles ($n_A(t \to \infty) \propto t^{-\gamma_{AB}/\gamma_{BC}}$).

For the stochastic description we have adapted the stochastic model that has been described in a previous section. We will not go into the details of the model but only exploit the results of our previous work. The general result for n_A, the density of A particles, is

$$\frac{d}{dt}n_A(t) = -\gamma_{AB} \, n_A(t) \, n_B(t) + \int_0^t dt' C(t - t') \, n_A(t') \, n_B(t'), \tag{32}$$

where $C(t) = \alpha \, \gamma_{AB}\left[(\pi t)^{-1/2} - \alpha \exp(\alpha^2 t) \, \mathrm{erfc}(\alpha \sqrt{t})\right]$ with $\alpha = \gamma_{AB}/\sqrt{4(D_A + D_B)}$; D_A and D_B are the diffusivities of particles A and B respectively and $n_B(t)$ is the (variable) density of B particles. Even though this equation was originally derived for a simple annihilation process, it can be proved to be correct for the general case of variable $n_B(t)$. For our particular cases this density comes from a trapping (case (a)) or an annihilation (case (b)) process.

The indicated integro-differential equation must be solved numerically. However, the asymptotic analysis can be done analytically by means of Laplace transformation procedures yielding for the double trapping case the final value reached for A density

$$n_{Af} = n_A(t \to \infty) = n_{A0}\left(1 + \frac{n_{B0}}{n_C}\sqrt{\frac{D_A + D_B}{D_B + D_C}}\right)^{-1}. \tag{33}$$

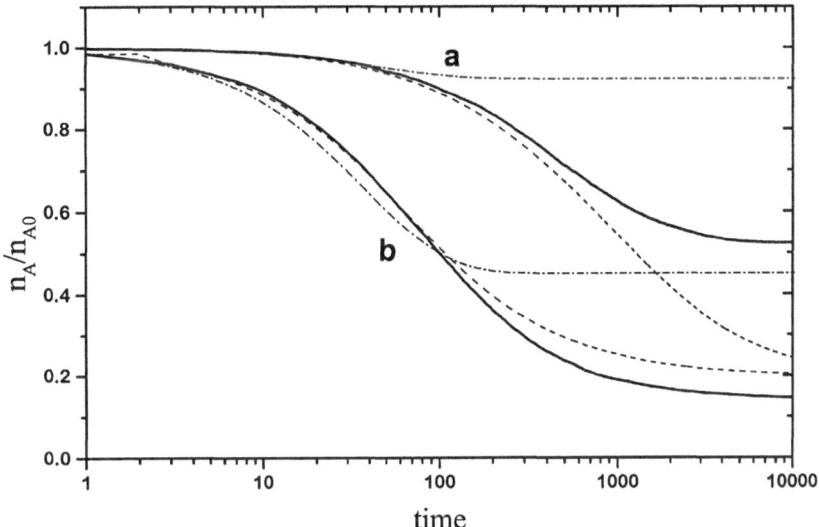

FIGURE 1. Temporal evolution of the density of A particles for a double trapping system. The solid line corresponds to simulations, the dashed one to the Galanin model (numerical integration) and the dashed-dot one to mean field results. The simulation (10 realizations) was performed in a 1000 sites lattice with periodic boundary conditions. The common parameters are: $n_{A0} = 1000$, $n_{B0} = 200$, $n_C = 50$, $q_A = q_B = q_C = .1$, $\gamma_{BC} = .4$, while $\gamma_{AB} = .008$ for (a) and $.08$ for (b). The final value $n_{Af}/nA0$ for the Galanin model is $.2$ in both cases.

This, again, is a finite value that contrasts with the second case where we have a potential decay to zero $n_A(t \to \infty) \propto t^{-1/4}$. In the most general way, the asymptotic analysis within this model predicts that, if we assume a long time behaviour $n_B(t) \propto t^\beta$, for $-1/2 < \beta \leq 0$, n_A will have a potential decay to zero, with an exponent $-(1/2+\beta)$. On the contrary, as indicated in the double trapping case, when $\beta = -1/2$, n_A reaches a finite value. Hence, we can expect that for $\beta < -1/2$, n_A will also reach a finite asymptotic value.

In figure 1 we show the result of simulations (see [37] for details), our model and the mean field for the density of A particles for two different values of γ_{AB}. Note that in order to keep a constant macroscopic rate γ we must change the microscopic absorption probability p whenever we change the jump probability q. It is clear that the stochastic model offers a better description of the problem than mean field approach although both give a qualitative good description for short times.

Figures 2 and 3 show a study of the asymptotic value dependence with the diffusivities and the macroscopic absorption rate respectively. The error bars indicate the standard deviation of values. For computing the final value we waited until $n_b = 0$ in each

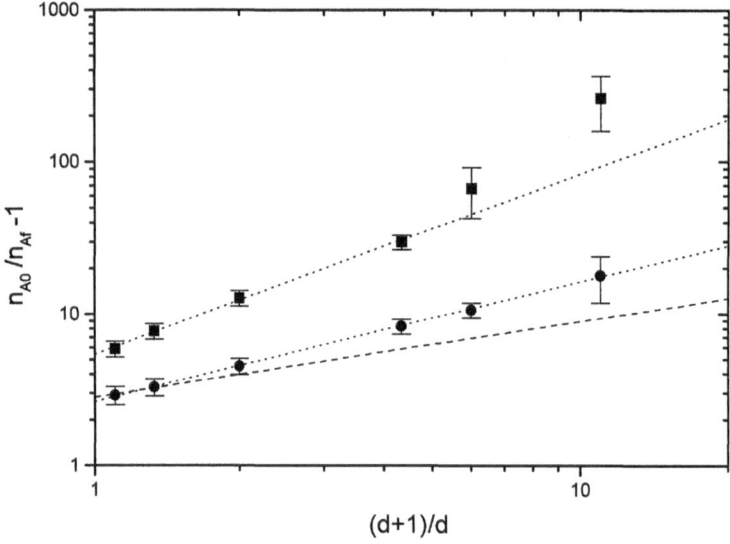

FIGURE 2. Variation of the final value n_{Af} as function of $d = D_B/D_A$. Here we choose the same initial densities and lattice size as in FIG. 1. The diffusivities are $D_B = D_C = .03$ in all the cases. The parameters are $\gamma_{AB} = .2$ (squares) and $.02$ (circles), and $\gamma_{BC} = .2$ for both cases. The dashed line indicates the Galanin result (Eq.(7)). The corresponding (constant) values for the mean field model are 53.6 (squares) and .5 (circles). The dotted lines correspond to a fitting of the simulation results.

realization. In fig. 2 we include the theoretical expression (33) in order to compare with simulations. In both figures we restricted ourselves to the case where $D_B = D_C$. We can see that the dependence of the asymptotic value of n_A on the diffusivity is qualitatively well described by the Galanin model; while it does not depend at all on the diffusivities for the mean field description. However, the stochastic model Galanin expression do not show any dependence on the γ's though it is qualitatively well described by the mean field. Matching both theoretical results we expect that the diffusivity dependence will appear only on the ratio $d = D_B/D_A$ while the reaction rates dependence will appear as the ratio γ_{BC}/γ_{AB}. We have computed n_{Af} in some simulations, with results that confirm (at least approximately) this guess.

In figure 4 we show results for the annihilation case. It is worth remarking here that the result of n_A reaching a zero value is only valid for an infinite lattice. For a finite one we can, for example, reduce γ_{AB} until the trapping reaction becomes so slow that all the traps can be annihilated before they can trap all A particles. This is shown in the insert of Fig. 4.

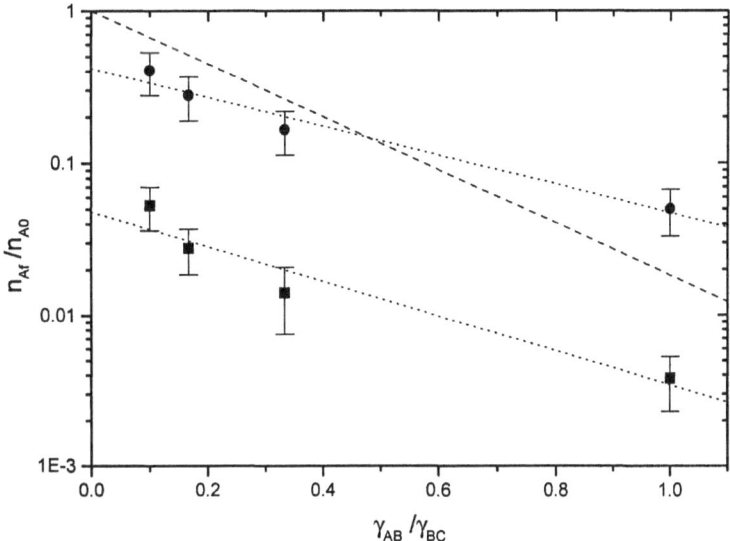

FIGURE 3. Variation of final value n_{Af} as function of γ_{BC}/γ_{AB}. Here we have chosen the same initial densities and lattice size as in Fig 1. The parameters are $D_B = .3$ (circles) and .03 (squares) while $\gamma_{BC} = .2$, $D_A = .3$ in both cases. The dashed line correspond to mean field result (Eq. (5)). The (constant) Galanin values are .2 (circles) and .96 (squares). The dotted lines correspond to a fitting of the simulation results.

FINAL DISCUSSION

The recently introduced stochastic scheme to describe the reaction $A + B \to B$ [13-15], extended to the case of two species annihilation reaction $A + B \to C$ [16,17], has been applied to two problems: a) the problem of initially separated reactants in the reaction $A + B \to C$, b) trapping with a time-dependent number of traps.

In the first case, the system we have considered is one-dimensional and we have assumed that the evolution between collisions is diffusive (with equal diffusion coefficients for both reactants), and that there is only a finite probability of annihilation in any collision of unlike particles. This method allows us to obtain, within certain approximations, a closed equation for the quantity δn (17) that can be treated through scaling procedures. In such a framework, we have obtained scaling exponents that agree with those obtained in the most recent simulations.

In the second case, we have shown that, for the reaction $A + B \to B$, the time evolution of the number of traps (B) can strongly influence the time evolution of the trapping process. Even more, there is a critical exponent ($n_B(t \to \infty) \propto t^\beta$, $\beta_c = -1/2$) separating different qualitative asymptotic behaviours for n_A: for $-1/2 < \beta \le 0$ we have complete extinction, while for $\beta \le -1/2$ we obtain the asymptotic survival of the A particles.

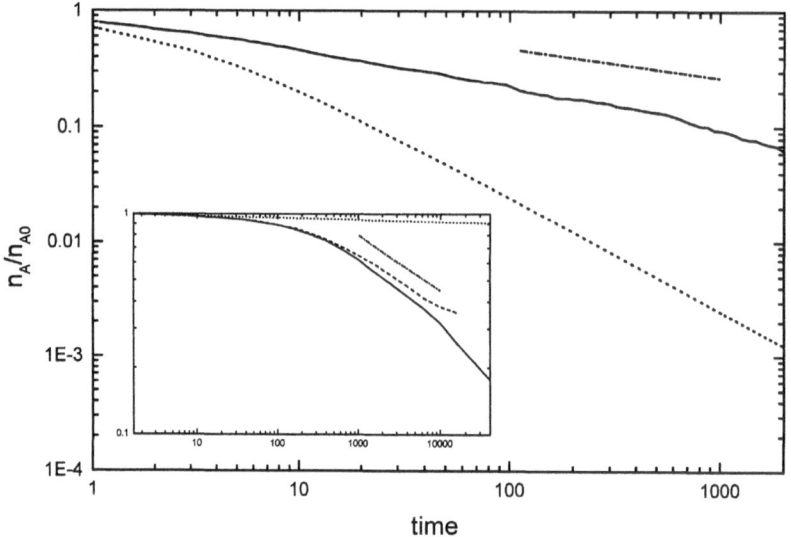

FIGURE 4. Temporal evolution of the density of A particles for a trapping with annihilated traps. The solid line correspond to simulations, and the dots to mean field results. The dashed-dot line indicates the $t^{-1/4}$ slope. The simulations (10 realizations or more) were performed in a lattice of 100 sites with periodic boundary conditions. The parameters are: $n_{A0} = 100$, $n_{B0} = n_{C0} = 50$, $q_A = q_B = q_C = .1$, $\gamma_{BC} = \gamma_{AB} = .8$. The inset shows the dependence with the lattice size: a 100-site lattice (dashed, $n_{Af}/n_{A0} = .35 \pm .06$) and a 200-site lattice (solid, $n_{Af}/n_{A0} = .16 \pm .06$); keeping the same density and the same parameters except that $\gamma_{AB} = .008$.

In the double trapping case, matching mean field and stochastic results, we have seen that the n_{Af} dependence on diffusivities and reaction rates comes through the ratios D_B/D_A and γ_{BC}/γ_{AB}. The adapted stochastic model gives a much better agreement with simulations than mean field, with a better qualitative prediction of the parameter dependence.

The adequacy of the present method to treat reaction kinetic processes is apparent from the present results as well as previous ones for the case of homogeneous systems and other results corresponding to extensions of the stochastic (Galanin-like) approach for: traps executing a non-Markovian motion [18], the inclusion of sources [19], competition between coupled reactions [37].

Finally, it is worth to remark that, for the first case, the analysis of the form of the the scaling function will require further investigation in order to provide a more complete picture of nonhomogeneous situations within the present model. For instance, the analysis of the related problem of a point like source injected into a homogeneous substrate, as discussed in [40], was discussed in [41]. For the second case, the study needs further improvement in order to obtain a still better quantitative agreement with simulations. Also, an analysis for the double annihilation process ($A + B \rightarrow 0$ and $B + C \rightarrow 0$) is also requiered and will be the subject of further work.

ACKNOWLEDGEMENTS

The authors want to thank G. Abramson, D.H. Zanette, D. ben Avraham, P. Alemany, A. Bru, Ernesto Nicola, for fruitful discussions and or collaborations, S.J. Cornell for providing information related with the present work prior to publication and V. Grunfeld for the revision of the manuscript. Financial support from Ministerio de Educación y Ciencia, Spain, through its program of Cooperación with Iberoamérica and Project DGICyT No.PB93-0054-C02-02, and from CONICET, Argentina through Project PIP-4953/96 are greatly acknowledged.

REFERENCES

1. See the recent review by S.Redner and F.Leyvraz, in *Fractals in Science*, A.Bunde and S.Havlin editors (Springer–Verlag, (1994)), pg. 197, nd the references included therein. See also several articles on this subject in the Raoul Kopelman Festschrift issue of *The Journal of Physical Chemistry* vol.**98**, July 28th (1994).

2. S. Havlin and D. Ben-Avraham; Adv.Phys. **36**, 695 (1987);
   Y.B. Zeldovich and A.S. Mikhailov; Usp.Fiz.Nauk. **153**, 469 (1987);
   K. Lindenberg, B.J.West and R. Kopelman, in *Noise and Chaos in Nonlinear Dynamical Systems*, edited by F. Moss, L. Lugiato and W. Schleich (Cambridge U.P., Cambridge, England, 1990), p.142;
   D. Ben-Avraham, M.A. Burschka and C.R. Doering; J.Stat.Phys. **60**, 695 (1990).

3. L.W. Anacker and R. Kopelman; Phys.Rev.Lett. **58**, 289 (1987).

4. K. Lindenberg, B.J. West and R. Kopelman; Phys.Rev.Lett. **60**, 1777(1988).

5. S. Kanno; Prog.Theoret.Phys. **79**, 721(1988).

6. B.J. West, R. Kopelman and K. Lindenberg; J.Stat.Phys. **54**, 1429 (1989).

7. C.R. Doering and D. Ben-Avraham; Phys.Rev. **A38**, 3035(1988).

8. R. Schoonover, D. Ben-Avraham, S. Havlin, R. Kopelman and G.H. Weiss; Physica **A171**, 232 (1991).
   P.K. Datta and A.M. Jayannavar; Physica **A184**,135 (1992).

9. H. Taitelbaum; Phys.Rev. **A43**, 6592 (1991).

10. H.S. Wio, M.A. Rodriguez, C.B. Briozzo and L. Pesquera; Phys.Rev. **A44**, R813 (1991).

11. W. Horsthemke, C.R. Doering, T.S. Roy and M.A. Burschka; Phys.Rev. **A45**, 5492(1992).

12. H.S.Wio, in *Instabilities and Nonequilibrium Structures IV*, E.Tirapegui and W.Zeller Eds.(Kluwer, 1993).

13. M.A. Rodriguez, G.Abramson, H.S. Wio and A.Bru Espino; Phys. Rev. E **48**, 829 (1993).

14. H.S.Wio, G.Abramson, M.A.Rodriguez and A.Bru Espino; Chaos, Solitons and Fract. **6**, 575 (1995).

15. G.Abramson and H.S.Wio; Chaos, Solitons and Fract. **6**, 1 (1995).

16. G.Abramson, A.Bru Espino, M.A. Rodriguez and H.S. Wio; Phys.Rev.E **50**, 4319 (1994);

17. G.Abramson, Ph.D. thesis, Instituto Balseiro, Universidad Nacional de Cuyo, Argentina, 1995; A. Bru, Ph.D. thesis, Universidad Complutense de Madrid, Spain, 1995.

18. G.Abramson and H.S.Wio, Phys.Rev.E. **53**, 2265 (1996).

19. A.D. Sanchez and H.S. Wio, Physica A **237**, 452 (1997).

20. A.D.Galanin, *Thermal Reactor Theory*, (Pergamon, N.Y., 1960, 2nd.Ed.);
    M.M.R. Williams,*Random Processes in Nuclear Reactors*, (Pergamon, Oxford, 1974).

21. I. Martinez and M.A. Rodriguez; Ann.Nucl.Energy **12**, 113 (1985).

22. L.Galfi and Z.Racz; Phys.Rev. A **38**, 3151 (1988).

384

23. Z.Jiang and C.Ebner; Phys.Rev. A **42**, 7483 (1990).
24. B.Chopard and M.Droz, Europhys.Lett. **15**, 459 (1991); S.Cornell, M.Droz and B.Chopard, Phys. Rev. A **44**, 4826 (1991); S.Cornell, M.Droz and B.Chopard, Physica A **188**, 322 (1992).
25. Y.-E.Lee Koo and R.Kopelman; J.Stat.Phys. **65**, 893 (1991).
26. H.Taitelbaum, S.Havlin, J.Kiefer, B.Trus and G.H.Weiss; J.Stat.Phys. **65**, 873 (1991).
27. H.Taitelbaum, Y.-E.Lee Koo, S.Havlin, R.Kopelman and G.H.Weiss; Phys.Rev. A **46**, 2151 (1992).
28. E.Ben-Naim and S.Redner; J.Phys.A: Math.Gen. **25**, L575 (1992).
29. a) H.Larralde, M.Araujo, S.Havlin and H.E.Stanley; Phys.Rev. A **46**, 855 (1993); b) H.Larralde, M.Araujo, S.Havlin and H.E.Stanley; Phys.Rev. A **46**, R6121 (1992); c) M.Araujo, S.Havlin, H.Larralde and H.E.Stanley; Phys. Rev. Lett. **68**, 1791 (1992).
30. B.Chopard and M.Droz, T.Karapiperis and Z.Racz; Phys.Rev. E **47**, R40 (1993).
31. S.J. Cornell, Phys.Rev.Lett. **75**, (1995); S.J. Cornell, Phys.Rev.E **51**, 4055 (1995);
32. F.G.Nicolini, M.A.Rodriguez, H.S.Wio, *Time behaviour of the reaction front in the catalytic $A + B \to B + C$ reaction–diffusion processes*, ICTP IC/94/162 (1994); M.A.Rodriguez and H.S.Wio, *Behaviour of the reaction front for $A + B \to 0$ reaction–diffusion system: an analytic approach*, ICTP IC/94/161 (1994).
33. B.P. Lee and J. Cardy, Phys. Rev. E **50**, R3287 (1994); B.P. Lee and J. Cardy, J. Stat. Phys. **80**, 971 (1995); M. Howard and J. Cardy, J. Phys. A **28**, 3599 (1995).
34. D.Avnir and Kagan; Nature **307**, 717 (1984), G.T.Dee; Phys. Rev. Lett. **57**, 275 (1986), R.E.Liesegang; Naturwiss. Wochensch. **11**, 353 (1896), T.A.Witten and L.M.Sander; Phys. Rev. Lett. **47**, 1400 (1981), K.F.Mueller; Science **225**, 1021 (1984).
35. M.A.Rodriguez and H.S.Wio, Phys.Rev. E **56**, 1724 (1997).
36. A.Schenkel, P. Wittwer and J. Stubbe, Physica D **69**, 135 (1993).
37. A.D. Sánchez, E.M. Nicola and H.S. Wio, Phys. Rev. Lett. **78**, 2244 (1997).
38. A. Sánchez, M.A. Rodriguez and H.S. Wio, Phys. Rev. E **57**, 6390-6397 (1998).
39. H. Larralde, Y. Lereah, P. Trunfio, J. Dror, S. Havlin, R. Rosenbaum and H. E. Stanley, Phys. Rev. Lett. 70 (1993) 1461
40. S. Bouzat and H. S. Wio, Anales AFA **8**, 14 (1997).
41. S. Bouzat, A. Sánchez and H.S.Wio, Physical Review E **60**, 2677 (1999).

Subject Index

Author Index

Nonlinear Phenomena and Complex Systems

KLUWER ACADEMIC PUBLISHERS – DORDRECHT / BOSTON / LONDON